T0213647

Springer Series in the Data Sciences

Springer Series in the Data Sciences focuses primarily on monographs and graduate level textbooks. The target audience includes students and researchers working in and across the fields of mathematics, theoretical computer science, and statistics. Data Analysis and Interpretation is a broad field encompassing some of the fastest-growing subjects in interdisciplinary statistics, mathematics and computer science. It encompasses a process of inspecting, cleaning, transforming, and modeling data with the goal of discovering useful information, suggesting conclusions, and supporting decision making. Data analysis has multiple facets and approaches, including diverse techniques under a variety of names, in different business, science, and social science domains. Springer Series in the Data Sciences addresses the needs of a broad spectrum of scientists and students who are utilizing quantitative methods in their daily research. The series is broad but structured, including topics within all core areas of the data sciences. The breadth of the series reflects the variation of scholarly projects currently underway in the field of machine learning.

More information about this series at http://www.springer.com/series/13852

Nickolay Trendafilov · Michele Gallo

Multivariate Data Analysis on Matrix Manifolds

(with Manopt)

 Springer

Nickolay Trendafilov🆔
School of Mathematics and Statistics
Open University
Milton Keynes, Buckinghamshire, UK

Michele Gallo🆔
Department of Human and Social Sciences
University of Naples "L'Orientale"
Naples, Italy

ISSN 2365-5674   ISSN 2365-5682   (electronic)
Springer Series in the Data Sciences
ISBN 978-3-030-76976-5   ISBN 978-3-030-76974-1   (eBook)
https://doi.org/10.1007/978-3-030-76974-1

Mathematics Subject Classification: 58A05, 58C05, 62–07, 62H12, 62H25, 62H30, 62H99, 65C60, 65Fxx, 65K99, 90C26, 90C51

This Springer imprint is published by the registered company Springer Nature Switzerland AG
The registered company address is: Gewerbestrasse 11, 6330 Cham, Switzerland

To the memory of

my mother, Zdravka, and my father, Trendafil

and to my family

wife, Irina, son, Iassen,
and grandchildren, Veronica and Christian

Preface

We want to start with few remarks predating considerably the emerging of the idea for writing this book and our collaboration in general. They are related to the first author's own experience which made him explore matrix manifolds in data analysis problems.

NTT was in an early stage of his research career, after his Ph.D. was completed on a completely different topic. He was assigned to study factor analysis (FA) and do some programming for a particular software product. While working on FA, NTT realized that the most interesting part for him is the FA interpretation and the so-called rotation methods (see Section 4.5). NTT recognized that the main goal of FA is to produce simple for interpretation results, which is achieved by orthogonal rotation of some initial, usually difficult to interpret, FA solution Λ (known as factor loadings matrix). However, how we can define what is "simple for interpretation results"? The problem is really tough, especially if you try to capture its meaning in a single mathematical expression/formula. Because of that, a huge number of different formulas were proposed each of them claiming to approximate in some sense the idea for "simple for interpretation results". In FA, these formulas are called simple structure criteria. They are supposed to measure/quantify the simplicity of a certain FA solution. Let f be such a criterion. Then, if Λ_2 is simpler than Λ_1, and if Λ_3 is even simpler than Λ_2, then we should have $f(\Lambda_1) > f(\Lambda_2) > f(\Lambda_3)$, or $f(\Lambda_1) < f(\Lambda_2) < f(\Lambda_3)$ depending on the sign of f. Clearly, one would be interested to find the solution Λ, for which f reaches its smallest/largest value, and the problem smells of optimization.

For reasons to become clear in Section 4.5, the rotation methods transform the initial (difficult to interpret) solution Λ_0 into another, simpler, one $\Lambda_0 Q$, where Q is an orthogonal matrix, which gives the name "rotation methods". Thus, in formal terms, the rotation methods look for an orthogonal matrix Q, such that $f(\Lambda_0) > f(\Lambda_0 Q)$. The problem stays the same no matter what rotation criterion f is considered. Thus, one needs to solve

$$\min_Q f(\Lambda_0 Q), \tag{1}$$

where Q is orthogonal and f is an arbitrary criterion.

NTT was surprised to see in the psychometric literature, e.g. Harman (1976); Mulaik (1972), that the introduction of any new simplicity criterion, f, was inevitably connected to a new method for solving (1). At that moment his knowledge in optimization was quite limited, but intuitively NTT felt that this cannot be right: a new optimization method for every new function f. It sounded ridiculous!

And then, NTT started to look for a general approach that can work for every f, as far as the variable Q is an orthogonal matrix. In this way, NTT realized that the rotational problem can be naturally defined as optimization on the matrix manifold of all orthogonal matrices. The understanding that, in fact, all MDA problems can be seen defined on specific matrix manifold(s) followed then straight away.

Milton Keynes, UK   Nickolay Trendafilov
Naples, Italy   Michele Gallo

Contents

Chapter 1

Introduction

N. Trendafilov and M. Gallo, *Multivariate Data Analysis on Matrix Manifolds*,
Springer Series in the Data Sciences, https://doi.org/10.1007/978-3-030-76974-1_1

The main idea of the book is to give a unified presentation, treatment and solution of several frequently used techniques for multivariate data analysis (MDA). The book differs from the existing MDA books, because it treats the data analysis problems as optimization problems on matrix manifolds.

To some extent, the book resembles (Helmke and Moore, 1994), where a number of signal processing and control problems are treated as optimization problems on matrix manifolds. Besides the difference in the subject areas (signal processing versus data analysis), both books differ in the way they approach the problems: (gradient) dynamical systems and ordinary differential equations (ODE) versus iterative schemes on matrix manifolds, which is adopted in this book. Of course, all MDA problems considered in this book can also be expressed as dynamical systems on matrix manifolds and solved by numerical ODE solvers. However, such an approach may be impractical for large data analysis applications. Nevertheless, it is worth noting that the theoretical study of dynamical systems (and their behaviour) emerging from data analysis problems is a pretty much untouched area, except principal component analysis (PCA), and is likely to bring new interesting and challenging questions.

This is a graduate-level textbook and the targeted audience are researchers, postgraduate/Ph.D. students and postdocs in and across the fields of mathematics, theoretical computer science, machine learning, computational statistics, data analysis and more broadly data science and analytics. Thus, it is assumed that the reader is mathematically literate and familiar to some extent with matrix analysis (Halmosh, 1987; Horn and Johnson, 2013), matrix computing (Eldén, 2007; Golub and Van Loan, 2013), MDA (Izenman, 2008; Mardia et al., 1979) and optimization (Bertsekas, 1999; Boyd and Vandenberghe, 2009). It probably makes sense to suggest also some entry-level MDA books as (Everitt and Dunn, 2001), and particularly (Adachi, 2016), where matrix language is thoroughly adopted to present the MDA models and solutions.

First of all, what is MDA and its goal? There are different types of data, but MDA is a collection of methods for analysing data which can be presented in a form of a matrix (or several matrices). Depending on the specific data type, MDA builds a particular model which proposes a simplified representation of the data, reflecting our understanding of the problem. The different types of data call for different models. The common feature of all MDA models is that each of them is given by a specific matrix function of matrix unknown parameters. The main goal of MDA is to find these parameter matrices such that the model matrix fits the data matrix as close as possible in some sense. Such type of problems are called in statistics parameter estimation. In fact, they require the solution of certain optimization problem. In this book, most frequently the goodness of the model fit (to the data) will be measured in terms of least squares (LS). In some methods, the maximum likelihood (ML) will be used as well as a goodness-of-fit measure.

Second, why we need matrix manifolds for MDA? This is because the unknown parameter matrices in the MDA model usually have certain structure which has to be preserved in the solution. For example, an unknown matrix may be required to be orthogonal matrix or rectangular matrix with prescribed rank, etc. The sets of such matrices form (smooth) matrix manifolds, which specific geometry is worth utilizing when solving numerically the particular MDA problem.

Thus, from mathematical point of view, many MDA problems require solution of optimization problems on matrix manifolds. A general treatment of such problems and methods for their numerical solution is given in (Absil et al., 2008; Boumal, 2020). The purpose of this book is to concentrate on a number of MDA problems and their formulation as constrained optimization problems on matrix manifolds defined by the MDA model parameters. As a result, they can be readily solved by Manopt, a free software for optimization on a great variety of matrix manifolds (Boumal et al., 2014). Initially, Manopt was developed as a MATLAB-based software, but Python and Julia versions are already available through the Manopt website. Julia is relatively new, and an idea how it works with matrices and numerical problems can be obtained by taking a look in (Boyd and Vandenberghe, 2018).

To be more specific about the nature of the book, let us consider an example. The well-known ordinary LS regression of $y \in \mathbb{R}^n$ on $X \in \mathbb{R}^{n \times p}$ requires

$$\min_{b}(y - Xb)^{\top}(y - Xb) \ ,$$

where y and X are assumed with zero column sums (centred). It is helpful if X has full rank. The matrix generalization of this problem is straightforward and is defined as follows. Let $Y \in \mathbb{R}^{n \times q}$ and $X \in \mathbb{R}^{n \times p}$ be given centred matrices ($n \geq p \geq q$). We are interested to find a matrix $B \in \mathbb{R}^{p \times q}$ such that XB is as close as possible to Y in LS sense. Formally, this can be expressed as

$$\min_{B} \|Y - XB\|_F^2 \ , \tag{1.1}$$

where $\| \ \|_F$ denotes the Frobenius matrix norm of the argument, which is $\|X\|_F^2 = \text{trace} X^\top X$. Problem (1.1) is well known as the multivariate regression (Izenman, 2008, Ch 6). The unknown B in (1.1) is not restricted in any way, though in many applications B can be required to obey certain linear constraints (Izenman, 2008, 6.2.4), i.e. B should stay in some (linear) subspace of $\mathbb{R}^{p \times q}$. In both situations, the multivariate regression (1.1) can be solved analytically, provided X is a full rank matrix.

Now, suppose that the unknown in (1.1) is required to be an orthonormal matrix $Q \in \mathbb{R}^{p \times q}$, i.e. $Q^\top Q = I_q$ (identity matrix). The set of all such matrices forms a smooth manifold known as the Stiefel manifold. This is a very special multivariate regression problem known as the Procrustes problem (Gower and Dijksterhuis, 2004). It also has a closed-form solution, based on the singular value decomposition (SVD) of $X^\top Y$.

However, for given $Y \in \mathbb{R}^{n \times r}$, $X \in \mathbb{R}^{n \times p}$ and $Z \in \mathbb{R}^{q \times r}$ the generalized Procrustes problem

$$\min_{Q, \ Q^\top Q = I_q} \|Y - XQZ\|_F^2 \, , \tag{1.2}$$

also known as the Penrose regression, cannot be solved analytically. For such problems, one can rely on numerical methods for optimization on matrix manifolds. They can be solved efficiently by Manopt. For example, to solve (1.2) one simply needs to find the gradient of the objective function using standard matrix differentiation. Then, the objective function and its gradient are given to Manopt, together with the information that Q is required to be in the Stiefel manifold. After specifying your preferred numerical method (steepest descent, conjugate gradient, etc.), Manopt produces the solution of (1.2). The benefit from this approach is that if you change the constraint set (manifold) of the problem, you only need to update the manifold information given before to Manopt.

The book is about considering a number of such more complicated problems which go beyond the standard MDA methods. The common feature is that they can be defined as optimization problems on matrix manifolds, and thus can be solved by Manopt. Additionally, we provide the Manopt codes for solving several MDA problems. They can be applied straightaway or used as templates for solving other related problems.

The book is organized as follows. The first two chapters collect background mathematical information essential for studying MDA on matrix manifolds which may not be widely popular among data analysts. The book begins with Chapter 2, where some basic facts are provided including the differentiation of functions of matrix unknowns. Such skills are essential to calculate gradients of MDA models. As we mentioned already, they are matrix functions of matrix parameters. Chapter 3 considers first several useful sets of matrices in MDA and their basic topological properties. Then, a formal definition of differentiable manifold is given, followed by several examples of matrix manifolds and their tangent spaces. Then, we are ready to explain how goodness-of-fit (scalar) functions can be optimized on matrix manifolds. We briefly list several methods for optimization on manifolds which are implemented in Manopt. The remaining part of the chapter is dedicated to a brief explanation of how Manopt works and what it needs from the user.

The remaining chapters consider different MDA techniques. All of them except one (Chapter 6) are in fact techniques for reducing the dimensionality of the original data. This is an important strategy that results in fewer output variables. They are produced in an optimal way and give a gist of what is contained in the available data. The MDA solutions (output variables) are linear combinations of the original/input ones, and the weighting coefficients indicate their (negative or positive) importance. A zero weight simply means that the corresponding input variable can be neglected. Finding such zero weights is crucial when analysing very large data. For this reason, nearly all of the considered, in the book, MDA techniques are accompanied by corresponding modifications that enhance the classical solutions with zero weights. To distinguish such solutions from the classical ones they are called *sparse*.

For example, Chapter 4 is dedicated to principal component analysis (PCA). The beginning of the chapter deals with description of the main features and methods of the technique. This includes details on how the classical PCA interprets the analysis results and the objective methods to achieve it. The remaining part considers modifications that produce sparse solutions. This chapter is very important because the main principals for interpretation of MDA results are discussed and linked to the modern concept of sparseness of solutions. The main features of the classical PCA are revisited to consider a hybrid version of PCA that combines the classical interpretation principle with modern sparse solutions. Finally, we consider pure sparse PCA and several possible methods that target different aspects of the dimension reduction. In this section, the reader will see how Manopt is applied to obtain sparse PCA solutions: problems requiring optimization on a single Stiefel manifold. The PCA methodology is also used to perform independent component analysis (ICA). The chapter is concluded by a brief introduction to several new dimension reduction approaches.

Chapter 5 treats a closely related technique to PCA technique for dimension reduction, namely, exploratory factor analysis (EFA). First, we state the EFA model and outline its features. Then, we consider a re-parametrization of the model which is essential for the EFA estimation to be formulated as optimization on matrix manifolds. Further, the new parameterization is additionally useful to establish sparse version of EFA. In this section the reader will see how Manopt is applied to obtain the classical EFA solutions: problems requiring optimization on a product of three matrix manifolds, a Stiefel manifold and two Euclidean spaces of different dimensions. Then, it is demonstrated how improper EFA solutions (Heywood cases) can be avoided by replacing the Euclidean spaces with the set (cone) of all positive definite (PD) matrices. The final two sections consider sparse EFA.

Chapter 6 is about a class of multivariate regression problems better known as Procrustes problems. We already met such problems in (1.1) and (1.2). The chapter considers first the general, so-called Penrose regression problem (1.2) which involves orthonormal (rectangular) unknown, and thus boils down to optimization on a Stiefel manifold. Then, we switch to more simple Procrustes problems widely used in the classical methods for interpretation of PCA and EFA. They involve orthogonal and oblique rotations (square matrices), and thus optimization on the orthogonal group $\mathcal{O}()$ and the oblique manifold $\mathcal{OB}()$. We briefly discuss robust version of the Procrustes problems when the nearness is measured in the ℓ_1 matrix norm rather than the usual ℓ_2 (Frobenius) norm. Finally, the chapter considers briefly the relation and application of Procrustes analysis (PA) to EFA of data with more variables than observations.

Chapter 7 considers the linear discriminant analysis (LDA) which is another very popular dimension reduction technique designed to work with grouped data. First, we define the LDA problem and consider the classical canonical variates and their orthogonal version. The difficulties with their interpretation naturally suggest to look for sparse solutions. Then, we consider and compare several approaches to LDA that achieve sparse canonical variates. Our experience with sparse PCA from Chapter 4 is extended to develop new sparse LDA techniques. The second part of the chapter is dedicated to LDA of (horizontal) data with more variables than observations.

Chapter 8 presents the canonical correlation analysis (CCA) as a closely related technique to LDA. First, we define the classical CCA problem and its solution as generalized eigenvalue decomposition (EVD). Then, we outline the relationships of CCA with LDA and common principal components (CPC). Next, we consider sparse versions of CCA by utilizing techniques for sparse PCA and sparse generalized EVD. Finally, we present several possible generalizations of CCA for more than two groups of variables and discuss their properties.

The method of CPC is presented in Chapter 9. First, we discuss the origins of CPC and its relation to LDA. Then, we formulate the ML–CPC and LS–CPC estimation as optimization problems on matrix manifolds. Some simplified alternative algorithms are also considered, followed by viewing CPC as simultaneous dimension reduction of several covariance/correlation matrices. Finally, we investigate some relations between CPC and other techniques as ICA and individual scaling (INDSCAL).

Chapter 10 collects several methods for analysing one or several proximity matrices. First, we consider the basic identities relating a data matrix to a matrix of distances and squared distances among its observations. The first technique to consider is the metric multidimensional scaling (MDS) (of squared distances), by defining its main goal and algorithm. Then, MDS is generalized to analyse simultaneously several matrices of squared distances (INDSCAL). Further, it is shown that a better fit can be achieved by working directly with original distances (DINDSCAL). Methods for analysing asymmetric dis/similarity matrices are considered in Section 10.7. The remaining part of the chapter gives a brief introduction to tensor data analysis. It provides some basic definitions and notations, as well as a prelude to methods for dimension reduction of tensor data.

Chapter 11 considers two MDA techniques which at first glance look unrelated. What connects them is the fact that their parametric spaces are simplexes (or product of simplexes). We believe that both techniques can benefit from each other's tools. Section 11.1 proposes a new treatment of archetypal analysis (AA), a technique that attracted considerable interest in the recent years. After defining the AA problem, we consider two approaches for its solution. One of them requires optimization on a product of two oblique manifolds, while the other directly solves the AA problem on a product of simplexes. Finally, the AA problem is presented as dynamical system (interior point flows) on a product of simplexes. Section 11.2 gives a brief introduction to the analysis of compositional data (CoDa). First, we list several most frequently used ways to transform CoDa that facilitate their statistical analysis. Then, one of them is adopted to analyse a real dataset and demonstrate the specific features of sparse PCA applied to CoDa.

The exercises in the book are mainly used to complement and enhance the main text. Some of them even constitute of open research problems. In addition, we provide Manopt codes for solving several types of MDA problems. These codes should be considered as templates which the readers can and should improve and/or tailor with respect to their individual tastes and needs. The reader is strongly encouraged to use these codes and explore their performance by solving own data analysis problems and/or check the provided in the book examples. Moreover, on several occasions, the readers are encouraged to write their own (new) Manopt codes based on the already available ones.

The notations adopted in the book are as follows. Scalars (real or complex numbers) are usually denoted by small Greek letters, with exception for π which may denote a projection. Occasionally, capital Greek letters, e.g. Λ, Σ, Ψ, are used for diagonal and correlation matrices. In general, small Roman letters denote vectors, and the same letter with single indexes is used for its elements. For example, $v = (v_1, v_2, v_3)$. The letters m, n, p, q, r make an exception as they are typically used for matrix/space dimensions, as well as i, j, k, l usually used for indexes. Functions and mappings are normally denoted by f or g, while d is kept for distances. Capital Roman letters denote matrices and linear operators (including tensors), while the corresponding small Roman letters with two or more indexes are used for their elements, e.g. $A = \{a_{ij}\}, X = \{x_{ijk}\}$. On general, the data matrices are denoted by X (or Z if standardized), while A, B and C are used for parameter matrices. Blackboard letters denote different real Euclidean spaces, e.g. $\mathbb{R}^n$. The calligraphic letters are used for sets, spaces and manifolds, e.g. $\mathcal{U}, \mathcal{E}, \mathcal{GL}$.

Chapter 2

Matrix analysis and differentiation

N. Trendafilov and M. Gallo, *Multivariate Data Analysis on Matrix Manifolds*,
Springer Series in the Data Sciences, https://doi.org/10.1007/978-3-030-76974-1_2

2.1   Matrix algebra

Every point $a = (a_1, \ldots, a_n) \in \mathbb{R}^n$ can be seen as a *vector* connecting the origin $0_n = \underbrace{\{0, \ldots, 0\}}_{n}$ of the coordinate system of $\mathbb{R}^n$ with the point a. For this reason, sometimes it is denoted as $\vec{a}$. The *matrix* is a collection of vectors. For example, the $n \times p$ matrix A,

$$A = \begin{bmatrix} a_{11} & a_{12} & \cdots & a_{1p} \\ a_{21} & a_{22} & \cdots & a_{2p} \\ \vdots & \vdots & \ddots & \vdots \\ a_{n1} & a_{n2} & \cdots & a_{np} \end{bmatrix},$$

can be seen as a collection of p *column vectors* $a \in \mathbb{R}^n$, i.e. $[a_{1i}, a_{2i}, \ldots, a_{ni}]^\top$ for $i = 1, \ldots, p$, or as a collection of n *row vectors* $a \in \mathbb{R}^p$, i.e. $[a_{i1}, a_{i2}, \ldots, a_{ip}]$ for $i = 1, \ldots, n$. In many occasions it is convenient, and, in fact, it is generally accepted to think of the vectors in $\mathbb{R}^n$ as $n \times 1$ matrices. Usually capital (Roman) letters are used for matrices, while the small (Roman) letters are kept for vectors. Let $\mathcal{M}(n, p)$ denote the set of all $n \times p$ *real* matrices, and $O_{n \times p} \in \mathcal{M}(n, p)$ be the matrix of zeros. The basic matrix operations are

- for $A, B \in \mathcal{M}(n, p)$, $A \pm B = C \in \mathcal{M}(n, p)$, and the elements of C are $c_{ij} = a_{ij} \pm b_{ij}$, where $i = 1, \ldots, n, j = 1, \ldots, p$; particularly, $A \pm O_{n \times p} = A$;

- for $A \in \mathcal{M}(n, p)$ and $\alpha \in \mathbb{R}$, $\alpha A = C \in \mathcal{M}(n, p)$, and the elements of C are $c_{ij} = \alpha a_{ij}$;

- for $A \in \mathcal{M}(n, p)$ and $B \in \mathcal{M}(p, r)$, $AB = C \in \mathcal{M}(n, r)$, and the elements of C are $c_{ij} = \sum_{k=1}^{p} a_{ik} b_{kj}$.

Note that the *matrix product* in the last item is *associative*, i.e. $A(BC) = (AB)C$, but not *commutative*, i.e. $AB \neq BA$, provided all involved matrices are compatible.

There is a number of other matrix products to be used in the book. For example, the *Hadamard matrix product* is defined for matrices of the same sizes as follows: for $A, B \in \mathcal{M}(n, p)$, their Hadamard product is $A \odot B = C \in \mathcal{M}(n, p)$, and the elements of C are $c_{ij} = a_{ij} b_{ij}$. This is an element-wise matrix product and $A \odot B = B \odot A$, i.e. it is commutative.

For any $A \in \mathcal{M}(n,p)$, the matrix $A^\top \in \mathcal{M}(p,n)$ obtained by exchanging the roles of the rows and the columns is called *transposed*, e.g.

$$A = \begin{bmatrix} a_{11} & a_{12} \\ a_{21} & a_{22} \\ a_{31} & a_{32} \end{bmatrix} \in \mathcal{M}(3,2) \text{ and } A^\top = \begin{bmatrix} a_{11} & a_{21} & a_{31} \\ a_{12} & a_{22} & a_{32} \end{bmatrix} \in \mathcal{M}(2,3) .$$

One can check directly that

- $(A^\top)^\top = A$,

- $(AB)^\top = B^\top A^\top$.

Particularly, any (column) vector $a \in \mathbb{R}^n$ can be considered as a matrix $a \in \mathcal{M}(n,1)$. Then, $a^\top \in \mathcal{M}(1,n)$ is a row vector.

A matrix with equal number of rows and columns ($n = p$) is called square. The subset of square matrices is denoted by $\mathcal{M}(n)$ or $\mathbb{R}^{n \times n}$. A square matrix A for which $A^\top = A$, i.e. $a_{ij} = a_{ji} \; \forall \; i,j$, is called *symmetric*. The set of all symmetric matrices in $\mathcal{M}(n)$ is denoted by $\mathcal{S}(n)$. The elements $a_{ii}, i = 1,\ldots,n$ are called *main diagonal* of the square matrix A. The rest of the elements of A are called *off-diagonal*.

If all off-diagonal elements of A are zeros, then A is called *diagonal* matrix. The set of all diagonal matrices in $\mathcal{M}(n)$ is denoted by $\mathcal{D}(n)$. To simplify the notations, the operator diag() has two meanings depending on its argument. For $a \in \mathbb{R}^n$, we set $\text{diag}(a) \in \mathcal{D}(n)$, a diagonal matrix in which main diagonal is a. Particularly, $I_n = \text{diag}(1_n)$ is called the *identity* matrix, where 1_n is a $n \times 1$ vector of ones. For $A \in \mathcal{M}(n)$, we set $\text{diag}(A) \in \mathcal{D}(n)$, a diagonal matrix, in which main diagonal is the main diagonal of A, i.e. $\text{diag}(A) = A \odot I_n$. Similarly, we define $\text{off}(A) = A - \text{diag}(A)$.

A matrix $A \in \mathcal{M}(n)$ for which $A^\top = -A$, i.e. $a_{ij} = -a_{ji} \; \forall \; i,j$, is called *skew-symmetric*. The set of all skew-symmetric matrices in $\mathcal{M}(n)$ is denoted by $\mathcal{K}(n)$. Note that for every $A \in \mathcal{M}(n)$:

$$A = \frac{A + A^\top}{2} + \frac{A - A^\top}{2} = A_S + A_K , \quad A_S \in \mathcal{S}(n), \; A_K \in \mathcal{K}(n) . \quad (2.1)$$

A *permutation matrix* P is a square matrix each row and column of which contains one element equal to 1, and the rest are 0s. In other words, the matrix $P \in \mathcal{M}(n)$ of 0s and 1s is a permutation matrix, if $1_n^\top P = 1_n$ and $1_n P = 1_n^\top$ hold simultaneously. Particularly, I_n is a permutation matrix.

Determinant of a square matrix $A \in \mathcal{M}(n)$ is the number

$$\det(A) = \sum_1^n \text{sign}(P) a_{1j_1} a_{2j_2} \ldots a_{nj_n} ,$$

where the sequence of integers $\{j_1, j_2, \ldots, j_n\}$ is a *permutation* of the natural numbers $\{1, 2, \ldots, n\}$, i.e. there exists a permutation matrix P, such that $(j_1, j_2, \ldots, j_n) = P(1, 2, \ldots, n)$. The function $\text{sign}(P)$ takes the value 1, if even number of rearrangements is needed to reconstruct $\{1, 2, \ldots, n\}$ from $\{j_1, j_2, \ldots, j_n\}$. Otherwise, if the number of rearrangements is odd, then $\text{sign}(P) = -1$. The total number of permutations of n numbers is $n! = 1 \times 2 \times \ldots \times n$, called n-factorial. Thus, the calculation of $\det(A)$ requires summation of $n!$ products, each of which contains a single element from every row and column of A. For example, for $A \in \mathcal{M}(2)$, we have

$$\det(A) = a_{11} a_{22} - a_{12} a_{21} ,$$

and for $A \in \mathcal{M}(3)$, the determinant is

$$\det(A) = a_{11} a_{22} a_{33} + a_{12} a_{23} a_{31} + a_{13} a_{21} a_{32} - a_{13} a_{22} a_{31} - a_{11} a_{23} a_{32} - a_{12} a_{21} a_{33} .$$

For larger n, the determinant of $A \in \mathcal{M}(n)$ is given by the following nice recurrent *Laplace's formula*:

$$\det(A) = a_{11} \mu_{11} - a_{12} \mu_{12} + \ldots + (-1)^{n-1} a_{1n} \mu_{1n} ,$$

where μ_{ij} is the (i, j)−minor of A, i.e. the determinant of the $(n-1) \times (n-1)$ submatrix of A obtained after deleting its ith row and jth column. Unfortunately, the formula becomes impractical even for very moderate n.

The matrix A is called *singular*, when $\det(A) = 0$. If $\det(A) \neq 0$, then A is called non-singular or *invertible*, because there exists a matrix $A^{-1} \in \mathcal{M}(n)$, such that $AA^{-1} = A^{-1}A = I_n$. In this case, we say that A^{-1} is the *inverse* of A. If $\det(A) \neq 0$, then $\det(A^{-1}) = [\det(A)]^{-1}$.

One can check directly that if $A, B \in \mathcal{M}(n)$ are non-singular, as well as AB and BA, then

- $\det(\alpha A) = \alpha^n \det(A)$,

- $\det(A^\top) = \det(A)$,

- $\det(AB) = \det(A)\det(B)$,

- $\det(AB) = \det(BA)$.

- $(A^{-1})^{-1} = A$,

- $(AB)^{-1} = B^{-1}A^{-1}$.

In many occasions it is convenient to consider $A \in \mathcal{M}(m, n)$ as a *block-matrix*, i.e. partitioned into matrices of smaller sizes called blocks. The blocks of these matrices are treated as their elements and the standard matrix operations apply. For example, for a 2×2 block-matrix $A \in \mathcal{M}(n)$ with blocks $A_{ii} \in \mathcal{M}(n_i)$ and $n_1 + n_2 = n$, we have

$$A = \begin{bmatrix} A_{11} & A_{12} \\ A_{21} & A_{22} \end{bmatrix} \text{ and } A^\top = \begin{bmatrix} A_{11}^\top & A_{21}^\top \\ A_{12}^\top & A_{22}^\top \end{bmatrix} .$$

If two block matrices $A, B \in \mathcal{M}(m, n)$ have the same partitions, then sum and product are calculated using the standard matrix addition and multiplication considering the blocks as elements of A and B as follows:

$$A \pm B = \begin{bmatrix} A_{11} \pm B_{11} & A_{12} \pm B_{12} \\ A_{21} \pm B_{21} & A_{22} \pm B_{22} \end{bmatrix} ,$$

and

$$AB^\top = \begin{bmatrix} A_{11}B_{11}^\top + A_{12}B_{12}^\top & A_{11}B_{21}^\top + A_{12}B_{22}^\top \\ A_{21}B_{11}^\top + A_{22}B_{12}^\top & A_{21}B_{21}^\top + A_{22}B_{22}^\top \end{bmatrix} .$$

Manipulations with block matrices are widely used in matrix computing. For example, the determinant of a 2×2 block-matrix $A \in \mathcal{M}(n)$ is expressed as

$$\det(A) = \det(A_{11})\det(A_{22} - A_{21}A_{11}^{-1}A_{12}) ,$$

requiring determinants of smaller size matrices, provided that A_{11}^{-1} exists. It can be proven by calculating the determinants of both sides of the following identity, known as the *Schur factorization* of A:

$$\begin{bmatrix} A_{11} & A_{12} \\ A_{21} & A_{22} \end{bmatrix} = \begin{bmatrix} I_{n_1} & O \\ A_{21}A_{11}^{-1} & I_{n_2} \end{bmatrix} \begin{bmatrix} A_{11} & O \\ O & A_{22} - A_{21}A_{11}^{-1}A_{12} \end{bmatrix} \begin{bmatrix} I_{n_1} & A_{11}^{-1}A_{12} \\ O & I_{n_2} \end{bmatrix} ,$$

where $A_{22} - A_{21}A_{11}^{-1}A_{12}$ is called the *Schur complement* of A_{11} in A, and O denote zero matrices of suitable dimensions. This identity also implies that $\begin{bmatrix} A_{11} & A_{12} \\ A_{21} & A_{22} \end{bmatrix}$ is invertible, if both A_{11} and $A_{22} - A_{21}A_{11}^{-1}A_{12}$ are invertible.

Similarly, for $A \in \mathcal{M}(m, n)$ and $B \in \mathcal{M}(n, m)$ one can establish the *Weinstein–Aronszajn* identity:

$$\det(I_m + AB) = \det(I_n + BA) ,$$

by making use of the following block matrices:

$$\begin{bmatrix} I_m & -A \\ B & I_n \end{bmatrix} \text{ and } \begin{bmatrix} I_m & A \\ O_{n \times m} & I_n \end{bmatrix} .$$

For short, a *block diagonal* matrix $A = \begin{bmatrix} A_{11} & O_{n_1 \times n_2} \\ O_{n_2 \times n_1} & A_{22} \end{bmatrix} \in \mathcal{M}(n_1 + n_2)$ is denoted by $\text{diag}_B(A_{11}, A_{22})$.

The sum of the elements on the main diagonal of $A \in \mathcal{M}(n)$ is called the *trace* of A, i.e. $\text{trace}(A) = \sum_1^n a_{ii}$. The following properties hold:

- $\text{trace}(\alpha A \pm \beta B) = \alpha\text{trace}(A) \pm \beta\text{trace}(B) , \ \alpha, \beta \in \mathbb{R} ,$

- $\text{trace}(A^\top) = \text{trace}(A) ,$

- $\text{trace}(AB) = \text{trace}(BA) ,$

- $\text{trace}(AD) = \text{trace}[\text{diag}(A)D]$, if D is a diagonal matrix ,

- $\text{trace}[A^\top(B \odot C)] = \text{trace}[(A^\top \odot B^\top)C].$

Euclidean *scalar/inner product* of two vectors $a, b \in \mathbb{R}^n$ is defined as $\langle a, b \rangle = \sum_1^n a_i b_i$. It is clear that $\langle a, b \rangle = \langle b, a \rangle$ and $\langle \alpha a, b \rangle = \alpha \langle a, b \rangle$, for some $\alpha \in \mathbb{R}$. The value of $+\sqrt{\langle a, a \rangle}$ is called (Euclidean) length of $a \in \mathbb{R}^n$. Similarly, a scalar/inner product of two matrices $A, B \in \mathcal{M}(n, p)$ can be defined as $\langle A, B \rangle = \text{trace}(A^\top B)$. Further details can be found in Section 2.4.

The vectors $a, b \in \mathbb{R}^n$ are called *orthogonal* when $\langle a, b \rangle = 0$. A matrix $Q \in \mathcal{M}(n)$ is called *orthogonal* if its rows and columns have unit lengths and are mutually orthogonal. The subset of all orthogonal matrices in $\mathcal{M}(n)$ is denoted by $\mathcal{O}(n)$. For $Q \in \mathcal{O}(n)$, we have $Q^\top = Q^{-1}$ and $\det(Q) = \pm 1$. For example, all 2×2 orthogonal matrices have the form:

$$
Q_+ = \begin{bmatrix} \cos\alpha & -\sin\alpha \\ \sin\alpha & \cos\alpha \end{bmatrix} \text{ or } Q_- = \begin{bmatrix} \cos\alpha & \sin\alpha \\ \sin\alpha & -\cos\alpha \end{bmatrix} ,
$$

where Q_+ represents anticlockwise rotation of the coordinate system of $\mathbb{R}^2$ with angle α, while Q_- represents a reflection about a line through the origin making an angle $\alpha/2$ with the axis $e_1 = (1,0)$ (Artin, 2011, 5.1.17). Particularly, the permutation matrices are orthogonal.

A $n \times 1$ (non-zero) vector v is called *eigenvector* or *proper vector* of $A \in \mathcal{M}(n)$, when $Av = \lambda v$ for some scalar λ , i.e. it is proportional to its product with A. Respectively, λ is called *eigenvalue* of A. The expression $Av = \lambda v$ is in fact a system of n linear equations for the unknown n elements of v. According to the *Cramer's rule*, $Av = \lambda v$ has a solution if $\det(A - \lambda I_n) = 0$, which is an equation in λ for the eigenvalues of A, known as *characteristic polynomial* of A. By direct calculations, one can check that it has the form:

$$
\det(A - \lambda I_n) = \lambda^n - \text{trace}(A)\lambda^{n-1} + \ldots + (-1)^n \det(A) = 0 , \qquad (2.2)
$$

and can have at most n roots, simple and/or multiple. Thus, every $n \times n$ matrix A has n eigenvalues (possibly complex) including their multiplicity. The set of all eigenvalues of A, i.e. solutions of (2.2), is called its *spectrum*. Let v_i be an eigenvector corresponding to the ith eigenvalue λ_i and $V = [v_1, \ldots, v_n]$ be the $n \times n$ matrix containing them all. If V is invertible, then A is said to have a *complete set of eigenvectors* and we have $V^{-1}AV = \text{diag}(\lambda_1, \ldots, \lambda_n) = \Lambda$. In this case A is called *diagonalizable* or that A is *similar* to a diagonal matrix (Λ). Note that there is *no* guarantee that V can be chosen orthogonal. In general, two $n \times n$ matrices A and B are called similar, if there exists a non-singular matrix T, such that $T^{-1}AT = B$.

Let k be the number of distinct roots of the characteristic equation (2.2). Then, it can be written as

$$
\det(A - \lambda I_n) = (\lambda - \lambda_1)^{n_1} \ldots (\lambda - \lambda_k)^{n_k} = 0 ,
$$

where n_i denotes the *algebraic* multiplicity of the ith root such that $n = n_1 + \ldots + n_k$. If $n_i = 1$ for some i, then the ith eigenvalue is called *simple*. If $n_i > 1$ for some i, let $V_i = [v_{i_1}, \ldots, v_{i_{m_i}}] \in \mathbb{R}^{n \times m_i}$ collect the eigenvectors associated with λ_i. The maximal m_i for which $V_i^\top V_i$ is invertible is called the *geometric multiplicity* of λ_i. In general, $m_i \leq n_i$ for $i = 1, \ldots, k$. The matrix $A \in \mathcal{M}(n)$ is diagonalizable if and only if all of its eigenvalues have the same algebraic and geometric multiplicities or are simple.

Now, for the ith distinct eigenvalue of A consider the following $n_i \times n_i$ matrix:

$$J_i = J(\lambda_i) = \begin{bmatrix} \lambda_i & 1 & & & \\ & \lambda_i & 1 & & \\ & & \ddots & \ddots & \\ & & & \lambda_i & 1 \\ & & & & \lambda_i \end{bmatrix} = \lambda_i I_{n_i} + N_{n_i} , \text{ for } i = 1, \ldots, k,$$

where the empty spaces are filled with zeros. It is known as the *Jordan block* of λ_i. The $n_i \times n_i$ matrix N_{n_i} is *nilpotent* of order n_i (i.e. $N_{n_i}^{n_i} = O_{n_i}$ and $N_{n_i}^l \neq O_{n_i}$ for $1 \leq l < n_i$). Then, there exists a non-singular T such that A can be written as $A = T\text{diag}_B(J_1, ..., J_k)T^{-1}$, known as its *Jordan canonical form/decomposition*. If A is real and all $\lambda_i \in \mathbb{R}$, then T can be chosen real too (Horn and Johnson, 2013, Th 3.1.11).

Every *symmetric* $A \in \mathcal{S}(n)$ has n *real* eigenvalues including their multiplicity. Moreover, there exists an orthogonal matrix $Q = [q_1, \ldots, q_n] \in \mathcal{O}(n)$, such that $Q^\top A Q = \text{diag}(\lambda_1, \ldots, \lambda_n) = \Lambda$, where λ_i is the ith eigenvalue of A, and q_i is its corresponding eigenvector. If n_i is the multiplicity of λ_i, then there should exist n_i eigenvectors $Q_i = [q_1, \ldots, q_{n_i}]$, such that $\det(Q_i^\top Q_i) \neq 0$. Any other eigenvector q associated with λ_i is their *linear combination*, i.e. it can be written as $q = \sum_1^{n_i} \xi_j q_j$ for some $\xi_j \neq 0$.

The expression $A = Q\Lambda Q^\top = \sum_{i=1}^n \lambda_i q_i q_i^\top$ is called *spectral or eigenvalue decomposition (EVD)* of A. It follows that:

- $\text{trace}(A) = \lambda_1 + \ldots + \lambda_n$,

- $\det(A) = \lambda_1 \ldots \lambda_n$.

In contrast, the eigenvalues of every *skew-symmetric* $K \in \mathcal{K}(n)$, for even n, are only *pure imaginary* numbers. For example, the characteristic polynomial of $\begin{bmatrix} 0 & 1 \\ -1 & 0 \end{bmatrix} \in \mathcal{K}(2)$ is $\lambda^2 + 1 = 0$ and the roots/eigenvalues are $\pm i \in \mathbb{C}$.

Nevertheless, there is a way to express the spectral features of a wide class of matrices in terms of real numbers, which is convenient for the purposes of MDA. Consider the set of $A \in \mathcal{M}(n)$, for which $A^\top A = AA^\top$. Such matrices are called *normal*. For example, all symmetric, skew-symmetric or orthogonal matrices are normal. The following important result holds: $A \in \mathcal{M}(n)$ is normal, if and only if there exists $Q \in \mathcal{O}(n)$, such that $Q^\top A Q = \text{Diag}_B(A_1, \ldots, A_n)$, where A_i is either a real number or $A_i = \begin{bmatrix} \alpha_i & \beta_i \\ -\beta_i & \alpha_i \end{bmatrix}$, with $\alpha_i, \beta_i \in \mathbb{R}$ (Horn and Johnson, 2013, Th. 2.5.8).

The matrix $A \in \mathcal{S}(n)$ is called *positive definite* (PD) or *positive semi-definite* (PSD) if $\lambda_i > 0$, or, respectively, $\lambda_i \geq 0$, and it is denoted by $A > (\geq) O_n$, where O_n is an $n \times n$ matrix of zeros to match the notation for a general $n \times p$ matrix of zeros $O_{n \times p}$. Similarly, for short E_n denotes an $n \times n$ matrix of ones, while the general $n \times p$ matrix of ones is denoted by $E_{n \times p}(= 1_n 1_p^\top)$.

The set of all PD matrices in $\mathcal{S}(n)$ is denoted by $\mathcal{S}_+(n)$. If A is PD, then A is invertible, as well as $a^\top A a > 0$ for all non-zero $a \in \mathbb{R}^n$. A *power* $\alpha(> 0)$ of a PSD matrix A is defined as $A^\alpha = Q \Lambda^\alpha Q^\top$. Particularly, for $\alpha = 1/2$ we have a *square root* of a PSD matrix A as $A^{1/2} = Q \Lambda^{1/2} Q^\top$. Similarly, for $\alpha = -1$ we have an expression for the inverse of a PD $A \in \mathcal{S}_+(n)$, $A^{-1} = Q \Lambda^{-1} Q^\top$, which implies that $\text{trace}(A^{-1}) = \frac{1}{\lambda_1} + \ldots + \frac{1}{\lambda_n}$.

The number of the *positive* eigenvalues of a PSD $A \in \mathcal{S}(n)$ is called *rank* of A. Obviously, any PD matrix $A \in \mathcal{S}_+(n)$ has rank exactly n and it is invertible.

Let $n > p$. Then, for any $A \in \mathcal{M}(n, p)$, the product $A^\top A \in \mathcal{S}(p)$ is PSD, and the rank of A can be defined as the rank of $A^\top A$. On general, the rank r of $A \in \mathcal{M}(n, p)$ is defined as the size of its largest invertible (square) submatrix and $r \leq \min\{n, p\}$. As $n > p$, then $A \in \mathcal{M}(n, p)$ can have rank at most p. If A does have rank p, it is called *full (column)rank* matrix and $A^\top A \in \mathcal{S}(p)_+$ (is PD). If its rank is $r < p$, then A is called *rank deficient*.

The vectors $u \in \mathbb{R}^n$ and $v \in \mathbb{R}^p$ are called (left and right) *singular vectors* of $A \in \mathcal{M}(n, p)$, when $Av = \sigma u$ and $A^\top u = \sigma v$, for some $\sigma \geq 0$ called *singular value* of A. Apparently, $A^\top A v = \sigma A^\top u = \sigma^2 v$ and $A A^\top u = \sigma A v = \sigma^2 u$. This implies that u is an eigenvector of $A A^\top \in \mathcal{M}(n)$ and that v is an eigenvector of $A^\top A \in \mathcal{M}(p)$. Moreover, they both have a common eigenvalue σ^2. Suppose, the eigenvalues of $A^\top A$ are $\lambda_1, \ldots, \lambda_p$. Then, the eigenvalues of $A A^\top$ are $\lambda_1, \ldots, \lambda_p, \underbrace{0, \ldots, 0}_{n-p}$, and $\lambda_i = \sigma_i^2$ for $i = 1, \ldots, p$.

In this term, the EVD of $A^\top A$ can be expressed as $A^\top A = V \Lambda V^\top$, where $V \in \mathcal{O}(p)$ and $\Lambda = \text{diag}(\lambda_1, \ldots, \lambda_p)$. Now, as the last $n - p$ eigenvalues of

$AA^\top$ are zeros, the corresponding eigenvectors can be omitted. Thus, the EVD of $AA^\top$ can be written as $AA^\top = U_p \Lambda U_p^\top$, where U_p contains only the first p columns (eigenvectors) of $U \in \mathcal{O}(n)$ corresponding to the non-zero eigenvalues. The expression $A = U_p \Sigma V^\top$ is known as the (economic) *singular value decomposition (SVD)* of A, where $\Sigma = \Lambda^{1/2}$.

In general, the rank $r(\leq p \leq n)$ of $A \in \mathcal{M}(n, p)$ is equal to the number of its non-zero singular values. Then, the economic SVD of A takes the form $A = U_r \Sigma_r V_r^\top$, where U_r and V_r contain, respectively, the first r columns of $U \in \mathcal{O}(n)$ and $V \in \mathcal{O}(p)$, and $\Sigma_r \in \mathcal{D}(r)$. The subspace spanned by $\{u_1, \ldots, u_r\}$ is called the *range* of A. The subspace spanned by $\{v_{r+1}, \ldots, v_p\}$ is called the *null space* of A.

2.2   Vector spaces, bases, and linear maps

The set $\mathcal{E}$ is called *real* vector (or linear) space, if

1. $\mathcal{E}$ is a commutative (Abelian) group with respect to the (internal) group addition, i.e. if $a, b \in \mathcal{E}$, then $a + b \in \mathcal{E}$, and it has the following properties:

    (a) $(a + b) + c = a + (b + c)$—associative property;

    (b) $a + 0 = 0 + a = a$—neutral element;

    (c) $a + b = b + a$—commutative property;

    (d) $a + (-a) = 0$—symmetric element;

2. It is defined (external) multiplication of elements from $\mathcal{E}$ with real numbers, i.e. if $a \in \mathcal{E}$, $\alpha \in \mathbb{R}$, then $\alpha a \in \mathcal{E}$ and it has the following properties:

    (a) $\alpha(\beta a) = (\alpha \beta) a$—associative property;

    (b) $\alpha(a + b) = \alpha a + \alpha b$, $(\alpha + \beta) a = \alpha a + \beta a$—distributive property;

    (c) $1a = a$—unit element of $\mathbb{R}$.

Usually, $0_\mathcal{E}$ and $1_\mathcal{E}$ are called *neutral elements* of $\mathcal{E}$ with respect to the addition and multiplication, respectively. In general, those two operations can be any "internal" and "external" operations defined in $\mathcal{E}$ and fulfilling the above list of conditions.

For $a_1, \ldots, a_n \in \mathcal{E}$ and $\alpha_1, \ldots, \alpha_n \in \mathbb{R}$, the new vector

$$\alpha_1 a_1 + \ldots + \alpha_n a_n \in \mathcal{E}$$

is called linear combination of $a_1, \ldots, a_n$. We say that the set $\mathcal{A}$ formed by all possible such linear combinations is spanned/generated by $a_1, \ldots, a_n$. Sometimes, $\mathcal{A}$ is called the *span* of $a_1, \ldots, a_n$, or *linear manifold*. The span $\mathcal{A}$ forms a *vector subspace* in $\mathcal{E}$, i.e. $\mathcal{A} \subset \mathcal{E}$ and $\alpha a + \beta b \in \mathcal{A}$, for $a, b \in \mathcal{A}$ and $\alpha, \beta \in \mathbb{R}$. Note that always $0_{\mathcal{E}} \in \mathcal{A}$.

The vectors $a_1, \ldots, a_n \in \mathcal{E}$ are called *independent*, if $\alpha_1 a_1 + \ldots + \alpha_n a_n = 0_{\mathcal{E}}$, implies $\alpha_1 = \ldots = \alpha_n = 0$. Otherwise, they are called dependent. In this case, not all α_i in $\alpha_1 a_1 + \ldots + \alpha_n a_n$ are zero. If, say, $\alpha_1 \neq 0$, then

$$a_1 = -\frac{\alpha_2}{\alpha_1} a_2 - \ldots - \frac{\alpha_n}{\alpha_1} a_n \ ,$$

i.e. a_1 is a linear combination of the rest of the vectors. Any subset of vectors from $a_1, \ldots, a_n$ are also independent. However, if there are no $n+1$ vectors in $\mathcal{E}$ which are also independent, then we say that the *dimension* of $\mathcal{E}$ is $n(< \infty)$, which is denoted by $\dim(\mathcal{E}) = n$. In other words, $\mathcal{E}$ is *finite-dimensional* space. If, for any n we can find n independent vectors in $\mathcal{E}$, then $\mathcal{E}$ is called *infinite-dimensional* vector space. The set of independent $a_1, \ldots, a_n$ is called *basis* of $\mathcal{E}$. Any other basis of $\mathcal{E}$ is also composed of n vectors. Note that 0_n cannot be part of any set of independent vectors.

For example, $\mathbb{R}^n$ is considered a vector space spanned by $e_1, \ldots, e_n$, where $e_i \in \mathbb{R}^n$ has unit entry at the ith position, and zeros elsewhere. This basis is called standard or canonical. Any $a \in \mathbb{R}^n$ can be represented uniquely as:

$$a = \alpha_1 e_1 + \ldots + \alpha_n e_n \ ,$$

and α_i are called coordinates of a. One notes that $e_i^\top e_i = 1$ for $i = 1, \ldots, n$ and $e_i^\top e_j = 0$ for $i < j$. Such a basis is called *orthonormal*. For short, we write $e_i^\top e_j = \delta_{ij}$, where δ_{ij} is known as the *Kronecker delta/symbol*. In MDA, we mainly work with finite-dimensional vector spaces, and particularly with different exemplars of $\mathbb{R}^n$.

If $m(< n)$ vectors $a_1, \ldots, a_m \in \mathcal{E}$ are independent in $\mathcal{E}$, there can be found $n - m$ supplementary vectors $a_{m+1}, \ldots, a_n \in \mathcal{E}$, such that $a_1, \ldots, a_n \in \mathcal{E}$ is a basis of $\mathcal{E}$. The span of $a_1, \ldots, a_m$ forms a m-dimensional subspace $\mathcal{F}$ of $\mathcal{E}$. The subspace $\mathcal{G}$ generated by $a_{m+1}, \ldots, a_n$ is called supplementary subspace to $\mathcal{F}$ in $\mathcal{E}$, and is also denoted by $\mathcal{F}^\perp$. Then, every $a \in \mathcal{E}$ can be written as:

$$a = \alpha_1 a_1 + \ldots + \alpha_m a_m + \alpha_{m+1} a_{m+1} + \ldots + \alpha_n a_n \ , \ \alpha_i \in \mathbb{R} \ ,$$

which means $a = f + g$ for $f \in \mathcal{F}$ and $g \in \mathcal{G}$. Note that $\mathcal{F} \cap \mathcal{G} = 0_n$, i.e. $\mathcal{F}$ and $\mathcal{G}$ have no common vectors except 0_n. In this case, we say that $\mathcal{E}$ is a *direct sum* of $\mathcal{F}$ and $\mathcal{G}$ and write $\mathcal{E} = \mathcal{F} \oplus \mathcal{G}$. It is clear that $\dim(\mathcal{E}) = \dim(\mathcal{F}) + \dim(\mathcal{G})$. Moreover, $f \in \mathcal{F}$ and $g \in \mathcal{G}$ have the form:

$$f = (\alpha_1, \ldots, \alpha_m, 0, \ldots, 0) \ , \ g = (0, \ldots, 0, \alpha_{m+1}, \ldots, \alpha_n) \ . \tag{2.3}$$

Let $\mathcal{E}$ and $\mathcal{F}$ be two real vector spaces. The map $A : \mathcal{E} \to \mathcal{F}$ is called linear if, for any $a, b \in \mathcal{E}$

$$A(a + b) = A(a) + A(b) \ ,$$

$$A(\alpha a) = \alpha A(a) \ ,$$

where $\alpha \in \mathbb{R}$. The set of all linear mappings from $\mathcal{E}$ to $\mathcal{F}$ form a real vector space which is denoted as $\mathcal{L}(\mathcal{E}, \mathcal{F})$ and if $\mathcal{E} = \mathcal{F}$, we write $\mathcal{L}(\mathcal{E})$, and A is called *endomorphism* or simply *linear operator*. If $\mathcal{F} = \mathbb{R}$, then A is called *linear form* or *linear functional* on $\mathcal{E}$. A is called *surjective*, if for every $b \in \mathcal{F}$ there exists is at least one element $a \in \mathcal{E}$, such that $A(a) = b$. It is called *injective*, if $A(a) \neq A(b)$, for any $a, b \in \mathcal{E}, a \neq b$. When A is both injective and surjective, it is called *bijective*. As A is linear it is also called *isomorphism*, because it preserves the operations defined in $\mathcal{E}$ and $\mathcal{F}$ (summation and multiplication by a scalar).

The linear mappings A from $\mathcal{L}(\mathcal{E}, \mathcal{F})$ have the following important properties:

- $A(\mathcal{E})$ is a vector subspace of $\mathcal{F}$, called *image* or *range* of A; if $\dim(\mathcal{E}) = n$, then $\dim(A(\mathcal{E})) = r \leq n$, and r is called *rank* of A.

- The zero element of $\mathcal{E}$, say $0_\mathcal{E}$, is mapped into the zero element of $\mathcal{F}$, say $0_\mathcal{F}$, i.e. $A(0_\mathcal{E}) = 0_\mathcal{F}$.

- $A^{-1}(0_\mathcal{F})$ is a vector subspace in $\mathcal{E}$ and is called *kernel* or *null space* of A and denoted by $\ker(A)$ or $\mathcal{N}(A)$.

- The necessary and sufficient condition for A to be injection from $\mathcal{E}$ to $\mathcal{F}$ is that its kernel contains a single element, $\{0_\mathcal{E}\}$, i.e. $A^{-1}(0_\mathcal{F}) = 0_\mathcal{E}$.

- A is surjective, if $A(\mathcal{E}) = \mathcal{F}$.

- Any injective A is bijective from $\mathcal{E}$ to $A(\mathcal{E})$.

- $\dim(A^{-1}(0_\mathcal{F})) = \dim(\mathcal{E}) - \dim(A(\mathcal{E}))$.

- If $a_1, \ldots, a_n$ are linearly independent in $\mathcal{E}$, their images $A(a_1), \ldots, A(a_n)$ may *not* be independent in $\mathcal{F}$; however, if the images $A(a_1), \ldots, A(a_n)$ of some $a_1, \ldots, a_n$ are independent in $\mathcal{F}$, then $a_1, \ldots, a_n$ are linearly independent in $\mathcal{E}$.

- If $\dim(\mathcal{E}) = \dim(\mathcal{F}) = n$, and A maps the basis of $\mathcal{E}$ into the basis of $\mathcal{F}$, then A is isomorphism; in this case, $A^{-1}(0_{\mathcal{F}}) = 0_{\mathcal{E}}$.

- Any two real vector spaces with equal dimensions n are isomorphic, and are isomorphic to $\mathbb{R}^n$.

- If $A \in \mathcal{L}(\mathcal{E})$ and $\dim(\mathcal{E}) = n$, the linear operator A is isomorphism, when its rank is n.

Let $\mathcal{E}, \mathcal{F}$ and $\mathcal{G}$ be three real vector spaces, and $A_1 \in \mathcal{L}(\mathcal{E}, \mathcal{F})$ and $A_2 \in \mathcal{L}(\mathcal{F}, \mathcal{G})$. Then, for any $a \in \mathcal{E}$, their product is defined as

$$(A_2 A_1)(a) = A_2(A_1(a)) \, ,$$

and $A_2 A_1 \in \mathcal{L}(\mathcal{E}, \mathcal{G})$. In general, $A_2 A_1$ and $A_1 A_2$ are different.

Let $\mathcal{E}$ and $\mathcal{F}$ be two real vector spaces with dimensions n and p, respectively. Let $e_1, \ldots, e_n$ and $f_1, \ldots, f_p$ be their bases. The vectors $A(e_i) \in \mathcal{F}$ can be represented as linear combinations of the basis in $\mathcal{F}$ as $A(e_i) = \sum_{j=1}^{p} a_{ij} f_j$. The coefficients of those linear combinations are collected in a table:

$$A = \begin{bmatrix} a_{11} & a_{12} & \cdots & a_{1p} \\ a_{21} & a_{22} & \cdots & a_{2p} \\ \vdots & \vdots & \ddots & \vdots \\ a_{n1} & a_{n2} & \cdots & a_{np} \end{bmatrix} \, ,$$

called the matrix of A. To shorten the notations, the linear mapping and its matrix are usually denoted with the same capital letter. The rank of the matrix A is equal to the rank of the linear mapping A. The matrix of the linear operator is a square matrix. The matrix of the identity operator is the identity matrix. If $\mathcal{E} = \mathcal{F} \oplus \mathcal{G}$, then the matrix of every linear operator $A \in \mathcal{L}(\mathcal{E})$ is a block-matrix with respect to the basis defining the direct sum and has the following form:

$$A = \begin{bmatrix} A_1 & O_{m \times (n-m)} \\ O_{(n-m) \times m} & A_2 \end{bmatrix} \, ,$$

where $A_1 \in \mathcal{M}(m)$ and $A_2 \in \mathcal{M}(n - m)$.

Further details on linear algebra and matrix analysis can be found in a number of excellent books (Bellman, 1960; Bhatia, 1997; Gantmacher, 1966; Gel'fand, 1961; Halmosh, 1987; Horn and Johnson, 2013; Lang, 1987; Lax, 2007; Mirsky, 1955). For additional interesting aspects, see (Artin, 2011; Bhatia, 2007; Dieudonné, 1969b; Halmosh, 1995; Kostrikin and Manin, 1997).

2.3   Metric, normed and inner product spaces

The set $\mathcal{E}$ is called metric space if with any two of its elements x and y there is associated a real number $d(x, y)$, called the *distance* between x and y, with the following properties:

1. $d(x, y) > 0$ if $x \neq y$ and $d(x, x) = 0$– non-negativity;

2. $d(x, y) = d(y, x)$– symmetry;

3. $d(x, z) \leq d(y, x) + d(y, z)$, for any $z \in \mathcal{E}$– *triangle inequality*.

Any function d with these properties is called a distance function or a *metric*. The elements of the metric space $\mathcal{E}$ are called points. Thus, the metric d is a function defined from the *Cartesian product* $\mathcal{E} \times \mathcal{E}$ to $\mathbb{R}_+$, where $\mathbb{R}_+ = [0, +\infty)$. In general, the Cartesian product $\mathcal{E} \times \mathcal{F}$ of two sets $\mathcal{E}$ and $\mathcal{F}$ is defined as the set of all (ordered) pairs (x, y), such that $x \in \mathcal{E}$, $y \in \mathcal{F}$ and $(x, y) \neq (y, x)$.

The geometrical meaning of the triangle inequality is that each side in a triangle is less than the sum of the other two. The triangle inequality also implies that

$$d(x_1, x_n) \leq d(x_1, x_2) + d(x_2, x_3) + \ldots + d(x_{n-1}, x_n) \,,$$

for any $x_1, x_2, \ldots, x_n \in \mathcal{E}$.

Examples:

- The real line $\mathbb{R}$ is a metric space with $d(x, y) = |x - y|$.

- The real Euclidean space $\mathbb{R}^n$ is Cartesian product of n real lines, i.e. $\mathbb{R}^n = \underbrace{\mathbb{R} \times \mathbb{R} \ldots \times \mathbb{R}}_{n}$. Then, each point $x \in \mathbb{R}^n$ is given as an ordered list of n real numbers, i.e. $x = (x_1, x_2, \ldots x_n)$. The Euclidean space $\mathbb{R}^n$ is a metric space with the standard *Euclidean distance* $d(x, y) = \sqrt{\sum_{i=1}^{n} |x_i - y_i|^2}$.

- Discrete metric can be defined in any set $\mathcal{E}$ by setting $d(x, y) = 1$, if $x \neq y$, and $d(x, x) = 0$.

The set of all $x \in \mathcal{E}$ for which $d(a, x) = r < \infty$ is called a sphere with centre at $a \in \mathcal{E}$ and radius r. The set of all $x \in \mathcal{E}$ for which $d(a, x) < r$ or $d(a, x) \leq r$ is called an open, respectively, closed ball with centre at $a \in \mathcal{E}$ and (finite) radius r. It is denoted by $\mathcal{B}(a, r)$, respectively, $\overline{\mathcal{B}(a, r)}$, or simply by $\mathcal{B}(a)/\overline{\mathcal{B}(a)}$ if the radius r can be omitted. If $r = 0$, then the open ball is empty, and the closed ball collapses to its centre. So, it is always assumed that $r > 0$. A subset $\mathcal{U} \subset \mathcal{E}$ is called *open* if for every $x \in \mathcal{U}$ there exists an open ball $\mathcal{B}(x)$ with centre at x such that $\mathcal{B}(x) \subset \mathcal{U}$.

The open subsets in $\mathcal{E}$ have the following properties:

- $\mathcal{E}$ and $\emptyset$ are open;

- the intersection of a finite number of open sets is open;

- the union of a finite or infinite number of open sets is open;

- for any $x, y \in \mathcal{E}$ there exist two open sets $\mathcal{U}, \mathcal{V} \subset \mathcal{E}$ such that $x \in \mathcal{U}, y \in \mathcal{V}$ and $\mathcal{U} \cap \mathcal{V} = \emptyset$ (Hausdorff separation axiom).

The necessary and sufficient condition for $\mathcal{U} \subset \mathcal{E}$ to be open is that it is an union of open balls. A *neighbourhood* of $x \in \mathcal{E}$ is any set $\mathcal{U} \subset \mathcal{E}$, for which there exists at least one open ball $\mathcal{B}(x) \subset \mathcal{U}$.

A subset $\mathcal{U} \subset \mathcal{E}$ is called *closed* if its *complement* $\complement \, \mathcal{U}$ is an open set in $\mathcal{E}$, i.e. the set of all $x \in \mathcal{E}$ and $x \notin \mathcal{U}$ is open. Thus, the closed sets in $\mathcal{E}$ have, respectively, the following properties:

- $\mathcal{E}$ and $\emptyset$ are closed;

- the union of a finite number of closed sets is closed;

- the intersection of a finite or infinite number of closed sets is closed.

Interior of $\mathcal{U}$ is the set of points $x \in \mathcal{E}$ for which there exist an open ball $\mathcal{B}(x) \subset \mathcal{U}$. The interior of $\mathcal{U}$ sometimes is denoted by $\overset{\circ}{\mathcal{U}}$. *Exterior* of $\mathcal{U}$ is the interior of its complement in $\mathcal{E}$. The set of points that do not belong to neither the interior of $\mathcal{U}$ nor to its exterior is called *boundary* of $\mathcal{U}$, sometimes denoted by $\partial\mathcal{U}$. The union of the interior of $\mathcal{U}$ and its boundary is called the *closure* of $\mathcal{U}$, sometimes denoted by $\overline{\mathcal{U}}$. A subset $\mathcal{U} \subset \mathcal{E}$ is closed if it coincides with its closure. A subset $\mathcal{U} \subset \mathcal{E}$ is called *bounded* if it is contained in a ball with finite radius.

The infinite sequence of points $x_1, \ldots, x_n, \ldots \in \mathcal{E}$ is called *convergent* to $x_0 \in \mathcal{E}$ if for any $\epsilon > 0$ there exists an index n_0, such that $d(x_n, x_0) \leq \epsilon$ for any $n \geq n_0$. The same can be reworded as follows: for any open $\mathcal{U}$ containing x_0, there exists an index n_0, such that $x_n \in \mathcal{U}$ for any $n \geq n_0$. This means that the convergence of a sequence in $\mathcal{E}$ does not really need the existence of the metric d. The set $\mathcal{E}$ in which there exists a collection of open subsets fulfilling the above properties is called *topological space*. The collection of those open subsets is called *topology*. Clearly every metric space is topological, but not all topological spaces are metric.

If $\mathcal{U} \subset \mathcal{E}$ is closed then for every point $x \in \mathcal{U}$ one can find a sequence of points $x_1, \ldots, x_n, \ldots \in \mathcal{U}$ converging to x. Indeed, x belongs to the closure of $\mathcal{U}$, and one can always find a non-empty ball with centre in x and radius $\frac{1}{n}$ contained in $\mathcal{U}$, i.e. $x_n \in \mathcal{B}(x, \frac{1}{n}) \subset \mathcal{U}$. This can be repeated to obtain an infinite sequence of points $x_1, \ldots, x_n, \ldots \in \mathcal{U}$ converging to x.

A *mapping* f is defined on $\mathcal{E}$ with values in $\mathcal{F}$ when it maps every $x \in \mathcal{E}$ to some element of $\mathcal{F}$, which is denoted by $f(x)$. For example, the sequence $x_1, \ldots, x_n, \ldots \in \mathcal{U}$ is a mapping from the set of *natural numbers* $\mathbb{N} = \{1, 2, \ldots\}$ to $\mathcal{U}$. Thus, $\mathcal{X} = \{x_i\}_{i=1}^{\infty}$ is also an example of a *countable/enumerable* subset in $\mathcal{U}$, i.e. which is bijective to $\mathbb{N}$. As above, one defines the continuity of a mapping $f : \mathcal{E} \to \mathcal{F}$, where $\mathcal{E}$ and $\mathcal{F}$ are metric spaces. The mapping f is *continuous* at $x_0 \in \mathcal{E}$ if for any $\epsilon > 0$ there exists $\delta > 0$, such that $d(x, x_0) \leq \delta$ implies $d(f(x), f(x_0)) \leq \epsilon$. Its topological counterpart states that f is continuous at $x_0 \in \mathcal{E}$ if for any open ball $\mathcal{B}(f(x_0)) \subset \mathcal{F}$ there exists an open ball $\mathcal{B}(x_0) \subset \mathcal{E}$ such that $f(\mathcal{B}(x_0)) \subset \mathcal{B}(f(x_0))$. The mapping f is called continuous, when it is continuous at every point of $\mathcal{E}$. Moreover, $f : \mathcal{E} \to \mathcal{F}$ is continuous if and only if $f^{-1}(\mathcal{U}) \subset \mathcal{E}$ is open/closed for any open/closed $\mathcal{U} \subset \mathcal{F}$. If f is also bijective and its inverse f^{-1} is continuous, then f is called *homeomorphism*. Two metric spaces $\mathcal{E}$ and $\mathcal{F}$ are homeomorphic if there exists at least one homeomorphism between them. A necessary and sufficient condition for a bijective and continuous $f : \mathcal{E} \to \mathcal{F}$ to be a homeomorphism is that $f(U)$ be open/closed in $\mathcal{F}$, if U is open/-closed in $\mathcal{E}$ (i.e. requires the continuity of f^{-1}). If $f : \mathcal{E} \to \mathcal{F}$ and $g : \mathcal{F} \to \mathcal{G}$ are continuous, then their composition $g \circ f : \mathcal{E} \to \mathcal{G}$ is also continuous.

Topological vector space is a vector space $\mathcal{E}$ with such a topology that the addition $(x, y) \to x + y$ is a continuous function $\mathcal{E} \times \mathcal{E} \to \mathcal{E}$, as well as the multiplication by a real number $(\alpha, x) \to \alpha x$, mapping $\mathbb{R} \times \mathcal{E} \to \mathcal{E}$.

The sequence $x_1, \ldots, x_n, \ldots$ in a metric space $\mathcal{E}$ is called *Cauchy* or *fundamental* if for any $\epsilon > 0$ there exists an index l, such that $d(x_m, x_n) \leq \epsilon$ for any $m, n \geq l$. Every convergent sequence is a Cauchy sequence. The metric space $\mathcal{E}$ is called *complete*, if any Cauchy sequence in $\mathcal{E}$ is convergent. Not all metric spaces are complete, but $\mathbb{R}^n$ is complete.

The vector space $\mathcal{E}$ is called *normed vector space* when a function $\| \ \| : \mathcal{E} \to [0, \infty)$ is defined with the following properties:

1. $\|x\| \geq 0$ if $x \neq 0$, and $\|0\| = 0$;

2. $\|\alpha x\| = |\alpha| \|x\|$, $\alpha \in \mathbb{R}$;

3. $\|x + y\| \leq \|x\| + \|y\|$.

A. N. Kolmogorov proved that a (Hausdorff) topological vector space $\mathcal{E}$ is *normable* (i.e. there exists a norm in $\mathcal{E}$ such that the open balls of the norm generate its topology) if and only if there exists a bounded convex neighbourhood of the origin of $\mathcal{E}$ (Kantorovich and Akilov, 1982, p. 120).

For example, $\mathbb{R}^n$ is a normed space with $\|x\| = \sqrt{\sum_{i=1}^{n} x_i^2}$, which is called the *Euclidean norm*. Any normed vector space $\mathcal{E}$ is a metric space with the distance $d(x, y) = \|x - y\|$. Such a (induced) metric space has two additional features: $d(x + z, y + z) = d(x, y)$ and $d(\alpha x, \alpha y) = |\alpha| d(x, y)$. Clearly, in every metric space one can define norm by setting: $\|x\| = d(x, 0)$. However, not every metric can be generated by a norm. The norm $\| \ \|$ considered as a function from $\mathcal{E}$ to $\mathbb{R}$ is continuous. A complete normed vector space is called *Banach space*. Of course, $\mathbb{R}^n$ is a Banach space.

The vector space $\mathcal{E}$ is called *Hilbert space* when a function $\langle, \rangle : \mathcal{E} \times \mathcal{E} \to \mathbb{R}$ can be defined with the following properties, for any two $x, y \in \mathcal{E}$:

1. $\langle x, x \rangle \geq 0$ if $x \neq 0$, and $\langle 0, 0 \rangle = 0$;

2. $\langle x, y \rangle = \langle y, x \rangle$;

3. $\langle \alpha x, y \rangle = \alpha \langle x, y \rangle$;

4. $\langle x, y + z \rangle = \langle x, y \rangle + \langle x, z \rangle$.

Function with such properties is called *inner (scalar) product*. The inner product is called *positive definite* if $\langle x, x \rangle > 0$ for any $x \neq 0$. The Hilbert space becomes a normed space with $\|x\| = \sqrt{\langle x, x \rangle}$. The space ℓ^2 of all (infinite) real sequences $x = \{x_1, x_2, \ldots\}$ with $\sqrt{\sum_{i=1}^{\infty} x_i^2} < \infty$ is a classical example of a Hilbert space. A finite-dimensional Hilbert space is called *Euclidean space*. In fact, historically we start with the Euclidean space and generalize the spaces in opposite direction by keeping as many of the useful features of the Euclidean space as possible.

The following *Cauchy–Schwartz inequality*,

$$|\langle x, y \rangle|^2 \leq \langle x, x \rangle \langle y, y \rangle = \|x\|^2 \|y\|^2 , \qquad (2.4)$$

is a very important tool in every inner product space. If $\langle x, y \rangle = 0$, there is nothing to prove. Then, assume that $\langle x, y \rangle \neq 0$, and consider:

$$0 \leq \langle \alpha x + \beta y, \alpha x + \beta y \rangle = \alpha^2 \langle x, x \rangle + \beta^2 \langle y, y \rangle + 2\alpha\beta \langle x, y \rangle ,$$

which is true for any $x, y \in \mathcal{E}$ and $\alpha, \beta \in \mathbb{R}$. Let us take $\alpha = \langle x, y \rangle / |\langle x, y \rangle|$, i.e. be the sign of $\langle x, y \rangle$. After substitution of this α, we end up with a quadratic inequality with respect to β, which can be true for any β, only if

$$|\langle x, y \rangle|^2 - \langle x, x \rangle \langle y, y \rangle \leq 0 .$$

The Cauchy–Schwartz inequality (2.4) immediately implies the *triangle inequality* for norms, already known from the distance properties. Indeed:

$$\begin{aligned} \|x + y\|^2 &= \langle x + y, x + y \rangle = \langle x, x \rangle + 2\langle x, y \rangle + \langle y, y \rangle \\ &\leq \|x\|^2 + 2\|x\|\|y\| + \|y\|^2 \\ &= (\|x\| + \|y\|)^2 , \end{aligned} \qquad (2.5)$$

i.e. implying

$$\|x + y\| \leq \|x\| + \|y\| . \qquad (2.6)$$

It can be further generalized to

$$\left\| \sum_{i=1}^{n} x_i \right\| \leq \sum_{i=1}^{n} \|x\| .$$

Now, let $\| \ \|$ be the norm derived from the above inner product. By direct calculations, one can check that the following *parallelogram identity*,

$$\|x + y\|^2 + \|x - y\|^2 = 2(\|x\|^2 + \|y\|^2) , \tag{2.7}$$

is valid (in any Hilbert space). However, if it is not valid for a particular norm, then we can conclude that this norm *cannot* be obtained from an inner product. The identity (2.7) implies the *parallelogram inequality*:

$$\|x + y\|^2 \leq 2(\|x\|^2 + \|y\|^2) , \tag{2.8}$$

which sometimes might be more useful than the triangle inequality (2.6).

A clear and brief reference to such topics can be found in (Rudin, 1976, Ch 1,2,4) or more advance sources, e.g. (Dieudonné, 1969a, Ch I–VI) and (Kantorovich and Akilov, 1982, Part I).

2.4 Euclidean spaces of matrices and their norms

Let $x, y \in \mathbb{R}^n$ be written in coordinate form as $x = (x_1, \ldots, x_n)$ and $y = (y_1, \ldots, y_n)$, respectively. Then $\mathbb{R}^n$ becomes an Euclidean space with the following (Euclidean) inner product of any two $x, y \in \mathbb{R}^n$:

$$\langle x, y \rangle = \sum_{i=1}^{n} x_i y_i . \tag{2.9}$$

The vectors x and y are called *orthogonal* when $\langle x, y \rangle = 0$. With $x = y$, (2.9) becomes $\langle x, x \rangle = \sum_{i=1}^{n} x_i^2$, which defines the Euclidean norm of $x \in \mathbb{R}^n$:

$$\|x\|_2 = \sqrt{\sum_{i=1}^{n} x_i^2} . \tag{2.10}$$

The matrices in $\mathcal{M}(n)$ and $\mathcal{M}(n,p)$ can be considered vectors (points) in $\mathbb{R}^{n^2}$ and $\mathbb{R}^{np}$, respectively, with the corresponding definitions of scalar product and norm. Thus, $\mathcal{M}(n)$ and $\mathcal{M}(n,p)$ possess all of the nice properties of the Euclidean space. For example, they are complete normed spaces (as any other finite-dimensional normed space). Also, every bounded and closed set $\mathcal{U} \subset \mathbb{R}^n$ is *compact*, i.e. every infinite sequence of points $x_1, \ldots, x_n, \ldots \in \mathcal{U}$ contains a subsequence converging in $\mathcal{U}$. The compact sets are very useful. Indeed, consider compact $\mathcal{U} \subset \mathbb{R}^n$ and continuous $f : \mathbb{R}^n \to \mathbb{R}^p$, then

- $f(\mathcal{U}) \subset \mathbb{R}^p$ is compact;

- for $p = 1$, $f(\mathcal{U})$ is a closed interval $[a,b] \in \mathbb{R}$, and thus it reaches its minimal and maximal values, a and b, respectively;

- f is uniformly continuous on $\mathcal{U}$.

There are a plenty of other norms that can be defined in $\mathbb{R}^n$. The most popular ones are

- ℓ_1-norm, also known as the *sum-norm, Manhattan* or *taxicab norm*:

$$\|x\|_1 = \sum_{i=1}^{n} |x_i| \; ; \qquad (2.11)$$

- $\ell_\infty-norm$, also known as the *max-norm*:

$$\|x\|_\infty = \max_{1 \le i \le n} |x_i| \; ; \qquad (2.12)$$

- ℓ_p-norm, for $1 < p < \infty$:

$$\|x\|_p = \left(\sum_{i=1}^{n} |x_i|^p \right)^{1/p} . \qquad (2.13)$$

The following *Hölder inequality* for ℓ_p-norms is very important:

$$|\langle x, y \rangle| \le \|x\|_p \|y\|_q \; , \quad \text{for } \frac{1}{p} + \frac{1}{q} = 1 \; , \qquad (2.14)$$

and gives the Cauchy–Schwartz inequality (2.4) for $p = q = 2$.

The good news is that all norms in $\mathbb{R}^n$ (and in any finite-dimensional space) are equivalent in the following sense. If $\| \ \|_\dagger$ and $\| \ \|_\ddagger$ are two norms, then there exist constants α and β, such that

$$\alpha\| \ \|_\dagger \leq \| \ \|_\ddagger \leq \beta\| \ \|_\dagger \ .$$

For example, if $x \in \mathbb{R}^n$, then

$$\|x\|_2 \leq \|x\|_1 \leq \sqrt{n}\|x\|_2 \ ,$$

$$\|x\|_\infty \leq \|x\|_2 \leq \sqrt{n}\|x\|_\infty \ ,$$

$$\|x\|_\infty \leq \|x\|_1 \leq n\|x\|_\infty \ .$$

This implies that convergence and continuity in $\mathbb{R}^n$ do not depend on the adopted norm. For example, proving the continuity of the linear mapping $A \in \mathcal{L}(\mathbb{R}^n, \mathbb{R}^p)$ is particularly simple with respect to the max-norm (2.12) in $\mathbb{R}^n$. In fact, the continuity of A at $0_n \in \mathbb{R}^n$ is sufficient for its continuity in every point. Indeed, for any $x \in \mathbb{R}^n$, i.e. $x = \sum_1^n x_i e_i$, we have

$$\|Ax\| = \left\| A\left(\sum_{i=1}^n x_i e_i\right) \right\| = \left\| \sum_{i=1}^n x_i A e_i \right\|$$

$$\leq \left(\sum_{i=1}^n \|Ae_i\|\right) \max_{1 \leq i \leq n} |x_i| \leq \lambda\|x\|_\infty \ ,$$

which implies continuity at 0_n. Now, for any $y \in \mathbb{R}^p$, substitute above $x := x - y$, which leads to

$$\|Ax - Ay\| \leq \lambda\|x - y\|_\infty \ ,$$

implying continuity at y. The inequality is valid for all $x, y \in \mathbb{R}^n$, and thus it implies *uniform continuity*. As the constant λ is the same for all $x, y \in \mathbb{R}^n$, then A is also *Lipschitz continuous* with *Lipschitz constant* λ. The inequality $\|Ax\| \leq \lambda\|x\|_\infty$ (for all $x \in \mathbb{R}^n$) also shows that $\|A\| \leq \lambda$, i.e. A is *bounded*. The opposite is also true: if A is linear and bounded, then it is continuous at 0_n.

The case $A \in \mathcal{L}(\mathbb{R}^n, \mathbb{R})$, when A is linear functional, is of particular interest. Let $x \in \mathbb{R}^n$ and $x = \sum_1^n x_i e_i$. Then, making use of the inner product (2.9), we find

$$Ax = A\left(\sum_{i=1}^n x_i e_i\right) = \sum_{i=1}^n x_i A(e_i) = \sum_{i=1}^n x_i a_i = \langle a, x \rangle \ ,$$

which gives the general form of any linear functional $A \in \mathcal{L}(\mathbb{R}^n, \mathbb{R})$. The space of all linear functionals $\mathcal{L}(\mathbb{R}^n, \mathbb{R})$ is called the *dual or conjugate space* of $\mathbb{R}^n$ and is denoted by $(\mathbb{R}^n)^*$. In general, $\mathcal{L}(\mathcal{E}, \mathbb{R})$ is denoted by $\mathcal{E}^*$. If $\mathbb{R}^n$ is Euclidean with (2.9), then $(\mathbb{R}^n)^*$ is also $n-$dimensional Euclidean space. The vector $a = (Ae_1, \ldots, Ae_n)$ is specific for every A, and can be considered as a vector from $(\mathbb{R}^n)^*$. In tensor algebra, vectors from both $\mathbb{R}^n$ and $(\mathbb{R}^n)^*$ are considered simultaneously. In order to distinguish them, the vectors from $(\mathbb{R}^n)^*$ are called *covariant*, while the "standard" vectors of $\mathbb{R}^n$ are called *contravariant*. Consider the following Cartesian product $\mathcal{F}$ of p copies of the real vector space $\mathcal{E}$ and q copies of its dual $\mathcal{E}^*$:

$$\mathcal{F} = \underbrace{\mathcal{E} \times \ldots \times \mathcal{E}}_{p} \times \underbrace{\mathcal{E}^* \times \ldots \times \mathcal{E}^*}_{q} .$$

Then, the linear functional (some authors prefer bi-linear functional):

$$T : \mathcal{F} \to \mathbb{R} \text{ (or } T : (\mathcal{E})^p \times (\mathcal{E}^*)^q \to \mathbb{R}) \tag{2.15}$$

is called $(p, q)-$tensor. All such tensors (linear functionals) form a vector space denoted by $\mathcal{T}_p^q(\mathcal{E})$. The tensor theory is important part of (multi)linear algebra, e.g. (Bourbaki, 1974, Ch3) or (Kostrikin and Manin, 1997, Ch4). Details related to modern applications of tensors in data analysis and their computational aspects can be found in (Hackbusch, 2012; Landsberg, 2012). Some elements of tensor decomposition applied to MDA problems are discussed briefly in Section 10.8.

The above definitions of inner product (2.9) and norm (2.10) can be elaborated to treat $\mathcal{M}(n)$ and $\mathcal{M}(n, p)$ directly as metric or normed spaces without going through vectorization. For any $A, B \in \mathcal{M}(m, n)$, one can define

$$\langle A, B \rangle_F = \text{trace}(A^\top B) , \tag{2.16}$$

which is known as the *Frobenius inner product*. Then, the corresponding norm in $\mathcal{M}(n)$ and $\mathcal{M}(m, n)$ is the *Frobenius norm* defined as

$$\|A\|_F = \sqrt{\text{trace}(A^\top A)} , \tag{2.17}$$

and one can check that it fulfils the three conditions for a norm.

Then, the corresponding metric in $\mathcal{M}(n)$ and $\mathcal{M}(m, n)$ is defined by

$$d(A, B) = \|A - B\|_F = \sqrt{\text{trace}(A^\top B)} . \tag{2.18}$$

Let λ_i be the eigenvalues of $A \in \mathcal{S}(n)$. It follows from (2.17) that $\|A\|_F = \sqrt{\sum_i \lambda_i^2}$. This implies that $|\lambda| \leq \|A\|_F$ for any eigenvalue λ of A, as well as $|\text{trace}(A)| \leq \sqrt{n}\,\|A\|_F$. Moreover, if $\|A\|_F < 1$, then $I_n + A > 0$ (is PD).

Such type of norms (with matrix argument) are called *matrix norms* (Horn and Johnson, 2013, 5.6). Other popular choices, apart from (2.17), are

- 1-norm of $A \in \mathcal{M}(m, n)$:

$$\|A\|_1 = \sum_{i,j} |a_{ij}| \; ; \tag{2.19}$$

- ∞-norm of $A \in \mathcal{M}(m, n)$:

$$\|A\|_\infty = \max_{i,j} |a_{ij}| \; ; \tag{2.20}$$

- p-norm of $A \in \mathcal{M}(m, n)$, for $1 < p < \infty$:

$$\|A\|_p = \max_{\|x\|_p=1} \|Ax\|_p \; , \tag{2.21}$$

where $\| \;\|_p$ on the right-hand side of (2.21) is the corresponding ℓ_p-norm of $n \times 1$ vectors. Note that $\|A\|_2 \leq \sqrt{\|A\|_1\|A\|_\infty}$. The special case of $p = 2$ is very important and is also well known as the *spectral norm* of $A \in \mathcal{M}(m, n)$, because $\|A\|_2 = \sqrt{\lambda_{\max}(A^\top A)}$ or

$$\|A\|_2 = \sigma_{\max}(A), \text{ the largest singular value of } A \; . \tag{2.22}$$

The equivalence of the different matrix norms is quantified by inequalities, many of which are listed in (Horn and Johnson, 2013, 5.6.P23). Several other useful inequalities between matrix norms are provided in (Golub and Van Loan, 2013, Ch 2.3). For example, if $A \in \mathcal{M}(m, n)$, then

$$\|A\|_2 \leq \|A\|_F \leq \sqrt{\text{rank}(A)}\|A\|_2 \leq \sqrt{\min\{m, n\}}\|A\|_2 \; .$$

Further details about norms of matrices and matrix inequalities can be found in (Bhatia, 1997, Ch IV), (Horn and Johnson, 2013, Ch 5), (Magnus and Neudecker, 1988, Ch 11), as well as in (Marshall et al., 2011).

2.5   Matrix differentials and gradients

Let $\mathcal{E}$ and $\mathcal{F}$ be two normed vector spaces over $\mathbb{R}$ and $\mathcal{U}$ is an open set in $\mathcal{E}$. The mapping $f : \mathcal{E} \to \mathcal{F}$ is Fréchet differentiable at $x \in \mathcal{U}$, if there exists a continuous linear mapping $L_x : \mathcal{E} \to \mathcal{F}$, such that for any $h \in \mathcal{E}$ that $x + h \in \mathcal{U}$,

$$\delta f = f(x + h) - f(x) = L_x h + \omega(h) , \tag{2.23}$$

where $\frac{\omega(h)}{\|h\|} \to 0$, when $h \to 0$ in the norm of $\mathcal{F}$. The main part of δf is a linear mapping of h called *Fréchet or total derivative* of f at x, or F-derivative for short, and is denoted by $f'(x)$. Equivalently, the Fréchet differentiability of f at x requires the existence of the following limit:

$$\lim_{h \to 0} \frac{\|f(x + h) - f(x) - f'(x)h\|}{\|h\|} = 0 . \tag{2.24}$$

If f is differentiable at every $x \in \mathcal{U}$, then f is called differentiable on $\mathcal{U}$. In this case its derivative is denoted simply by df.

In the same conditions, one can define another derivative known as the *Gâteaux* or *directional derivative*, or G-derivative for short. Let $\mathcal{E}$ and $\mathcal{F}$ be two normed vector spaces over $\mathbb{R}$ and $\mathcal{U}$ is an open set in $\mathcal{E}$. The mapping $f : \mathcal{U} \to \mathcal{F}$ is Gâteaux differentiable at $x \in \mathcal{U}$ along $h \in \mathcal{E}$, if the following limit exists

$$\lim_{\epsilon \to 0} \frac{f(x + \epsilon h) - f(x)}{\epsilon} , \tag{2.25}$$

where h is an unit vector defining the direction of differentiation. If the limit (2.25) exists, it is denoted as $f'_h(x)$. Note that $f'_h(x) \in \mathcal{F}$, while the Fréchet derivative $f'(x)$ is a linear mapping, i.e. $f'(x) \in \mathcal{L}(\mathcal{E}, \mathcal{F})$.

If f is Fréchet differentiable at x, then the derivative is unique and it is continuous at x. Moreover, f is Gâteaux differentiable at x along any direction h. However, the opposite implication is not correct: the existence of G-derivative (even in any direction) does not imply F-differentiability. However, if the G-derivative exists in $x \in \mathcal{U}$ (in any direction h), is linear in h, and $f'_h(x)$ is bounded when regarded as linear operator, and is continuous in x in the topology of $\mathcal{L}(\mathcal{E}, \mathcal{F})$, then f is F-differentiable (in $\mathcal{U}$) and the G and F derivatives are the same.

Note that $f(x) + L_x h$ in (2.23) is a linear mapping of h, and defines a *tangent hyperplane* to f at x. The set of all tangent vectors at x form a vector space in $\mathcal{E} \times \mathcal{F}$ which is called the *tangent space* of f at x. In general, the set of points $(x, f(x)) \in \mathcal{E} \times \mathcal{F}$ called *graph* of f is an example of a *differentiable manifold* defined by the equation $y = f(x)$, which is a high-dimensional generalization of curves $(\mathcal{E} = \mathbb{R}^2, \mathcal{F} = \mathbb{R})$ and surfaces $(\mathcal{E} = \mathbb{R}, \mathcal{F} = \mathbb{R}^2)$.

If $\{e_1, \ldots, e_n\}$ is the basis of $\mathcal{E}$, then $f'(x)e_i = \frac{\partial f}{\partial x_i}(x) = \partial_i f$ for $i = 1, \ldots, n$ are called *partial derivatives*. If $\dim(\mathcal{E}) = \dim(\mathcal{F}) = n$, the determinant of the matrix of $f'(x)$ is called *Jacobian* of f at x.

Consider the case $\mathcal{F} = \mathbb{R}$. Then, $f'(x)$ is a linear functional from $\mathcal{E}$ to $\mathbb{R}$, for which there exists an unique element $a \in \mathcal{E}$, such that $f'(x)h = \langle a, h \rangle$, which is called *gradient* of f at x and is denoted by $\mathrm{grad} f(x)$. If the basis $\{e_1, \ldots, e_n\}$ of $\mathcal{E}$ is *orthonormal*, i.e. $\langle e_i, e_j \rangle = \delta_{ij}$, where $\delta_{ii} = 1$ and 0 otherwise, then $\mathrm{grad} f(x)_i = \frac{\partial f}{\partial x_i} = \partial_i f$, i.e. the coordinates of the gradient coincide with the partial derivatives of f in orthonormal basis.

Differentiation is a linear operation. As the standard derivatives of functions of a single variable, the F (and G) derivatives possess the following well-known properties:

1. If df and dg exist, then $d(\alpha f + \beta g)$ also exists and $d(\alpha f + \beta g) = \alpha df + \beta dg$ for $\alpha, \beta \in \mathbb{R}$.

2. If $h(x) = g(f(x))$ is a composition of two mappings, then $h'(x) = g'(f(x))f'(x)$, when the involved derivatives exist.

3. If f has an inverse f^{-1}, i.e. $f^{-1}(f(x)) = I_\mathcal{E}$ and $f(f^{-1}(y)) = I_\mathcal{F}$, and if f is differentiable at x and f^{-1} is differentiable at $y = f(x)$, then its derivative is given by $(f^{-1})'(y) = (f'(x))^{-1}$, provided $\det(f'(x)) \neq 0$.

4. If df and dg exist, then $d(fg)$ exists and $d(fg) = df g(x) + f(x) dg$. More generally, if $B(x, y)$ is a *bi-linear mapping* from $\mathcal{E} \times \mathcal{E}$ to $\mathcal{F}$, i.e. $B(\alpha x + \beta y, z) = \alpha B(x, z) + \beta B(y, z)$ and $B(x, \alpha y + \beta z) = \alpha B(x, y) + \beta B(x, z)$ for any $x, y, z \in \mathcal{E}$. Then, $dB(x, y) = B(dx, y) + B(x, dy)$.

5. Let $\mathcal{U} \subset \mathbb{R}^m \times \mathbb{R}^n$ be open and let $f : \mathcal{U} \to \mathbb{R}^n$ be differentiable, such that $f(a, b) = 0_n$ for some $(a, b) \in \mathcal{U}$. Suppose that for fixed $x \in \mathbb{R}^m$, the partial derivative $\partial_y f$ at (a, b) is a linear homeomorphism in $\mathcal{L}(\mathbb{R}^n)$, i.e. the $n \times n$ (partial) Jacobian is invertible. Then, there exists sufficiently small open ball $\mathcal{B}(a, \epsilon) \subset \mathbb{R}^m$, such that for every $x \in \mathcal{B}(a, \epsilon)$, the equation $f(x, y) = 0_n$ has an unique solution $y \in \mathbb{R}^n$ and $(x, y) \in \mathcal{U}$. This defines a differentiable map $g : \mathcal{B}(a, \epsilon) \to \mathbb{R}^n$,

called the *implicit function*, such that $b = g(a)$ and $f(x, g(x)) = 0_n$ for every $x \in \mathcal{B}(a, \epsilon)$. Moreover, its derivative at a is given by

$$g'(a) = -[\partial_y f(a, g(a))]^{-1} \partial_x f(a, g(a)) .$$

For proofs, see (Dieudonné, 1969a, Ch X, 2), (Lang, 1993, XIV, §2) or (Rudin, 1976, Th 9.27).

6. Let $\mathcal{GL}(n) \subset \mathcal{M}(n)$ be the subset of all invertible $n \times n$ matrices; then, the (bijective) mapping $X \to X^{-1}$ defined on $\mathcal{GL}(n)$ is differentiable and its Fréchet derivative (2.23) is given by the matrix

$$d(X^{-1}) = -X^{-1}(dX)X^{-1} , \qquad (2.26)$$

which follows from $(X + dX)^{-1} - X^{-1} \approx -X^{-1}(dX)X^{-1}$ for small $\|dX\|(< \|X^{-1}\|^{-1})$ and from the fact that $dX \to -X^{-1}(dX)X^{-1}$ is a linear mapping from $\mathcal{M}(n)$ to $\mathcal{M}(n)$; for $n = 1$, it reduces to

$$\left(\frac{1}{x}\right)' = -\frac{x'}{x^2} .$$

Let $\mathcal{U} \subset \mathcal{E}$ be open and *convex* (i.e. if $x, y \in \mathcal{U}$, then $\alpha x + (1 - \alpha)y \in \mathcal{U}$ for $\forall \alpha \in [0, 1]$), and f be $F-$differentiable in $\mathcal{U}$. Then, for any $x, x + h \in \mathcal{U}$ there exists $\tau \in [0, 1]$, such that

$$f(x + x) - f(x) = \langle \mathrm{grad} f(x + \tau h), h \rangle ,$$

which is also known as the *mean value theorem*.

We already know that the $F-$derivative $f'(x)$ is a mapping from $\mathcal{U} \subset \mathcal{E}$ to $\mathcal{L}(\mathcal{E}, \mathcal{F})$. If this mapping is also differentiable, then it is denoted by $f''(x)$ and is a function from $\mathcal{U} \subset \mathcal{E}$ to $\mathcal{L}(\mathcal{E}, \mathcal{L}(\mathcal{E}, \mathcal{F}))$. It can be proven that $f''(x)$ is, in fact, a *symmetric* bi-linear mapping from $\mathcal{E} \times \mathcal{E}$ to $\mathcal{F}$, i.e. $B(x, y) = B(y, x)$. For $\mathcal{E} = \mathcal{F} = \mathbb{R}^n$, the matrix of $f''(x)$ is called *Hessian* of f at x. It is symmetric and contains the second-order partial derivatives: $\partial_{ij} f(x) = \partial_{ji} f(x)$.

See for details (Nashed, 1966; Sagle and Walde, 1973; Spivak, 1993, Ch 2) and (Kantorovich and Akilov, 1982, Ch XVII), and, in fact, any good text discussing differentiation of functions of several variables, e.g. (Lang, 1993, Ch XIII) and (Dieudonné, 1969a, Ch VIII). Many practical tools needed in MDA are collected in (Athans and Schweppe, 1965; Magnus and Neudecker, 1988). Further generalizations of differentiation can be found, for example, in (Michal, 1940; Schechter, 1984).

2.6 Conclusion

This chapter presents an essential prerequisite for reading and working with the book. Its purpose is to serve as a quick reference and introduction to the book notations. Nevertheless, the reader is advised to visit the provided references for further and deeper education. Section 2.1 collects a great number of useful definitions and facts from matrix algebra and analysis. Section 2.2 gives a brief intro to linear algebra and linear operators in finite-dimensional spaces. Section 2.3 lists the main properties of the topological, metric and normed spaces, as well as the basic concepts of continuity, convergence, compactness, etc. Section 2.4 concentrates on the Euclidean spaces of matrices. The matrix differentiation of functions and functionals depending on matrix variable(s) is refreshed in Section 2.5.

2.7 Exercises

1. Let $A \in \mathcal{M}(n)$ be a 2×2 block-matrix with $A_{ii} \in \mathcal{M}(n_i), n_1 + n_2 = n$. Assuming that A^{-1} and the other involved inverses exist, check that

$$A^{-1} = \begin{bmatrix} (A_{11} - A_{12}A_{22}^{-1}A_{21})^{-1} & A_{11}^{-1}A_{12}(A_{21}A_{11}^{-1}A_{12} - A_{22})^{-1} \\ A_{22}^{-1}A_{21}(A_{12}A_{22}^{-1}A_{21} - A_{11})^{-1} & (A_{22} - A_{21}A_{11}^{-1}A_{12})^{-1} \end{bmatrix}.$$

2. For $A_{11} \in \mathcal{S}(n_1)$ and $A_{22} \in \mathcal{S}(n_2)$, prove that $\begin{bmatrix} A_{11} & B \\ B^{\top} & A_{22} \end{bmatrix} > 0$, if and only if $A_{11} > 0$ and $A_{22} > B^{\top}A_{11}^{-1}B$.

3. Prove that O_n always belongs to the span $\mathcal{A}$.

4. Check that:

   (a) the non-zero real numbers $(\mathbb{R} - \{0\})$ under multiplication form a commutative group;

   (b) the invertible matrices in $\mathcal{M}(n)(n > 1)$ form a *non*-commutative group with respect to the matrix multiplication.

5. Check that $\mathcal{M}(n, p)$ is a real vector space with respect to the matrix summation and the multiplication by a real number. What about the subset of all singular matrices $A \in \mathcal{M}(n)$, i.e. having $\det(A) = 0$? Do they also form vector subspace?

6. In MDA it is very common to centre the data matrix X prior to further analysis, such that the variables/columns have zero sums. Let $X \in \mathcal{M}(n,p)$ and define the *centring operator/matrix* $J_n = I_n - 1_n 1_n^\top / n = I_n - E_n^\top / n$. Then the matrix $X_0 = J_n X$ is the *centred data matrix* with zero column sums.

   (a) Check that the set of all centred matrices $\mathcal{M}_0(n,p)$ forms a vector subspace in $\mathcal{M}(n,p)$ with dimension $(n-1)p$.

   (b) Check that the centring operator/matrix $J_n : \mathcal{M}(n,p) \to \mathcal{M}_0(n,p)$ is a *projector*, i.e. the range of J_n is $\mathcal{M}_0(n,p)$ and $J_n^2 = J_n = J_n^\top$. Moreover, if $X \in \mathcal{M}_0(n,p)$, then $J_n X = X$.

7. Let $\mathcal{M}_{00}(n)$ be the subset of all $n \times n$ matrices in which columns and rows sum to zeros.

   (a) Check that $\mathcal{M}_{00}(n)$ is a vector subspace of $\mathcal{M}(n)$ with dimension $(n-1)^2$;

   (b) Check that the subset $\mathcal{S}_{00}(n)$ of all symmetric matrices in $\mathcal{M}_{00}(n)$ is a vector subspace of $\mathcal{M}_{00}(n)$ with dimension $\frac{n(n-1)}{2}$;

   (c) Check that the subset $\mathcal{K}_{00}(n)$ of all skew-symmetric matrices in $\mathcal{M}_{00}(n)$ is a vector subspace with dimension $\frac{(n-1)(n-2)}{2}$.

8. Find the dimension of the subset of all traceless matrices $A \in \mathcal{M}(n)$, i.e. having trace$(A) = 0$.

9. Check that the definition of rank of linear function A implies that the rank of the corresponding matrix A is equal to the number of its independent columns.

10. $\mathcal{E}$ and $\mathcal{F}$ are real vector spaces and $A : \mathcal{E} \to \mathcal{F}$ is a linear function. Suppose that $\dim(\mathcal{E}) = n$ and $\dim(A(\mathcal{E})) = r$ (the rank of A). Prove that the dimension of the kernel $A^{-1}(0_\mathcal{F}) = \{a \in \mathcal{E} : A(a) = 0\}$ is $\dim(A^{-1}(0_\mathcal{F})) = \dim(\mathcal{E}) - \dim(A(\mathcal{E})) = n - r$.

11. Check that the representation of every normal matrix $A \in \mathcal{M}(n)$ as $Q^\top A Q = \mathrm{Diag}_B(A_1, \ldots, A_n)$, for some $Q \in \mathcal{O}(n)$, takes the following specific forms (Horn and Johnson, 1985, 2.5.14):

   (a) if $A \in \mathcal{S}(n)$, then all A_i are simply the (real) eigenvalues of A;

   (b) if $A \in \mathcal{K}(n)$, then all $A_i = \beta_i \begin{bmatrix} 0 & 1 \\ -1 & 0 \end{bmatrix}$ and $\beta_i > 0$;

(c) if $A \in \mathcal{O}(n)$, then all $A_i = \begin{bmatrix} \cos \alpha_i & \sin \alpha_i \\ -\sin \alpha_i & \cos \alpha_i \end{bmatrix}$ and $\alpha_i \in (0, \pi)$.

12. A matrix $K \in \mathcal{M}(n)$ is called *skew-symmetric* if $K^\top = -K$.

    (a) Check that the subset $\mathcal{K}(n)$ of all skew-symmetric matrices in $\mathcal{M}(n)$ forms a real vector (sub)space.

    (b) Check that the following matrices form a basis in $\mathcal{K}(3)$:

$$\begin{bmatrix} 0 & 1 & 0 \\ -1 & 0 & 0 \\ 0 & 0 & 0 \end{bmatrix}, \begin{bmatrix} 0 & 0 & 1 \\ 0 & 0 & 0 \\ -1 & 0 & 0 \end{bmatrix}, \begin{bmatrix} 0 & 0 & 0 \\ 0 & 0 & 1 \\ 0 & -1 & 0 \end{bmatrix}.$$

    (c) If $K \in \mathcal{K}(n)$, then K is normal.

    (d) Prove that the eigenvalues of $K \in \mathcal{K}(n)$ are either pure imaginary numbers or zero(s).

    (e) If $K \in \mathcal{K}(n)$ is invertible, then n is even. Hint: Use $\det(\alpha K) = \alpha^n \det(K)$ for any $\alpha \in \mathbb{R}$.

    (f) If $K \in \mathcal{K}(2n)$ is non-singular, then there exists $A \in \mathcal{GL}(2n)$, such that $AKA^\top = \begin{bmatrix} O_n & I_n \\ -I_n & O_n \end{bmatrix}$.

    (g) For $K \in \mathcal{K}(n)$, prove that $\det(K) \geq 0$. Moreover, $\det(K)$ is the square of an integer, if K has integer entries (Artin, 2011, p. 258).

13. Strassen's algorithm for matrix multiplication (Golub and Van Loan, 2013, 1.3.11), (Kostrikin and Manin, 1997, p. 33).

    (a) We are given the following two matrices $A, B \in \mathcal{M}(2)$:

$$A = \begin{bmatrix} a_{11} & a_{12} \\ a_{21} & a_{22} \end{bmatrix}, \quad B = \begin{bmatrix} b_{11} & b_{12} \\ b_{21} & b_{22} \end{bmatrix},$$

and let us consider their matrix product $C = AB$, with elements:

$$c_{11} = a_{11}b_{11} + a_{12}b_{21},$$
$$c_{12} = a_{11}b_{12} + a_{12}b_{22},$$
$$c_{21} = a_{21}b_{11} + a_{22}b_{21},$$
$$c_{22} = a_{21}b_{12} + a_{22}b_{22}.$$

One can count that to find C we need 8 multiplications and 4 additions. Can you prove that in case of $A, B \in \mathcal{M}(n)$, the number of multiplications and additions needed is n^3 and $n^3 - n^2$, respectively?

(b) Consider the following expressions:

$$a = (a_{11} + a_{22})(b_{11} + b_{22}) \ , \quad b = (a_{21} + a_{22})b_{11} \ ,$$

$$c = a_{11}(b_{12} - b_{22}) \ , \quad d = a_{22}(b_{21} - b_{11}) \ ,$$

$$e = (a_{11} + a_{12})b_{22} \ , \quad f = (a_{21} - a_{11})(b_{11} + b_{12}) \ ,$$

$$g = (a_{12} - a_{22})(b_{21} + b_{22}) \ .$$

Check that the elements of the matrix product $C = AB$ can also be calculated as:

$$c_{11} = a + d - e + g \ ,$$

$$c_{12} = c + e \ ,$$

$$c_{21} = b + d \ ,$$

$$c_{22} = a + c - b + f \ .$$

Clearly, we need 7 multiplications and 18 additions to find the product C. This is known as Strassen's algorithm.

(c) Applying Strassen's algorithm to matrices of order 2^n, partitioned into four $2^{n-1} \times 2^{n-1}$ blocks, show that one can multiply them using 7^n multiplications (instead of the usual $(2^n)^3 = 8^n$) and $6(7^n - 4^n)$ additions. Hint: Suppose that $A, B \in \mathcal{M}(4)$ are composed of 2×2 blocks. Then, Strassen's algorithm can be applied to get the blocks of $C = AB$ in terms of the blocks of A, B using 7 multiplications of 2×2 matrices, i.e. one needs $7^2 = 49$ multiplications instead of the usual $4^3 = 64$.

(d) Extend matrices of order N to the nearest matrix of order 2^n by filling with zeros and show that approximately $N^{\log_2 7} \approx N^{2.81}$ operations are sufficient for multiplying them. Look for some recent developments in (Landsberg, 2012, 1.1). Can you still think of something better?

14. Prove that the triangle inequality implies

$$d(x, z) \geq |d(y, x) - d(y, z)| \ ,$$

i.e. that each side in a triangle cannot be less than the difference between the other two.

15. Check that the identity map $i : \mathcal{E} \to \mathcal{F}$ is bijective and continuous, but it is *not* homeomorphism, if $\mathcal{E} = \mathbb{R}$ with the following discrete metric:

$$d(x, y) = \begin{cases} 1 \text{ if } x \neq y \\ 0 \text{ if } x = y \end{cases} ,$$

and $\mathcal{F} = \mathbb{R}$ with the standard metric $d(x, y) = |x - y|$.

16. Prove that the normed vector space is a topological vector space. Hint: Prove that $x_n \to x_0, y_n \to y_0$ and $\alpha_n \to \alpha_0$ imply $x_n + y_n \to x_0 + y_0$ and $\alpha_n x_n \to \alpha_0 x_0$.

17. (*Riesz's representation theorem*) Let $\mathcal{M}^*(n)$ be the vector space of all linear functionals $f : \mathcal{M}(n) \to \mathbb{R}$. Prove that for every $f \in \mathcal{M}^*(n)$, there exists an unique matrix $F \in \mathcal{M}(n)$, such that $f(A) = \text{trace}(F^\top A)$ for $\forall A \in \mathcal{M}(n)$.

18. Prove that for any two matrices A and B of appropriate sizes:

    (a)
    $$|\text{trace}(A^\top B)| \leq \frac{\text{trace}(A^\top A) + \text{trace}(B^\top B)}{2} ; \qquad (2.27)$$

    (b) a matrix version of the Cauchy–Schwartz inequality (2.4):
    $$\text{trace}(A^\top B)]^2 \leq \text{trace}(A^\top A)\text{trace}(B^\top B) , \qquad (2.28)$$

    with equality if and only if $A = \alpha B$.

19. Check that ℓ^2 is

    (a) A vector space with respect to $x + y = \{x_1 + y_1, x_2 + y_2, \ldots\}$ and $\alpha x = \{\alpha x_1, \alpha x_2, \ldots\}$. Hint: Use (2.27).

    (b) *Complete* with norm of $x \in \ell^2$ defined by $\|x\| = \sqrt{\sum_{i=1}^{\infty} x_i^2}$, and the corresponding inner product $\langle x, y \rangle = \sum_{i=1}^{\infty} x_i y_i$ is well defined. Hint: Use (2.8).

    (c) a *separable* space, i.e. it contains a *countable* subset $\mathcal{N} \subset \ell^2$, which is *dense* in ℓ^2, i.e. $\overline{\mathcal{N}} \equiv \ell^2$. Hint: Consider the subset of sequences $\{r_1, \ldots, r_n, 0, 0, \ldots\} \subset \ell^2$, where r_i are rational numbers.

20. Check that the parallelogram identity (2.7) does not hold for the ℓ_∞–norm (2.12), i.e. the max-norm cannot be generated by an inner product.

21. Prove that the following *mixed norm*:

$$\|x\|_\lambda = \sqrt{\lambda \|x\|_1^2 + (1 - \lambda)\|x\|_2^2}\ ,$$

is indeed a norm in $\mathbb{R}^n$ for any $\lambda \in [0, 1]$ (Qi et al., 2013). Is the parallelogram identity (2.7) fulfilled for the mixed norm $\|x\|_\lambda$?

22. Let $\mathcal{E} = \mathbb{R}^n$, $\mathcal{F} = \mathbb{R}$ and $f(x) = \|x\|$.

    (a) Prove that the G-derivative at $x = 0$ exists along any direction.
    (b) Prove that the F-derivative at $x = 0$ does not exist.

23. For $A \in \mathcal{M}(n)$, its *matrix exponential* is defined as

$$\exp(A) = I_n + A + \frac{A^2}{2!} + \frac{A^3}{3!} + \ldots \in \mathcal{M}(n). \qquad (2.29)$$

    (a) Check that the exponential of a diagonal matrix $D = \mathrm{Diag}(d_1, \ldots, d_n)$ is given by

$$
\begin{aligned}
\exp(D) \;&=\; I_n + \mathrm{diag}(d_1, \ldots, d_n) + \mathrm{diag}\left(\frac{d_1^2}{2!}, \ldots, \frac{d_n^2}{2!}\right) + \ldots \\
&=\; \mathrm{diag}\left(e^{d_1}, \ldots, e^{d_n}\right) .
\end{aligned}
$$

    (b) It follows from the properties of the Euclidean norm in $\mathbb{R}^{n \times n}$ that

$$\left\|\frac{A^k}{k!} + \ldots + \frac{A^{k+p}}{(k+p)!}\right\| \leq \left\|\frac{A^k}{k!}\right\| + \ldots + \left\|\frac{A^{k+p}}{(k+p)!}\right\| .$$

    Make use of it to show that the series in (2.29) is well defined (converges) for any A. Hint: Note that the series of the Euclidean norms of the terms in (2.29) is convergent.

    (c) Check that if $B \in \mathcal{M}(n)$ is invertible, then

$$\exp(BAB^{-1}) = B\exp(A)B^{-1} .$$

    (d) If A and B commute, i.e. $AB = BA$, then

$$\exp(A + B) = \exp(A)\exp(B) ;$$

    particularly, the matrix $\exp(A)$ is invertible for any A, and

$$[\exp(A)]^{-1} = \exp(-A) .$$

(e) $\| \exp(A) \| \leq \exp(\|A\|)$.

(f) If $A \in \mathcal{K}(n)$, then $\exp(A) \in \mathcal{O}(n)$, i.e. it is orthogonal matrix.

(g) Every $A \in \mathcal{K}(n)$ can be expressed as $A = \sqrt{B}$, for some *negative* definite matrix $B \in \mathcal{S}(n)$ (Rinehart, 1960).

24. Let $A \in \mathcal{M}(n)$ be diagonalizable, i.e. $A = Q \Lambda Q^{-1}$ for some non-singular Q and diagonal Λ, containing its eigenvalues (Horn and Johnson, 2013, Th 1.3.7.). Particularly, if $A \in \mathcal{S}(n)$, then Q can be chosen orthogonal, i.e. $Q^{-1} = Q^{\top}$. The term "function of a matrix" has different meanings. Here, function f of a diagonalizable matrix $A \in \mathcal{M}(n)$ is defined by $f(A) = Q f(\Lambda) Q^{-1}$ or, respectively, $f(A) = Q f(\Lambda) Q^{\top}$. It is assumed that f is defined on the spectre of A. Check that:

(a) $\det(A) = \prod_i \lambda_i$;

(b) $\text{trace}(A) = \sum_i \lambda_i$;

(c) $\exp(A) = Q \exp(\Lambda) Q^{\top}$;

(d) $\log(A) = Q \log(\Lambda) Q^{\top}$, assuming $A > 0$;

(e) $\det(\exp(A)) = \exp(\text{trace}(A))$ (Jacobi's formula).

25. Prove that every $A \in \mathcal{M}(n)$ can be written as a sum $A = S + N$ of commuting diagonalizable S and nilpotent N, i.e. $SN = NS$. Hint: Use the Jordan canonical form of A. Then, show that Jacobi's formula is true for an arbitrary $A \in \mathcal{M}(n)$ (Hall, 2015, Ch 2.2, A.3).

26. A function $f : (a, b) \to \mathbb{R}$ is called monotonically increasing (decreasing) on (a, b) if $a < x < y < b$ implies $f(x) \leq f(y) (f(x) \geq f(y))$. Prove that a differentiable f is increasing if and only if $f' \geq 0$ in (a, b).

27. A function $f : (a, b) \to \mathbb{R}$ is called convex ($-f$ is called concave) if

$$f(\alpha x + (1 - \alpha) y) \leq \alpha f(x) + (1 - \alpha) f(y) , \quad \forall \alpha \in (0, 1) .$$

If f is differentiable in (a, b), prove that f is convex, if and only if f' is monotonically increasing in (a, b), sometimes also expressed as

$$f(x_2) - f(x_1) \geq (x_2 - x_1) f'(x_1) , \quad a \leq x_1 \leq x_2 \leq b .$$

If f is twice differentiable, prove the condition is $f'' \geq 0$ in (a, b).

28. The function $f : (a,b) \to \mathbb{R}$ is called *matrix convex* of order n, if for any $A, B \in \mathcal{S}(n)$ with spectra in (a,b) and $\alpha \in (0,1)$, it follows

$$f(\alpha A + (1-\alpha)B) \le \alpha f(A) + (1-\alpha)f(B) \ .$$

If this is true for any n, then f is simply called matrix convex. Show that if f is also continuous, the following weaker condition is enough for convexity

$$f\left(\frac{A+B}{2}\right) \le \frac{f(A)+f(B)}{2} \ .$$

29. The function f is called *matrix monotone* of order n, if for any $A, B \in \mathcal{S}(n)$, the relation $A \ge B$ ($A - B$ is PSD) implies $f(A) \ge f(B)$. If this is true for any n, then f is called matrix monotone. Prove that x^2 is

   (a) matrix convex on $\mathbb{R}$,

   (b) *not* matrix monotone of order 2 on $[0, \infty)$. Give examples that there exist $A, B \ge 0$ and $A, B > 0$, such that $A^2 - B^2$ is not PSD. What more is needed (Horn and Johnson, 2013, Th 7.7.3(b))?

30. Prove that for $x > 0$ the function:

   (a) x^{-1} is matrix convex. Hint: $\left(\frac{1+x}{2}\right)^{-1} \le \frac{1+x^{-1}}{2}, x > 0$.

   (b) $-x^{-1}$ is matrix monotone (or, if $A, B \in \mathcal{S}_+(n)$, then $A \ge B > 0 \Rightarrow B^{-1} \ge A^{-1}$).

   (c) $x^p, p \in (0,1)$, is matrix monotone and concave on $\mathcal{S}_+(n)$.

31. If $A > 0$ and $B \ge 0$, then $\det(A + B) \ge \det(A)$, with equality if and only if $B = 0$. Hint: Let $A = Q\Lambda Q^\top, \Lambda > 0$ be the EVD of A, and express $A + B = Q\Lambda^{1/2}(I + \Lambda^{-1/2}Q^\top BQ\Lambda^{-1/2})\Lambda^{1/2}Q^\top$.

32. Let f be continuous on $\mathbb{R}$. Then if

   (a) f is monotone increasing, so is $A \to \text{trace}[f(A)]$ on $\mathcal{S}(n)$.

   (b) f is convex, so is $A \to \text{trace}[f(A)]$ on $\mathcal{S}(n)$.

   (c) $A \to \log\{\text{trace}[\exp(A)]\}$ is convex on $\mathcal{S}(n)$. Hint: For $A, B \in \mathcal{S}(n)$, consider $x \to \log\{\text{trace}[\exp(A + xB)]\}$ defined on $(0,1)$.

33. Prove that

   (a) $\mathcal{S}_+(n)$ is open and convex, and is a cone, i.e. if $X_1, X_2 \in \mathcal{S}_+(n)$, then $X_1 + \alpha X_2 \in \mathcal{S}_+(r)$ for any $\alpha > 0$;

(b) the function $X \to \log[\det(X)]$ is concave on $\mathcal{S}_+(n)$;

(c) for a fixed $A \in \mathcal{S}(n)$, the function $X \to \text{trace}[\exp(A + \log X)]$ is concave on $\mathcal{S}_+(n)$ (Tropp, 2015, 8.1.4).

34. Let $a = (a_1, \dots, a_n)^\top$ be a *probability vector*, i.e. $a_i \geq 0$ and $\mathbf{1}_n^\top a = 1$. Assuming $0 \log 0 = 0$, the *Boltzmann–Shannon entropy* of a is defined as

$$h(a) = -\sum_{i=1}^{n} a_i \log a_i . \tag{2.30}$$

A matrix analogue of (2.30) is the *von Neumann* entropy defined as

$$h(A) = -\text{trace}(A \log A) , \tag{2.31}$$

for any *density matrix* A, i.e. $A \geq O_n$ with $\text{trace}(A) = 1$. Prove that

(a) the set of all density matrices is convex;

(b) $h(A)$ in (2.31) depends on A only through its eigenvalues. Hint: let λ_i be the eigenvalues of A, check that $h(A) = -\sum_{i=1}^{n} \lambda_i \log \lambda_i$;

(c) $h(A) \leq \log(n)$, with equality if and only if $A = I_n/n$.

35. Let f be convex and differentiable on $\mathbb{R}$, and $A, B \in \mathcal{S}(n)$. Prove:

(a) (Bhatia, 1997, IX.8.12), (Bhatia, 2007, 4.3.2), Carlen (2010)

$$\text{trace}[f(A)] - \text{trace}[f(B)] \geq \text{trace}[(A - B)f'(B)] ;$$

(b) Klein's inequality for $A, B \geq O_n$ (apply $f(x) = x \log x$)

$$\text{trace}[A(\log A - \log B)] \geq \text{trace}(A - B) ;$$

(c) let $s(A|B) = \text{trace}[A(\log A - \log B)]$; prove a more precise result than Klein's inequality (Furuta, 2005, (4.6)):

$$s(A|B) \geq \text{trace}\left(A\{\log[\text{trace}(A)] - \log[\text{trace}(B)]\}\right) \geq \text{trace}(A{-}B) ;$$

(d) $s(A|B)$ is called the *(Umegaki) relative entropy* of the density matrices A and B. Prove that $s(A|B) \geq 0$. It generalizes the relative entropy of two probability vectors a and b defined as

$$s(a|b) = \sum_{i=1}^{n} a_i \log\left(\frac{a_i}{b_i}\right) = \sum_{i=1}^{n} a_i(\log a_i - \log b_i) , \tag{2.32}$$

and better known as the *Kullback–Leibler divergence*. For further related results, see (Hiai and Petz, 2014; Petz, 2008).

36. Prove that the differential of the determinant is given by

$$d(\det(A)) = \det(A)\text{trace}(A^{-1}dA) \ .$$

   Hint: For a non-singular A and for a "small" H, i.e. $\|H\| < \epsilon$, consider $\det(A+H) - \det(A)$ and make use of $\det(I + A^{-1}H) \approx 1 + \text{trace}(A^{-1}H)$.

37. (The Banach Contraction Principle) Let $\mathcal{X}$ be a complete metric space and the mapping $f : \mathcal{X} \to \mathcal{X}$ be a *contraction*, i.e. there exists a positive constant $\kappa < 1$, such that for all $x, y \in \mathcal{X}$ it holds that $d(f(x), f(y)) \le \kappa d(x, y)$. Show that:

   (a) f has a unique *fixed point* $x_0 \in \mathcal{X}$, defined by $f(x_0) = x_0$;

   (b) moreover, for any $x \in \mathcal{X}$, we have $x_0 = \lim f(\dots f(f(x)))$.

   Hint: Consider the sequence $x_2 = f(x_1), x_3 = f(x_2), \dots$ and prove that it is a Cauchy sequence for any starting $x_1 \in \mathcal{X}$.

Chapter 3

Matrix manifolds in MDA

© Springer Nature Switzerland AG 2021

N. Trendafilov and M. Gallo, *Multivariate Data Analysis on Matrix Manifolds*,

Springer Series in the Data Sciences, https://doi.org/10.1007/978-3-030-76974-1_3

3.1 Several useful matrix sets

3.1.1 $\mathcal{GL}(n)$

The set of all non-singular matrices in $\mathcal{M}(n)$ is denoted by $\mathcal{GL}(n)$. It is an open set in $\mathbb{R}^{n \times n}$. To see this, consider first its complement, the set of all *singular* matrices in $\mathcal{M}(n)$, i.e. having zero determinant. The determinant is a continuous function f as a sum of polynomials (Section 10.1). Thus, as $\{0\}$ is closed in $\mathbb{R}^{n \times n}$, then $f^{-1}(\{0\})$ is closed as well. This proves that the set of all singular matrices in $\mathcal{M}(n)$ is closed. Then, its complement $\mathcal{GL}(n)$ is *open* in $\mathbb{R}^{n \times n}$, and thus it is not compact. It is not even bounded, because $\alpha I_n \in \mathcal{GL}(n)$ for any $\alpha > 0$ (which also implies that it is open).

The determinant of an arbitrary $A \in \mathcal{GL}(n)$ can be ether positive or negative. Thus, one can define two non-overlapping open subsets in $\mathcal{GL}(n)$ as follows:

$$\mathcal{GL}(n)^+ = \{A \in \mathcal{M}(n) : \det(A) > 0\}$$

and

$$\mathcal{GL}(n)^- = \{A \in \mathcal{M}(n) : \det(A) < 0\} \ ,$$

such that $\mathcal{GL}(n)^+ \cup \mathcal{GL}(n)^- = \mathcal{GL}(n)$. The existence of $\mathcal{GL}(n)^+$ and $\mathcal{GL}(n)^-$ with such features shows that $\mathcal{GL}(n)$ is not a *connected set*, i.e. it can be presented as an union of two non-overlapping open (or closed) subsets. The image of a connected set through a continuous function is also connected. If $\mathcal{GL}(n)$ is connected, then $\det(\mathcal{GL}(n))$ should be connected too. But we found that the image $\det(\mathcal{GL}(n)) = (-\infty, 0) \cup (0, \infty)$, i.e. not connected. Thus, $\mathcal{GL}(n)^+$ and $\mathcal{GL}(n)^-$ are the two connected components $\mathcal{GL}(n)$.

Finally, note that $\mathcal{GL}(n)$ forms a (non-commutative) group with respect to the matrix multiplication, known as the *general linear group*, which explains its abbreviation.

3.1.2 $\mathcal{O}(n)$

The set of all orthogonal matrices is denoted by $\mathcal{O}(n)$ and is defined as

$$\mathcal{O}(n) = \{Q \in \mathcal{M}(n) : Q^\top Q = QQ^\top = I_n\} \ . \tag{3.1}$$

Clearly, $\mathcal{O}(n)$ is bounded in $\mathbb{R}^{n \times n}$. To see that $\mathcal{O}(n)$ is also closed, let $Q \notin \mathcal{O}(n)$. Then, $Q \in \mathbb{R}^{n \times n} - \mathcal{O}(n)$ and $Q^\top Q \neq I_n$. As the matrix multiplication is continuous, there exists an open ball $\mathcal{B}_Q$ around Q, such that for any matrix $P \in \mathcal{B}_Q$ we have $P^\top P \neq I_n$. Thus $P \notin \mathcal{O}(n)$, or $\mathcal{B}_Q \subset \mathbb{R}^{n \times n} - \mathcal{O}(n)$, which proves that the complement of $\mathcal{O}(n)$ in $\mathbb{R}^{n \times n}$ is open, or that $\mathcal{O}(n)$ is closed. Thus, $\mathcal{O}(n)$ is *compact* in $\mathbb{R}^{n \times n}$.

It follows from the definition (3.1) that every orthogonal matrix $Q \in \mathcal{O}(n)$ has an inverse $Q^{-1} = Q^\top$. Moreover, $\mathcal{O}(n)$ forms a group with respect to the matrix multiplication, known as the *(real) orthogonal group*. Thus, $\mathcal{O}(n) \subset \mathcal{GL}(n)$ and $\det(Q) \neq 0$ for every $Q \in \mathcal{O}(n)$. Then, we have for every $Q \in \mathcal{O}(n)$ that

$$\det(I_n) = \det(Q^\top Q) = \det(Q^\top)\det(Q) = (\det(Q))^2 = 1 \ ,$$

which implies that $\det(Q) = \pm 1$. Thus, one can define two non-overlapping closed subsets in $\mathcal{O}(n)$ composed of orthogonal matrices with determinants 1 and -1, respectively. Usually they are denoted by $\mathcal{O}(n)^+$ and $\mathcal{O}(n)^-$. An orthogonal matrix Q is called *rotation*, when $Q \in \mathcal{O}(n)^+$, and is called *reflection* (about a line) when $Q \in \mathcal{O}(n)^-$. Alternatively, the elements of $\mathcal{O}(n)^-$ can be seen as (orthogonal) symmetries about a plane (subspace) (Dieudonné, 1969b, (7.2.1)).

Obviously, every rotation Q can be expressed as a product of two reflections. Every element of $\mathcal{O}(n)$ can be presented as a product of at most n reflections. Note that $\mathcal{O}(n)^+$ is a (non-commutative) subgroup of $\mathcal{O}(n)$ with respect to the matrix multiplication. It is better known as the *special orthogonal group* $\mathcal{SO}(n)$, sometimes called the *rotation group*.

The existence of $\mathcal{O}(n)^+$ and $\mathcal{O}(n)^-$, such that $\mathcal{O}(n) = \mathcal{O}(n)^+ \cup \mathcal{O}(n)^-$ and $\mathcal{O}(n)^+ \cap \mathcal{O}(n)^- = \emptyset$, shows that $\mathcal{O}(n)$ is not a *connected set*, i.e. it can be presented as an union of two non-overlapping closed subsets. The image of a connected set through a continuous function is also connected. If $\mathcal{O}(n)$ is connected, then $\det(\mathcal{O}(n))$ should be connected too. But we found that $\det(\mathcal{O}(n)) = \{1, -1\}$, i.e. not connected. $\mathcal{O}(n)^+$ and $\mathcal{O}(n)^-$ are the two connected components of $\mathcal{O}(n)$.

Note that $\mathcal{O}(n)$ can be defined alternatively, as the subset of matrices $Q \in \mathcal{M}(n)$ satisfying (3.30) (Curtis, 1984, Ch 2).

There is an important connection between $\mathcal{GL}(n)$ and $\mathcal{O}(n)$: for any $A \in \mathcal{GL}(n)$, there exists an unique upper triangular matrix U with positive entries on its main diagonal such that $AU^\top \in \mathcal{O}(n)$. Moreover, the entries of U are smooth functions of the entries of A. For short, the set of such non-singular upper triangular matrices with positive diagonal entries is denoted by $\mathcal{U}_+(n)$. This makes $\mathcal{O}(n)$ a *retract* of $\mathcal{GL}(n)$, i.e. there exists a continuous map (*retraction*) $f : \mathcal{GL}(n) \to \mathcal{O}(n)$ such that $f(Q) = Q$ for every $Q \in \mathcal{O}(n)$ (restricting itself to the identity on $\mathcal{O}(n)$). More formally, the retraction is a continuous map $f : \mathcal{E} \to \mathcal{E}$ such that $f \circ f = f$, i.e. being identity on its image (Borsuk, 1967, Ch I, 2.). Thus, the retraction can be viewed as a generalization (possibly non-linear) of the projection operator/matrix.

These results are closely related to the *Gram–Schmidt orthogonalization* (Horn and Johnson, 2013, 0.6.4) and to the *Cholesky* matrix decomposition (Horn and Johnson, 2013, 7.2.9). The Gram–Schmidt orthogonalization is also in the heart of the *QR-decomposition* of an arbitrary rectangular matrix $A \in \mathcal{M}(n, p)$ into a product $Q[U^\top O_{p \times (n-p)}]^\top$ with $Q \in \mathcal{O}(n)$ and $U \in \mathcal{U}(p)$ (Horn and Johnson, 2013, 2.1.14). Here, $\mathcal{U}(p)$ denotes the subspace of all upper $p \times p$ triangular matrices. The number of the non-zero diagonal entries of $U \in \mathcal{U}(p)$ gives the rank of A, and thus such a decomposition is *rank revealing* (Golub and Van Loan, 2013, 5.4.5).

More generally, for any matrix $A \in \mathcal{GL}(n)$, there exists an unique symmetric PD matrix $S \in \mathcal{S}_+(n)$, such that $A = QS$ and $Q \in \mathcal{O}(n)$ (Kosinski, 1993, p. 228) or (Chevalley, 1946, Ch I, §5). The parallel of this result in numerical linear algebra is the *polar decomposition* (Horn and Johnson, 2013, 7.3.1).

These matrix decompositions and several others, as eigenvalue decomposition (EVD) and singular value decomposition (SVD), are widely used in MDA to facilitate the model definitions and, in many cases, to provide closed-form solutions. They are particularly important for MDA on matrix manifolds for unification of the parameters constraints. The matrix decompositions are also called *matrix factorizations*.

3.1.3 $\mathcal{O}(n, p)$

Along with $\mathcal{O}(n)$ we consider the set of all *orthonormal* $n \times p$ matrices Q known also as the *Stiefel manifold* (Stiefel, 1935). It is defined as

$$\mathcal{O}(n, p) = \{Q \in \mathcal{M}(n, p) : Q^\top Q = I_p\}, \tag{3.2}$$

i.e. it contains all full column rank $n \times p$ matrices Q with columns, orthogonal to each other and normalized to unit length. $\mathcal{O}(n, p)$ is compact in $\mathbb{R}^{n \times p}$, following the same argument used for the compactness of $\mathcal{O}(n)$.

The Stiefel manifold $\mathcal{O}(n, p)$ is connected. Note that $\mathcal{O}(n, n-1)$ is diffeomorphic (by expanding it to a full basis of $\mathbb{R}^n$) to $\mathcal{SO}(n)$, which is connected. For $p < n - 1$, all $\mathcal{O}(n, p)$ are continuous images of $\mathcal{O}(n, n - 1)$ and thus connected.

3.1.4 $\mathcal{O}_0(n, p)$

It is quite usual to centre the data matrix $X \in \mathbb{R}^{n \times p}$ before further analysis. The centred data matrix $X_0 = J_n X$ has zero column sums, where $J_n = I_n - 1_n 1_n^\top / n$ is known as the centring operator/matrix. The set of all such $n \times p$ matrices forms a vector subspace $\mathcal{M}_0(n, p)$ in $\mathbb{R}^{n \times p}$ (or $\mathcal{M}(n, p)$) of dimension $(n - 1)p$. Then, it is natural to consider orthonormal matrices which arise in $\mathcal{M}_0(n, p)(n > p)$ and also have columns, which sum to 0, i.e.

$$\mathcal{O}_0(n, p) := \{Q \in \mathcal{M}(n, p) | \ Q^\top Q = I_p \text{ and } 1_n Q = 0_n\} \ ,$$

which is called the *centred Stiefel manifold*. It is *not* possible to construct orthogonal matrix $Q \in \mathbb{R}^{n \times n}$, in which columns sum to 0_n.

For example, such centred orthonormal matrices arise in the economy SVD of $X \in \mathcal{M}_0(n, p)$, which is written as $X = QDP^\top$, but here $Q \in \mathcal{O}_0(n, p)$ and $P \in \mathcal{O}(p)$. The same is true for the economy QR-decomposition of $X \in \mathcal{M}_0(n, p)$, which is written as usual as $X = QR$, but with $Q \in \mathcal{O}_0(n, p)$ and R an upper $p \times p$ triangular matrix. Note that, if $Q \in \mathcal{O}_0(n, p)$, then $Q^\top J_n Q = I_p$ and $Q^\top 1_n 1_n^\top Q = O_p$, a $p \times p$ matrix of zeros.

3.1.5 $\mathcal{OB}(n)$

We already know that the orthogonal matrices $Q \in \mathcal{O}(n)$ have mutually orthogonal columns with unit lengths. Geometrically, they perform orthogonal rotations preserving segment lengths and angles. Historically, the initial goal of FA (and PCA) is to find uncorrelated factors underlying the correlation structure of the data. Of course, the correlation matrix of those initial factors is the identity. Frequently the factors are easier to interpret after a proper orthogonal rotation. Moreover, the orthogonally rotated factors remain uncorrelated.

However, it was recognized that the factors interpretation can be further improved if their orthogonality is sacrificed. Such factors can be achieved by *oblique* rotation. The set of all $n \times n$ oblique rotation matrices is denoted by $\mathcal{OB}(n)$. They also have unit length columns, but they are *not* mutually orthogonal. Their mutual (scalar) products give the cosines of the angles between the corresponding (oblique) columns. In other words, if $Q \in \mathcal{OB}(n)$, then $Q^\top Q$ has unit diagonal, but non-zero off-diagonal entries. In particular, the classical FA rotates the initial factors with a matrix $Q^{-\top}$ for some $Q \in \mathcal{OB}(n)$. This implies that the oblique rotations should be non-singular (Harman, 1976, 13.26). The resulting factors are correlated and their correlation matrix is given by $Q^\top Q$ (Harman, 1976, 13.16). Further details are given in Section 5 and Section 6.

Thus, a $n \times n$ matrix Q is called oblique if $Q^\top Q$ is a *correlation matrix*, i.e. if $Q^\top Q$ is PD with unit main diagonal. This can be written as

$$\mathcal{OB}(n) = \{Q \in \mathcal{M}(n) : Q^\top Q\text{--correlation matrix}\} \ . \tag{3.3}$$

It is well known that $\operatorname{rank}(Q^\top Q) \leq \min\{\operatorname{rank}(Q^\top), \operatorname{rank}(Q)\} = \operatorname{rank}(Q)$ (Horn and Johnson, 2013, 0.4.5(c)) However, if the $p \times p$ matrix Q is also not singular, then

$$\begin{aligned}
\operatorname{rank}(Q) &= \operatorname{rank}(Q^{-\top} Q^\top Q) \leq \min\{\operatorname{rank}(Q^{-\top}), \operatorname{rank}(Q^\top Q)\} \\
&= \min\{\operatorname{rank}(Q), \operatorname{rank}(Q^\top Q)\} = \operatorname{rank}(Q^\top Q) \ .
\end{aligned}$$

Combining the two rank inequalities shows that $\operatorname{rank}(Q^\top Q) = \operatorname{rank}(Q)$, i.e. $Q^\top Q$ is necessarily PD if $Q \in \mathcal{GL}(n)$. Then, the definition (3.3) can be transformed into

$$\mathcal{OB}(n) = \{Q \in \mathcal{GL}(n) : \operatorname{diag}(Q^\top Q) = I_n\} \ . \tag{3.4}$$

Clearly, $\mathcal{OB}(n)$ is bounded, because the columns of $Q \in \mathcal{OB}(n)$ have unit lengths, i.e. are bounded in $\mathbb{R}^{n \times n}$. However, $\mathcal{OB}(n)$ is open by the same argument that $\mathcal{GL}(n)$ is open. Thus, $\mathcal{OB}(n)$ is not compact.

Example: A general oblique 2×2 matrix Q can be written as

$$Q(\alpha, \beta) = \begin{bmatrix} \cos\alpha & \cos\beta \\ \sin\alpha & \sin\beta \end{bmatrix} \ ,$$

for some $\alpha, \beta \in \mathbb{R}$, such that $\sin(\beta - \alpha) \neq 0$ or $\beta \neq \alpha \pm k\pi$ and $k = 0, 1, 2, \ldots$ It is clear that one can choose a sequence $\alpha_i (\neq \beta \pm k\pi)$ converging to β. As sin and cos are continuous functions, the corresponding sequence of oblique matrices $Q(\alpha_i, \beta)$ converges to $Q(\beta, \beta)$, which is singular. Thus, $Q(\beta, \beta)$ is a limit point of a sequence from $\mathcal{OB}(2)$ which does not belong to it, i.e. $\mathcal{OB}(2)$ is not closed.

Many authors relax the above definition of oblique matrix as follows: an $n \times n$ matrix $Q = [q_1, \ldots, q_n]$ is oblique when its columns have unit lengths, i.e. $q_1^\top q_1 = \ldots = q_n^\top q_n = 1$ (ten Berge and Nevels, 1977). In particular, it is also adopted in Manopt (http://www.manopt.org/tutorial.html). The set of such matrices is equivalent to the product $\prod_n \mathcal{S}^{n-1}$ of n unit spheres $\mathcal{S}^{n-1}$ in $\mathbb{R}^n$. It is compact as a finite product of compact sets: a product of p exemplars of the unit sphere $\mathcal{S}^{n-1}$, which is compact in $\mathbb{R}^n$.

To evaluate the difference between these two definitions, consider the set of all singular $n \times n$ matrices with unit length columns:

$$\mathcal{OB}_0(n) = \{Q \in \mathcal{M}(n)| \det(Q) = 0 \text{ and } \mathrm{diag}(Q^\top Q) = I_n\}, \qquad (3.5)$$

e.g. $\frac{1}{\sqrt{2}} \begin{bmatrix} 1 & 1 \\ 1 & 1 \end{bmatrix} = \frac{1}{\sqrt{2}} E_2 \in \mathcal{OB}_0(2)$. Clearly, $\mathcal{OB}_0(n)$ is bounded and closed in $\mathbb{R}^{n \times n}$.

By construction, we have that $\mathcal{OB}(n) \cap \mathcal{OB}_0(n) = \emptyset$ and $\mathcal{OB}(n) \cup \mathcal{OB}_0(n)$ is the set of all matrices in $\mathbb{R}^{n \times n}$ with unit length columns, which is another way to express $\prod_p \mathcal{S}^{p-1}$. Thus, $\mathcal{OB}(n)$ and $\mathcal{OB}_0(n)$ are, respectively, the interior and the boundary of $\prod_p \mathcal{S}^{p-1}$. Indeed, for $\mathcal{OB}_0(n)$ to be the boundary of $\prod_n \mathcal{S}^{n-1}$ means that if $Q \in \mathcal{OB}_0(n)$, then any ball $\mathcal{B}_Q$ contains matrices from both $\mathcal{OB}(n)$ and the complement of $\prod_n \mathcal{S}^{n-1}$. It follows from (3.5) that if $Q \notin \mathcal{OB}_0(n)$, then either $\det(Q) \neq 0$ or $\mathrm{diag}(Q^\top Q) \neq I_n$ or both, i.e. either $Q \in \mathcal{OB}(n)$ or Q belongs to the complement of $\prod_n \mathcal{S}^{n-1}$.

This shows that the difference between the two definitions of oblique matrices is the set $\mathcal{OB}_0(n)$, which has *measure zero*. This follows from the fact that the set of the roots of any polynomial (and particularly the determinant) has measure zero. It is said that $\mathcal{U}$ has measure zero in $\mathbb{R}^n$, when for any $\epsilon > 0$, it can be covered by a countable sequence of n-dimensional balls $B_i, i = 1, 2, \ldots$ with total volume less than ϵ. Loosely speaking, this means that the chance/probability to pick up a matrix $Q \in \mathcal{OB}_0(n)$ is 0.

3.1.6 $\mathcal{OB}(n,p)$

A $n \times p$ $(n > p)$ matrix Q of full column rank is called oblique if $Q^\top Q$ is a correlation matrix. Then, the set of all $n \times p$ oblique matrices is defined as

$$\mathcal{OB}(n,p) = \{Q \in \mathcal{M}(n,p) : \text{rank}(Q) = p, \ Q^\top Q\text{---correlation matrix}\} .$$
(3.6)

As Q has full column rank p, the product $Q^\top Q$ is necessarily positive definite (Horn and Johnson, 2013, 7.1.6). Thus, as in the squared case, the definition (3.6) can be modified to

$$\mathcal{OB}(n,p) = \{Q \in \mathcal{M}(n,p) : \text{rank}(Q) = p, \ \text{diag}(Q^\top Q) = I_p\} ,$$
(3.7)

i.e. the full column rank matrices with unit length columns are called oblique. It follows from the previous considerations that $\mathcal{OB}(n,p)$ is bounded and open in $\mathbb{R}^{n \times p}$, and that its closure is $\prod_p \mathcal{S}^{n-1}$. Particularly, $\prod_p \mathcal{S}^1$ is known as the p-dimensional *torus*.

3.1.7 $\mathcal{G}(n,p)$

Many MDA techniques, e.g. PCA and EFA, obtain matrix solutions which are not unique in a sense that if they are multiplied by a non-singular or orthogonal matrix, the resulting product is still a valid solution. Such situations naturally require their consideration on *Grassmann manifold*.

The set of all p-dimensional subspaces of $\mathbb{R}^n$ $(n > p)$ (all p-dimensional hyperplanes in $\mathbb{R}^n$ passing through 0) is known as the Grassmann manifold $\mathcal{G}(n,p)$. There are many alternative ways to define $\mathcal{G}(n,p)$ and prove that it is a smooth compact manifold without boundary with dimension $p(n-p)$, see (Lee, 2003, p. 22) and (Munkres, 1966, 5.2).

Every element of $\mathcal{G}(n,p)$ is generated by p linearly independent vectors $a_1, \ldots, a_p$ in $\mathbb{R}^n$. If they are arranged (as columns) in a matrix $A \in \mathcal{M}(n,p)$, it will be a full rank matrix. The subset $\mathcal{M}_p(n,p)$ of all full rank matrices in $\mathcal{M}(n,p)$ is known as the *non-compact Stiefel manifold*. Let $A \in \mathcal{M}_p(n,p)$ and $\mathcal{A}$ be the subspace spanned by its columns. Then, the surjective map $f : \mathcal{M}_p(n,p) \to \mathcal{G}(n,p)$ induces topology in $\mathcal{G}(n,p)$ as follows: $\mathcal{V} \subset \mathcal{G}(n,p)$ is open in $\mathcal{G}(n,p)$, if $f^{-1}(\mathcal{V}) = \mathcal{U}$ is open in $\mathcal{M}_p(n,p)$.

In a similar fashion, we can take the set of all orthonormal full rank matrices, i.e. the Stiefel manifold $\mathcal{O}(n,p) \subset \mathcal{M}_p(n,p)$. Note that $\mathcal{O}(n,p)$ and $\mathcal{M}_p(n,p)$ are homeomorphic: the identity map in the right direction, and the Gram–Schmidt orthogonalization in the opposite direction. Then, if $\mathcal{Q}$ is the subspace spanned by the columns of $Q \in \mathcal{O}(n,p)$, the map $f_{\mathcal{O}} : \mathcal{O}(n,p) \to \mathcal{G}(n,p)$ induces the same topology in $\mathcal{G}(n,p)$ as above. As $f_{\mathcal{O}}$ is continuous and $\mathcal{O}(n,p)$ is compact, its image $\mathcal{G}(n,p)$ is also compact.

3.2 Differentiable manifolds

The term differentiable manifold was already mentioned in Section 2.5 in relation to the graph of a differentiable function f, i.e. the set formed by the pairs $\{x, f(x)\}$ on some open $\mathcal{U}$. This section provides further information which will be a general basis to consider a number of parametric spaces appearing in MDA.

Let $\mathcal{U} \subset \mathbb{R}^n$ and $\mathcal{V} \subset \mathbb{R}^p$ be open. Then, the function $f : \mathcal{U} \to \mathcal{V}$ is called *differentiable of class C^k* , if it has continuous derivatives of order k, and we write $f \in C^k$. If $f \in C^k$ for all $k > 0$, then f is of class C^∞ and is called *smooth*. In particular, the continuous function f is C^0. Clearly, we have

$$C^0 \supset C^1 \supset \ldots \supset C^\infty .$$

If f is bijective and smooth (together with f^{-1}), then it is called *diffeomorphism*. If f is diffeomorphism, then $n = p$ and its Freshet derivative is a non-singular linear operator. We assume that $f \in C^\infty$, unless specified otherwise.

A subset $\mathcal{M} \subset \mathbb{R}^n$ is called p-dimensional manifold $(n \geq p)$, or simply p-manifold, if it "looks locally" as $\mathbb{R}^p$. Every vector space is a manifold, e.g. $\mathbb{R}^n$ is a smooth n-manifold. A n-(sub)manifold in $\mathbb{R}^n$ is simply an open set in $\mathbb{R}^n$ and vice versa. A set of isolated points in $\mathbb{R}^n$ is a 0-manifold. By convention, the empty set $\emptyset$ is a manifold with dimension -1. The curves and surfaces in $\mathbb{R}^3$ are standard examples of one- and two-dimensional manifolds.

Depending on f being of class C^0, C^k or C^∞, the set of points $\{x, f(x)\}$ forms a manifold, which is topological, $k-$differentiable or smooth, respectively. In particular, the one-dimensional C^k manifolds are called curves (of class C^k), the two-dimensional C^k manifolds are C^k surfaces, etc.

However, not all manifolds can be expressed in this form. For example, this is the case with the unit sphere $\mathcal{S}^2$ in $\mathbb{R}^3$, for which we have

$$\left\{x_1, x_2, \pm\sqrt{1 - x_1^2 - x_2^2}\right\} ,$$

i.e. we need *two* different functions in its definition.

A more general definition that covers such sets is as follows. A subset $\mathcal{M} \subset \mathbb{R}^n$ is called p-dimensional *smooth manifold* if every $x \in \mathcal{M}$ has a neighbourhood $\mathcal{U} \cap \mathcal{M}$ diffeomorphic to an open set $\mathcal{V} \subset \mathbb{R}^p$. Let $\phi : \mathcal{V} \to \mathcal{U} \cap \mathcal{M}$ denote this diffeomorphism. Then, ϕ is called *parametrization* of $\mathcal{U} \cap \mathcal{M}$, and $\phi^{-1} : \mathcal{U} \cap \mathcal{M} \to \mathcal{V}$ is called *system of coordinates* on $\mathcal{U} \cap \mathcal{M}$. If $\phi \in C^k$, then $\mathcal{M}$ is called *differentiable manifold* of class C^k. If $\phi \in C^0$, i.e. ϕ is only homeomorphism, then $\mathcal{M}$ is called *topological manifold*. The pair $(\mathcal{V}, \phi)$ is called *coordinate chart*. Two charts $(\mathcal{V}_1, \phi_1)$ and $(\mathcal{V}_2, \phi_2)$ are called *related* when either $\mathcal{V}_1 \cap \mathcal{V}_2 = \emptyset$ or $\phi_1(\phi_2^{-1})$ and $\phi_2(\phi_1^{-1})$ are diffeomorphisms. The collection of all related charts $(\mathcal{V}_i, \phi_i)$ of $\mathcal{M}$ such that $\mathcal{M} \subset \cup\phi_i(\mathcal{V}_i)$ is called *atlas* on $\mathcal{M}$. If $\mathcal{M}$ admits countable atlas, it is called *Riemannian manifold*, i.e. it has *Riemannian metric*, see for details Section 3.5.2. Of course, if $\mathcal{M}$ is compact, then it has a finite atlas. If $\mathcal{M}$ admits countable atlas, then any submanifold $\mathcal{N} \subset \mathcal{M}$ also admits one (Chevalley, 1946, Ch III, §IX). In particular, if $\mathcal{N}$ is an n-dimensional manifold in $\mathbb{R}^m (m \geq n)$, then $\mathcal{N}$ is Riemannian. A brief and clear contemporary coverage is given in (Do Carmo, 1992, Ch 1).

There exist sets with boundary which behave as manifolds, e.g. the unit ball in $\mathbb{R}^3$ with the unit sphere $\mathcal{S}^2$ as its boundary. They do not fit well in the standard manifold definition, which needs slight modification as follows: a subset $\mathcal{M} \subset \mathbb{R}^n$ is called $p-$dimensional *manifold with boundary* if every $x \in \mathcal{M}$ has a neighbourhood $\mathcal{U} \cap \mathcal{M}$ diffeomorphic to an open set $\mathcal{V} \cap \mathbb{H}^p$, where

$$\mathbb{H}^p = \{x \in \mathbb{R}^p : x_p \geq 0\}$$

is an Euclidean half-space or *hyperplane* with boundary $\partial\mathbb{H}^p = \mathbb{R}^{p-1} \times \{0\} \subset \mathbb{R}^p$. Sometimes, the (standard) manifolds without boundary are called *non-bounded*. The boundary $\partial\mathcal{M}$ is non-bounded manifold with dimension $p-1$. The interior $\mathcal{M} - \partial\mathcal{M}$ is also a manifold with dimension p. Hereafter, all manifolds are assumed non-bounded, unless explicitly stated.

Returning to $\mathcal{S}^2$, we can say that

$$(x_1, x_2) \to \left(x_1, x_2, \pm\sqrt{1 - x_1^2 - x_2^2}\right) ,$$

parametrize, respectively, the regions $x_3 > 0$ and $x_3 < 0$ of $\mathcal{S}^2$ and define two charts. By "rotating" x_1, x_2 and x_3 we see that there are altogether six ($= 3 \times 2$) such parametrizations (charts) of $\mathcal{S}^2$. This shows that $\mathcal{S}^2$ is a two-dimensional smooth manifold in $\mathbb{R}^3$. Following the same way of thinking one can conclude that the unit sphere $\mathcal{S}^{n-1}$ in $\mathbb{R}^n$ is $(n-1)$–dimensional manifold and needs $2n$ parametrizations, i.e. an atlas with $2n$ charts. However, by using the so-called stereographic projection of $\mathcal{S}^{n-1}$ one can see that only two parametrizations are enough, i.e. an atlas with only two charts exists. Indeed, the stereographic projection connects the north pole $(0,0,1)$ of $\mathcal{S}^{n-1}$ with every point on the plane passing through its equator, i.e. $\mathbb{R}^{n-1} \times \{0\}$, by a straight line. Thus, every point form the north hemisphere (where the line from $(0,0,1)$ crosses $\mathcal{S}^{n-1}$) is projected onto $\mathbb{R}^{n-1} \times \{0\}$. The same construction with the south pole gives the second parametrization.

We stress here on several essential facts that help to identify manifolds usually arising in MDA. The simplifying argument is that such MDA manifolds are (sub)manifolds in $\mathbb{R}^n$ for some n.

Let $\mathcal{N}$ and $\mathcal{P}$ be differential manifolds of dimensions n and p, respectively, and that $f : \mathcal{N} \to \mathcal{P}$ be smooth. We assume that $n \geq p$. Rank of f at $x \in \mathcal{N}$ is defined as the rank of its Freshet derivative at x, or more precisely of its matrix $d_x f$ (or its Jacobi matrix at x). If the rank of f is the same at every point of $\mathcal{N}$, then f is said to have *constant rank* on $\mathcal{N}$. The points $x \in \mathcal{N}$ where the rank of f is equal to $p = \min\{n, p\}$ are called *regular*. This means that the dimension of its image $d_x f(\mathcal{N})$ at these points is p, which implies that the derivative of f at x is surjective. The set of all regular points of f is open in $\mathcal{N}$ (possibly empty). If f is diffeomorphism, then it has regular points only, and its Jacobi matrix is non-singular. The values $y \in \mathcal{P}$ are called *regular values* of f, if $f^{-1}(y)$ contains *only* regular points.

The points $x \in \mathcal{N}$ where the rank of f is less than $\min\{n, p\}$ are called *critical* or *singular*, and the corresponding $f(x)$ are called *critical values*. So, note that y is a *regular value*, if all points in $f^{-1}(y)$ are regular, but it is critical value, if $f^{-1}(y)$ contains at least one critical point of f. The smooth maps from lower to higher dimensions have only critical points.

Thus, every point $y \in \mathcal{P}$ is either regular or critical of f. Sard's theo-

rem proves that the set of all critical values of f has measure zero in $\mathcal{P}$ (Milnor, 1965, §2, §3) or (Guillemin and Pollack, 1974, p. 39-40). Note that the assertion is about the critical values, *not* about the critical points. Also, if $f^{-1}(y) = \emptyset$, then it contains *no* critical points, i.e. y is a regular value without being a value of f at all!

If $y \in \mathcal{P}$ is a regular value of f, and $\mathcal{N}$ is compact, then $f^{-1}(y)$ being closed in $\mathcal{N}$ is also compact. Moreover, it is *discrete*, since f is local diffeomorphism at each $x \in f^{-1}(y)$, which follows from the *inverse function theorem* (Rudin, 1976, p. 221) or more general for mappings (Lang, 1962, Ch I, §5). Further, the compactness of $\mathcal{N}$ implies that $f^{-1}(y)$ is *finite*.

If f has constant rank p on $\mathcal{N}$ it is called *submersion*, i.e. if it has only regular points on $\mathcal{N}$. The *canonical submersion* is simply given by $f(x_1, \ldots, x_n) = (x_1, \ldots, x_p)$. Every submersion is locally equivalent to a canonical one. If f has constant rank n on $\mathcal{N}$ it is called *immersion*. If f is a submersion, then $n \geq p$; if f is a immersion, then $p \geq n$. If f is both submersion and immersion, then it is a diffeomorphism and $n = p$. An alternative definition is that the mapping f is immersion/submersion ar x if the linear mapping $d_x f$ is injective/surjective. If $f : \mathcal{M} \to \mathcal{N}$ and $g : \mathcal{N} \to \mathcal{P}$ are both immersions/submersions, then their composition $g \circ f : \mathcal{M} \to \mathcal{P}$ is immersion/submersion too.

If $f : \mathcal{N} \to \mathcal{P}$ is an *injective* immersion and is a *proper* function, i.e. the preimage $f^{-1}(C) \subset \mathcal{N}$ of every compact $C \subset \mathcal{P}$ is compact, then f is called *embedding*. Clearly, if $\mathcal{N}$ is a compact manifold, then every injective immersion is embedding. The embeddings are very useful for identifying manifolds. Indeed, if $f : \mathcal{N} \to \mathcal{P}$ is an embedding, then $f(\mathcal{N})$ is a (sub)manifold in $\mathcal{P}$.

As before, let $\mathcal{N}$ and $\mathcal{P}$ be two manifolds with dimensions n and p, and $f : \mathcal{N} \to \mathcal{P}$ be smooth. If $n \geq p$, and if $y \in \mathcal{P}$ is a regular value, then $f^{-1}(y) \subset \mathcal{N}$ is a smooth manifold of dimension $n - p$. With this in mind, it is much easier to show that $\mathcal{S}^{n-1}$ is a manifold, than constructing specific parametrizations. Indeed, consider the function $f : \mathbb{R}^n \to \mathbb{R}$ defined by $f(x) = x_1^2 + \ldots + x_n^2$. One can check that $y = 1$ is a regular value, and thus $f^{-1}(1) = \mathcal{S}^{n-1}$ is a smooth manifold of dimension $n - 1$. This construction can be extended to manifolds which are (locally) regular level sets of smooth maps. If the set of regular values of f is a k-manifold $\mathcal{K} \subset \mathcal{P}$, then $f^{-1}(\mathcal{K})$ is a manifold of dimension $n - p + k$ (or empty).

Another related result is provided by the constant rank theorem (Rudin,

1976, 9.32) or (Conlon, 2001, Th 2.5.3). Let $\mathcal{U} \subset \mathbb{R}^n$ and $\mathcal{V} \subset \mathbb{R}^p$ be open, and the smooth function $f : \mathcal{U} \to \mathcal{V}$ has constant rank k. Then, for $v \in \mathcal{V}$, $f^{-1}(v)$ is a smooth (sub)manifold of $\mathcal{U}$ of dimension $n - k$.

The following result is very convenient for "generation" of manifold with boundary. Let $\mathcal{M}$ be a non-bounded manifold ($\partial \mathcal{M} = \emptyset$) and 0 be a regular value of $g : \mathcal{M} \to \mathbb{R}$. Then $\{x \in \mathcal{M} : g(x) \geq 0\}$ is a smooth manifold with boundary defined by $g^{-1}(0)$.

Now, consider a smooth map $f : \mathcal{N} \to \mathcal{P}$ between manifolds with dimensions $n > p$, and assume that $\mathcal{N}$ has boundary, i.e. $\partial \mathcal{N} \neq \emptyset$. If $y \in \mathcal{P}$ is a regular value for both f and its restriction on $\partial \mathcal{N}$, then $f^{-1}(y) \subset \mathcal{N}$ is a smooth $(n - p)$-manifold with boundary $\partial(f^{-1}(y)) = f^{-1}(y) \cap \partial \mathcal{N}$ (Milnor, 1965, p. 13). In particular, this result helps to prove that any continuous function from the closed unit ball $\overline{\mathcal{B}(0, 1)} \subset \mathbb{R}^n$ to itself has a fixed point, i.e. $f(x) = x$ for some $x \in \overline{\mathcal{B}(0, 1)}$. This is known as the Brouwer fixed point theorem (Milnor, 1965, p. 14).

Standard references on the topic are (Guillemin and Pollack, 1974; Lang, 1962; Milnor, 1965). Further details can be found in (Conlon, 2001; Hirsh, 1976; Kosinski, 1993; Lee, 2003; Tu, 2010).

3.3 Examples of matrix manifolds in MDA

3.3.1 $\mathcal{GL}(n)$

The set of all non-singular matrices $\mathcal{GL}(n)$ in $\mathcal{M}(n)$ forms a smooth manifold of dimension n^2 in $\mathbb{R}^{n \times n}$. Indeed, we already know that $\mathcal{GL}(n)$ is an open subset of a smooth manifold ($\mathbb{R}^{n \times n}$), and thus it is a smooth manifold itself.

3.3.2 $\mathcal{O}(n)$

The set of all $n \times n$ orthogonal matrices $\mathcal{O}(n)$ forms a smooth manifold of dimension $n(n - 1)/2$ in $\mathbb{R}^{n \times n}$. To see this, let us consider the quadratic matrix mapping $f(Q) = Q^\top Q$ from $\mathbb{R}^{n \times n}$ to $\mathbb{R}^{\frac{n(n+1)}{2}}$, because f can take only $\frac{n(n+1)}{2}$ independent values as $Q^\top Q$ is symmetric. If we can prove that the identity I_n is a regular value for f, then we can conclude that $f^{-1}(I_n) = \mathcal{O}(n)$

is a smooth manifold in $\mathbb{R}^{n \times n}$ with dimension $n^2 - \frac{n(n+1)}{2} = \frac{n(n-1)}{2}$.

The derivative of f at Q is given by $d_Q f = (dQ)^\top Q + Q^\top (dQ)$. This is a linear function of dQ, mapping the tangent space around Q to the tangent space around $f(Q)$. As f is a mapping between two vector spaces, the tangent spaces at any of their points coincide with the vector spaces themselves. In other words, this bowls down to showing that the linear map $d_Q f : \mathbb{R}^{n \times n} \rightarrow \mathbb{R}^{\frac{n(n+1)}{2}}$ should be surjective. This also means that $d_Q f(\mathbb{R}^{n \times n}) = \mathbb{R}^{\frac{n(n+1)}{2}}$, and that the rank of f should be $\frac{n(n+1)}{2}$. Consider the symmetric matrix $S = \frac{S}{2} + \frac{S^\top}{2} = (dQ)^\top Q + Q^\top (dQ)$. Then we can find from $\frac{S}{2} = (dQ)^\top Q$ that $\frac{S}{2} Q^\top = (dQ)^\top$, which implies that for any $S \in \mathbb{R}^{\frac{n(n+1)}{2}}$ we can find $dQ \in \mathbb{R}^{n \times n}$, suggesting that $d_Q f$ is surjection.

Another way of proving this is to consider $Q \in \mathcal{O}(n)$ as $Q = [q_1, \ldots, q_n]$. Then, the mapping $f : \mathbb{R}^{n^2} \rightarrow \mathbb{R}^{\frac{n(n+1)}{2}}$ can be defined by $\frac{n(n+1)}{2}$ coordinate real functions $f_{ij} = q_i^\top q_j - \delta_{ij}$ for $1 \leq i \leq j \leq n$, where δ_{ij} is the usual Kronecker symbol. Then, one needs to prove that $d_Q f_{ij}$ are linearly independent (Guillemin and Pollack, 1974, p. 23). This can also be proven by seeing that the $n^2 \times \frac{n(n+1)}{2}$ matrix of partial derivatives $\{\partial_k f_{ij}\}$ has (full) rank $\frac{n(n+1)}{2}$. Indeed, for $1 \leq i \leq j \leq p$ and $1 \leq k \leq p$ we have $\frac{\partial f_{ij}}{\partial q_k} = 0_p$ if $k \neq i, j$, and non-zero otherwise. This shows again that the dimension of $Q \in \mathcal{O}(n)$ is $n^2 - \frac{n(n+1)}{2} = \frac{n(n-1)}{2}$.

We can mention only that both $\mathcal{GL}(n)$ and $\mathcal{O}(n)$ are also *Lie groups*. Indeed, they are smooth manifolds, they are groups with respect to the matrix multiplication, and the group operations $(X, Y) \rightarrow XY$ and $X \rightarrow X^{-1}$ are also smooth (Chevalley, 1946, Ch 4). If the operations are only continuous, then the groups are called *topological*.

Let $\mathcal{G}$ be a matrix Lie group. Then $\mathfrak{g} = \{A : \exp(\alpha A) \in \mathcal{G} \text{ for all } \alpha \in \mathbb{R}\}$ is called its *Lie algebra*. For example, the Lie algebra $\mathfrak{gl}(n)$ of $\mathcal{GL}(n)$ is $\mathcal{M}(n)$. What is the Lie algebra $\mathfrak{o}(n)$ of $\mathcal{O}(n)$?

The Lie group theory is a vast mathematical subject. In spite of being traditionally theoretical, in the last few decades it started to relate to certain numerical areas, e.g. Engø et al. (1999); Iserles et al. (2000). Here, we are not going to use the Lie group formalism because most of the manifolds appearing in MDA are not matrix groups and can simply be treated as Riemannian manifolds.

3.3.3 $\mathcal{O}(n, p)$

The set of all $n \times p$ $(n \geq p)$ orthonormal matrices $\mathcal{O}(n, p)$ is compact and forms a smooth manifold of dimension $np - p(p + 1)/2$ in $\mathbb{R}^{n \times p}$. This follows from the same arguments as for the case of $\mathcal{O}(n)$. It is known as the (compact) Stiefel manifold (Stiefel, 1935).

3.3.4 $\mathcal{O}_0(n, p)$

The set of all $n \times p$ centred orthonormal matrices $\mathcal{O}_0(n, p)$ forms a smooth manifold of dimension $(n - 1)p - p(p + 1)/2$ in $\mathcal{M}_0(n, p)$. This follows from the same arguments as for the case of $\mathcal{O}(n)$ and $\mathcal{O}(n, p)$, and by taking into account that the nth row of every $Q \in \mathcal{O}_0(n, p)$ can be expressed as a linear combination of the other $n - 1$ ones.

3.3.5 $\mathcal{OB}(n)$

The unit sphere $\mathcal{S}^{n-1} \subset \mathbb{R}^n$ is a (smooth) manifold of dimension $n - 1$. Thus, $\prod_n \mathcal{S}^{n-1}$ and $\prod_p \mathcal{S}^{n-1}$ are smooth manifolds of dimension $n(n - 1)$ and $p(n - 1)$, respectively. We already know that $\mathcal{OB}(n)$ and $\mathcal{OB}(n, p)$ are their interiors, and thus they are also smooth manifolds with dimensions $n(n - 1)$ and $p(n - 1)$, respectively (Milnor, 1965, p. 12).

An independent proof that $\mathcal{OB}(n)$ forms a smooth manifold in $R^{n \times n}$ is similar to that one for $\mathcal{O}(n)$. Consider $Q \in \mathcal{O}(n)$ as $Q = [q_1, \dots, q_n]$. Then, define the mapping $f : \mathbb{R}^{n^2} \to \mathbb{R}^n$ by n coordinate real functions $f_i = q_i^\top q_i - 1$ for $1 \leq i \leq n$. The $n^2 \times n$ matrix of partial derivatives $\{\partial_k f_{ij}\}$ has rank n. Indeed, for $1 \leq i \leq j \leq n$ we have $\partial_j f_i = 0_n$, if $i \neq j$, and non-zero otherwise. This shows that the dimension of $Q \in \mathcal{OB}(n)$ is $n^2 - n = n(n - 1)$.

Alternatively, one can follow the construction from Section 3.3.2. Consider the mapping $f : \mathbb{R}^{n \times n} \to \mathbb{R}^{n^2 - n}$ defined by $Q \to Q^\top Q$, because the number of independent values taken by f is $n(n - 1)$ as $\text{diag}(Q^\top Q) = I_n$. Now, we need to prove that 1_n is a regular value of f, i.e. that $d_Q f$ is surjective. The derivative of f at Q is given by $d_Q f = \text{diag}[(dQ)^\top Q + Q^\top (dQ)]$ and defines a linear map $d_Q f : \mathbb{R}^{n \times n} \to \mathbb{R}^n$, which should be surjective, i.e. that for every $v \in \mathbb{R}^n$ there should exists at least one dQ. To show this, consider the symmetric matrix with zero diagonal $S = S - S \odot I_n = (dQ)^\top Q + Q^\top (dQ)$. Then, from $\frac{S}{2} = (dQ)^\top Q$, we can find that $\frac{S}{2} Q^\top = (dQ)^\top$, i.e. for any symmetric matrix with zero diagonal $S \in \mathbb{R}^{\frac{n(n+1)}{2}}$ we have corresponding $dQ \in \mathbb{R}^{n \times n}$. This suggests that $d_Q f$ is surjection, which implies that $f^{-1}(1_n) = \mathcal{OB}(n)$ is a smooth manifold of dimension $n^2 - n = n(n-1)$.

The same arguments apply to prove that $\mathcal{OB}(n,p)$ forms a smooth manifold of dimension $np - p = p(n-1)$.

3.3.6 $\mathcal{G}(n,p)$

Now, we want to show that locally $\mathcal{G}(n,p)$ looks like $\mathbb{R}^{p(n-p)}$, i.e. that $\mathcal{G}(n,p)$ is a smooth manifold of dimension $p(n-p)$. To do this, we construct a manifold structure of $\mathcal{M}_p(n,p)$ and induce it to $\mathcal{G}(n,p)$. First, we note that $A_1, A_2 \in \mathcal{M}_p(n,p)$ span the same $\mathcal{A}$, if and only if their columns are linear combinations of each other, i.e. there exists a non-singular $X \in \mathcal{GL}(p)$, such that $A_2 = A_1 X$. In this sense, A_1 and A_2 are considered equivalent. All such equivalent matrices form the elements (called *equivalence classes*) of the *quotient set* $\mathcal{M}_p(n,p)/\mathcal{GL}(p)$. In this sense, $\mathcal{G}(n,p)$ can be identified with $\mathcal{M}_p(n,p)/\mathcal{GL}(p)$.

Similarly, we can identify $\mathcal{G}(n,p)$ with $\mathcal{O}(n,p)/\mathcal{O}(p)$, in which elements (equivalence classes) are the sets of the form $\mathcal{V} = \{VQ : Q \in \mathcal{O}(p)\}$ for all $V \in \mathcal{O}(n,p)$. These two $\mathcal{G}(n,p)$ interpretations make it important for a number of MDA techniques.

Let $A \in \mathcal{M}_p(n,p)$. Then, there exists $X_A \in \mathcal{GL}(p)$, such that $AX_A = \begin{bmatrix} A_1 \\ A_2 \end{bmatrix}$ and $A_1 \in \mathcal{GL}(p)$. We define a neighbourhood $\mathcal{U}_A$ of $A \in \mathcal{U}_A \subset \mathcal{M}_p(n,p)$ as

$$\mathcal{U}_A = \left\{ B \in \mathcal{M}_p(n,p) : BX_A = \begin{bmatrix} B_1 \\ B_2 \end{bmatrix} : B_1 \in \mathcal{GL}(p) \right\} ,$$

and a smooth function $g : \mathcal{U}_A \to \mathcal{M}(n-p,p)$ by the rule $A \to B_2 B_1^{-1}$. Clearly, $g(A) = g(AX)$ for any $X \in \mathcal{GL}(p)$, and in this sense all such matrices AX are equivalent. This helps to define $\hat{g} : \mathcal{U}_{AX} \to \mathcal{M}(n-p,p)$, which gives the required homeomorphism between $\mathcal{M}_p(n,p)/\mathcal{GL}(p)$ and $\mathcal{M}(n-p,p)$. Indeed, it is injective, because for $B, C \in \mathcal{U}_A \subset \mathcal{M}_p(n,p)$, we have $BX_A = \begin{bmatrix} B_1 \\ B_2 \end{bmatrix}$ and $CX_A = \begin{bmatrix} C_1 \\ C_2 \end{bmatrix}$, with $B_2 B_1^{-1} = C_2 C_1^{-1}$, or $B_2 = C_2 C_1^{-1} B_1$. This shows that there exists $Y = C_1^{-1} B_1 \in \mathcal{GL}(p)$, such that $B_1 = C_1 Y$ and $B_2 = C_2 Y$, which gives $BX_A = (CX_A)Y$, i.e. they are equivalent. Now, to show it is surjective, let $W \in \mathcal{M}(n-p, p)$. Then, for $Y = \begin{bmatrix} X_A^{-1} \\ WX_A^{-1} \end{bmatrix} \in \mathcal{M}_p(n, p)$, we have

$YX_A = \begin{bmatrix} I_p \\ W \end{bmatrix} \in \mathcal{U}_A$ and $g(Y) = W$. Thus, $\{\mathcal{U}_A, \hat{g}^{-1}\}$ defines coordinate chart at every point $A \in \mathcal{M}_p(n, p)/\mathcal{GL}(p)$. Moreover, for $\mathcal{U}_{A_1} \cap \mathcal{U}_{A_2} \neq \emptyset$, we have

$\hat{g}_2\left(\hat{g}_1^{-1}(W)\right) = \hat{g}_2\left(\begin{bmatrix} X_{A_1}^{-1} \\ WX_{A_1}^{-1} \end{bmatrix}\right) = \hat{g}_2\left(\begin{bmatrix} X_{A_1}^{-1}X_{A_2} \\ WX_{A_1}^{-1}X_{A_2} \end{bmatrix}\right) = W$ and all involved

matrix operations are smooth.

The equivalence of $\mathcal{G}(n, p)$ with both $\mathcal{M}_p(n, p)/\mathcal{GL}(p)$ and $\mathcal{O}(n, p)/\mathcal{O}(p)$ describes its structure well, but working with equivalence classes (sets) is not convenient for numerical purposes. The following explicit $\mathcal{G}(n, p)$ definition in terms of matrices,

$$\mathcal{G}(n, p) = \{Q \in \mathcal{O}(n) : Q^\top = Q \text{ and } \operatorname{trace}(Q) = 2p - n\} , \qquad (3.8)$$

is suggested in (Bhatia, 2001, p. 305) and extensively studied in (Lai et al., 2020). Indeed, let $\mathcal{U}$ and $\mathcal{U}_\perp$ be any complementary subspaces of $\mathbb{R}^n$ of dimensions p and $n - p$, respectively, i.e. $\mathbb{R}^n = \mathcal{U} \oplus \mathcal{U}_\perp$. Their basis vectors can be ordered as columns of $U \in \mathcal{O}(n, p)$ and $U_\perp \in \mathcal{O}(n, n - p)$, such that $[U \ U_\perp] \in \mathcal{O}(n)$. Then, $Q = UU^\top - U_\perp U_\perp^\top$ belongs to (3.8) and gives the difference between the orthogonal projectors on the subspaces $\mathcal{U}$ and $\mathcal{U}_\perp$ (spanned by the columns of U and $U_\perp$, respectively).

3.4 Tangent spaces

The term tangent space was mention already in Section 2.5 for the particular case of graph of a differentiable function. We know from (2.24) that for any smooth map $f : \mathcal{U} \to \mathcal{V}$ defined for open $\mathcal{U} \subset \mathbb{R}^n$ and $\mathcal{V} \subset \mathbb{R}^p$, the derivative $d_x f : \mathbb{R}^n \to \mathbb{R}^p$ is a linear mapping in which $p \times n$ (Jacobian) matrix is composed of its partial derivatives at $x \in \mathcal{U}$. Then, the tangent space of $\mathcal{U} \subset \mathbb{R}^p$ at x is defined as the entire vector space $\mathbb{R}^p$ and is denoted by $\mathcal{T}_x\mathcal{U}$.

In general, let $\mathcal{M} \subset \mathbb{R}^p$ be a smooth manifold and $\mathcal{U} \subset \mathbb{R}^n$ be open, such that $g : \mathcal{U} \to \mathcal{M}$ be parametrization of a neighbourhood $g(\mathcal{U})$ of $x \in \mathcal{M}$ and $g(u) = x$. Then, the derivative at u is defined as the linear mapping $d_u g : \mathbb{R}^n \to \mathbb{R}^p$. Its image $d_u g(\mathbb{R}^n) \subset \mathbb{R}^p$ is called *tangent space* of $\mathcal{M}$ at x and is denoted by $\mathcal{T}_x\mathcal{M}$. Thanks to the chain rule of the derivative, this definition/construction does not depend on the parametrization g (Milnor, 1965, p. 4). One can check that $\mathcal{T}_x\mathcal{M} \subset \mathbb{R}^p$ is a vector (sub)space and its elements are called *tangent vectors* of $\mathcal{M}$ at x. If x is not specified, we simply write $\mathcal{T}_{\mathcal{M}}$. The dimension of the manifold $\mathcal{M}$ is given by the dimension of its tangent space $\mathcal{T}_x\mathcal{M}$, i.e. $\dim(\mathcal{M}) = \dim(\mathcal{T}_x\mathcal{M})$. In many occasions, this is a convenient way to find the dimension of a manifold. If $\mathcal{M}$ is a surface in $\mathbb{R}^p$ defined by $k(< p)$ equations, then its dimension is $p - k$, or the number of parameters defining the surface.

Another useful definition of the tangent space $\mathcal{T}_x\mathcal{M}$ uses the tangent (velocity) vectors of all smooth parametrized curves on $\mathcal{M}$ passing through $x \in \mathcal{M}$. A tangent vector of $\mathcal{M}$ at x is the tangent vector at $t = 0$ of a curve $c : (-\alpha, \alpha) \to \mathcal{M}$ with $c(0) = x$ for some $\alpha \in \mathbb{R}$. Each particular tangent vector of $\mathcal{T}_x\mathcal{M}$ is identified by the equivalence of $c_1(t)$ and $c_2(t)$ with equal derivatives at 0, i.e. with $d_0(\phi^{-1}c_1) = d_0(\phi^{-1}c_2)$, for some local parametrization ϕ around $x \in \mathcal{U}$. The equivalence does not depend on the choice of the coordinate chart $(\mathcal{U}, \phi)$.

If $\mathcal{M}_1$ and $\mathcal{M}_2$ are m_1- and m_2-manifolds, and $f : \mathcal{M}_1 \to \mathcal{M}_2$ is diffeomorphism, then $d_x f : \mathcal{T}_x\mathcal{M}_1 \to \mathcal{T}_{f(x)}\mathcal{M}_2$ is an isomorphism of vector spaces and $m_1 = m_2$ (Milnor, 1965, p. 7).

Let $\dim(\mathcal{T}_x\mathcal{M}) = k$. According to (2.3), every $a \in \mathcal{T}_x\mathcal{M} \subset \mathbb{R}^p$ have the form $a = (a_1, \ldots, a_k, 0, \ldots, 0)$. Now, consider the set of vectors $b \in \mathbb{R}^p$ for which $b^\top a = 0$ for any $a \in \mathcal{T}_x\mathcal{M}$. Clearly, the general form of such vectors is $b = (0, \ldots, 0, b_{k+1}, \ldots, b_p)$. They also form a vector subspace in $\mathbb{R}^p$, which is called the *orthogonal complement* of $\mathcal{T}_x\mathcal{M}$ in $\mathbb{R}^p$ with respect to the Euclidean inner product (2.9). The orthogonal complements are denoted by $(\mathcal{T}_x\mathcal{M})^\perp$, but in the case of tangent space, it is usually called the *normal space* of $\mathcal{M}$ at x and is denoted by $\mathcal{N}_x\mathcal{M}$. We also say that $\mathcal{T}_x\mathcal{M}$ and $\mathcal{N}_x\mathcal{M}$ are *orthogonal*, because for every $x \in \mathcal{T}_x\mathcal{M}$ and $y \in \mathcal{N}_x\mathcal{M}$ we have $x^\top y = 0$, and write $\mathcal{T}_x\mathcal{M} \perp \mathcal{N}_x\mathcal{M}$.

In this relation, it should be recalled that a point $x \in \mathcal{M}$ is critical point of the smooth function $f : \mathcal{M} \to \mathbb{R}$ if and only if $\mathrm{grad}_x f = 0$. However, if $\mathrm{grad}_x f \neq 0$ and $y \in \mathbb{R}$ is a regular value of f such that $x \in f^{-1}(y)$, i.e. x is a regular point, then $\mathrm{grad}_x f \perp \mathcal{T}_x(f^{-1}(y))$, i.e. $\mathrm{grad}_x f \in \mathcal{N}_x(f^{-1}(y))$ and is orthogonal to all tangent vectors to $f^{-1}(y)$ at x.

By construction, $\mathcal{T}_x\mathcal{M} \cap \mathcal{N}_x\mathcal{M} = 0_p$. Thus, every point $c \in \mathbb{R}^p$ can be expressed as $c = a + b$, where $a \in \mathcal{T}_x\mathcal{M}$ and $b \in \mathcal{N}_x\mathcal{M}$. In other words, at every $x \in \mathcal{M} \subset \mathbb{R}^p$, the vector space $\mathbb{R}^p$ is a *direct sum* of the two subspaces $\mathcal{T}_x\mathcal{M}$ and $\mathcal{N}_x\mathcal{M}$, i.e. $\mathbb{R}^p = \mathcal{T}_x\mathcal{M} \oplus \mathcal{N}_x\mathcal{M}$. Moreover, $\dim(\mathcal{N}_x\mathcal{M}) = p - k$.

For $x_1, x_2 \in \mathcal{M}$ and $x_1 \neq x_2$, the tangent vector spaces $\mathcal{T}_{x_1}\mathcal{M}$ and $\mathcal{T}_{x_2}\mathcal{M}$ are different. The union of all $\mathcal{T}_x\mathcal{M}$ for all $x \in \mathcal{M}$ is called the *tangent bundle* of $\mathcal{M}$ and is denoted by $\mathcal{T}\mathcal{M}$, i.e. $\mathcal{T}\mathcal{M} = \{(x, y) \mid x \in \mathcal{M}, y \in \mathcal{T}_x\mathcal{M}\}$. Moving from x_1 to x_2 on $\mathcal{M}$ causes considerable complications in (numerical) analysis on manifolds because some features may change with the local coordinates. As a result, the introduction of additional (Riemannian) constructions is required, e.g. connections, parallel transport, etc.

It the sequel we consider the tangent spaces of several matrix manifolds regularly used in MDA. We derive formulas for the projection of a general vector (in fact, matrix) on particular tangent space. Thus, we will be able to find a "curve-linear" gradient on $\mathcal{M}$ by simply projecting the standard (Euclidean) gradient on $\mathcal{T}_\mathcal{M}$.

3.4.1 $\mathcal{GL}(n)$

The dimension of the manifold $\mathcal{GL}(n)$ is n^2 because its tangent space is $\mathcal{M}(n)$ (or $\mathbb{R}^{n \times n}$). Indeed, let $A \in \mathcal{M}(n)$ and define a curve $c : (-\alpha, \alpha) \to \mathcal{M}(n)$ by $c(t) = I_p + tA$. Then, there exists a neighbourhood of $c(0) = I_p \in \mathcal{GL}(n)$, such that $I_p + tA \in \mathcal{GL}(n)$ for small enough t (see Exercises 3.8). Then, we find $\left.\frac{d(I_p + tA)}{dt}\right|_0 = A$, which shows that the tangent vector of $\mathcal{GL}(n)$ is an element of $\mathcal{M}(n)$. The same conclusion simply follows from the fact that $\mathcal{GL}(n)$ is open in $\mathcal{M}(n)$.

3.4.2 $\mathcal{O}(n)$

For a while we assume that Q depends on the real parameter t, such that for all $t \geq 0$ the matrix $Q(t)$ is orthogonal, i.e. we have $Q(t)^\top Q(t) = Q(t)Q(t)^\top = I_p$. In other words, $Q(t)$ is a curve in $\mathcal{O}(n)$ passing through the identity I_p at $t = 0$. For brevity, we usually write simply Q. By definition, a tangent (matrix) H at Q is the velocity of the smooth path $Q(t) \in \mathcal{O}(n)$ at $t = 0$. By differentiating $Q^\top Q = I_p$ at $t = 0$, we obtain that the tangent space of $\mathcal{O}(n)$ at any Q is given by

$$
\begin{aligned}
\mathcal{T}_Q\mathcal{O}(n) &= \{H \in \mathcal{M}(n) \mid H^\top Q + Q^\top H = O_p\} \\
&= \{H \in \mathcal{M}(n) \mid Q^\top H \text{ is skew-symmetric}\} . \quad (3.9)
\end{aligned}
$$

This implies that any matrix H tangent to $\mathcal{O}(n)$ at $Q \in \mathcal{O}(n)$ is necessarily of the form $Q^{\top}H = K$, or $H = QK$, for some skew-symmetric matrix $K \in \mathcal{K}(n)$ (Chu and Driessel, 1990; Chu and Trendafilov, 1998a). Then, we can rewrite (3.9) as follows:

$$\mathcal{T}_Q\mathcal{O}(n) = \{H \in \mathcal{M}(n)|\ H = QK,\ K \text{ skew-symmetric}\} = Q\mathcal{K}(n)\ .$$

Therefore, we are able to define the following projection:

Theorem 3.4.1. *Let $A \in \mathbb{R}^{n \times n}$. Then*

$$\pi_{\mathcal{T}}(A) := Q\frac{Q^{\top}A - A^{\top}Q}{2} = A - Q\frac{Q^{\top}A + A^{\top}Q}{2} \qquad (3.10)$$

defines the projection of A onto the tangent space $\mathcal{T}_Q\mathcal{O}(n)$.

The result (3.10) follows directly from the basic fact that every $p \times p$ matrix can be presented as a sum of a symmetric and skew-symmetric matrix (Chu and Driessel, 1990; Chu and Trendafilov, 1998b). Alternative considerations can be found in (Edelman et al., 1998, 2.2.1).

3.4.3 $\mathcal{O}(n, p)$

The same derivations for the tangent space of $\mathcal{O}(n)$ can be extended for the Stiefel manifold $\mathcal{O}(n, p)$ (Chu et al., 2001), (Edelman et al., 1998, 2.2.1).

Again, we suppose that Q depends on the real parameter $t \geq$, such that the matrix $Q(t)$ is orthonormal, and $Q(t)$ forms a one-parameter family of $n \times p$ orthonormal matrices. Thus, we regard $Q(t)$ to as a curve evolving on $\mathcal{O}(n, p)$, but for short write Q. By definition, a tangent vector H of $\mathcal{O}(n, p)$ at Q is the velocity of the smooth curve $Q(t) \in \mathcal{O}(p, q)$ at $t = 0$. By differentiation of $Q^T Q = I_p$ at $t = 0$ we obtain that

$$\dot{Q}^T|_{t=0}Q + Q^T\dot{Q}|_{t=0} = H^TQ + Q^TH = O_p,$$

and thus its tangent space $\mathcal{T}_Q\mathcal{O}(n, p)$ at any orthonormal matrix $Q \in \mathcal{O}(n, p)$ is given by

$$\begin{aligned}\mathcal{T}_Q\mathcal{O}(n, p) &= \{H \in \mathcal{M}(n, p)|H^TQ + Q^TH = O_p\} \\ &= \{H \in \mathcal{M}(n, p)|Q^TH \text{ is skew-symmetric}\}\ . \qquad (3.11)\end{aligned}$$

To further characterize a tangent vector, we recall that a least squares solution $A \in \mathcal{M}(n,p)$ to the equation

$$MA = N$$

is given by

$$A = M^{\dagger}N + (I - M^{\dagger}M)W \ ,$$

where $M^{\dagger}$ is the *Moore–Penrose inverse* of M, I is an identity matrix, and W is an arbitrary matrix of proper dimensions. In our case, $M = Q^{T}$, where $Q \in \mathcal{O}(n,p)$, and $N = K \in \mathcal{K}(p)$ is skew-symmetric, we note that $(Q^{T})^{\dagger} = Q$. Therefore, this is summarized in the following:

Theorem 3.4.2. *Any tangent vector $H \in \mathcal{T}_{Q}\mathcal{O}(n,p)$ has the form*

$$H = QK + (I_{p} - QQ^{T})W \ ,$$

where $K \in \mathcal{K}(p)$ and $W \in \mathcal{M}(n,p)$ is arbitrary. When $n = p$, $H = QK$.

Theorem 3.4.2 can also be rewritten as the following decomposition of $\mathbb{R}^{n \times p}$:

Theorem 3.4.3. $\mathbb{R}^{n \times p}$ *can be written as the direct sum of three mutually perpendicular subspaces*

$$\mathbb{R}^{n \times p} = Q\mathcal{S}(p) \oplus Q\mathcal{K}(p) \oplus \{X \in \mathcal{M}(n,p) | Q^{T}X = O_{p}\} \ ,$$

where the last term is the null space of Q^{T}.

Therefore we are able to define the following projections.

Corollary 3.4.4. *Let $A \in \mathcal{M}(n,p)$. Then*

$$\pi_{\mathcal{T}}(A) := Q\frac{Q^{T}A - A^{T}Q}{2} + (I_{n} - QQ^{T})A = A - Q\frac{Q^{T}A + A^{T}Q}{2} \quad (3.12)$$

defines the projection of A onto the tangent space $\mathcal{T}_{Q}\mathcal{O}(n,p)$. Similarly,

$$\pi_{\mathcal{N}}(A) := Q\frac{Q^{T}A + A^{T}Q}{2} \quad (3.13)$$

defines the projection of A onto the normal space $\mathcal{N}_{Q}\mathcal{O}(n,p)$.

We note that the projection (3.12) on $\mathcal{O}(n,p)$ is the same as the projection (3.10) on $\mathcal{O}(n)$.

3.4.4 $\mathcal{O}_0(n,p)$

From the same construction, assume that Q depends on the real parameter t, such that for all $t \geq 0$ the matrix $Q(t)$ is orthonormal with zero column sums, i.e. $Q(t) \in \mathcal{O}_0(n,p)$ for $t \geq 0$. For brevity, we simply write Q. The tangent matrix H at Q is the velocity of the smooth path $Q(t) \in \mathcal{O}_0(n,p)$ at $t = 0$. After differentiation $Q^\top Q = Q^\top J_n Q = I_p$ at $t = 0$, we obtain that the tangent space at any $Q \in \mathcal{O}_0(n,p)$ is

$$
\begin{aligned}
\mathcal{T}_Q \mathcal{O}_0(n,p) &= \{H \in \mathcal{M}(n,p)|\ H^\top J_n Q + Q^\top J_n H = O_p\} \\
&= \{H \in \mathcal{M}(n,p)|\ Q^\top J_n H \in \mathcal{K}(p)\}\ .
\end{aligned}
$$

Thus, for every $Q \in \mathcal{O}_0(n,p)$ and some $K \in \mathcal{K}(p)$ one has $(Q^\top J_n)H = Q^\top H = Q^\top (J_n H) = K$, which suggests that H can be assumed centred, i.e. $H = J_n H \in \mathcal{M}_0(n,p)$. Then, every $H \in \mathcal{T}_Q \mathcal{O}_0(n,p)$ has the form $H = QK + (I_n - QQ^\top)W = QK + (J_n - QQ^\top)W$ for some $K \in \mathcal{K}(p)$ and arbitrary $W \in \mathcal{M}(n,p)$.

Therefore, the projection of a general $A \in \mathcal{M}(n,p)$ onto the tangent space $\mathcal{T}_Q \mathcal{O}_0(n,p)$ is defined by

$$
\begin{aligned}
\pi_{\mathcal{T}_Q \mathcal{O}_0(n,p)}(A) &= Q\frac{Q^\top A - A^\top Q}{2} + (J_n - QQ^\top)A \\
&= J_n A - Q\frac{Q^\top A + A^\top Q}{2}\ .
\end{aligned}
\tag{3.14}
$$

The projection formula (3.14) suggests that A can be centred first, and then be subject to the standard formula (3.12) for $\mathcal{O}(n,p)$, provided $Q \in \mathcal{M}_0(n,p)$.

3.4.5 $\mathcal{OB}(n)$

By differentiating $\mathrm{diag}(Q^\top Q) = I_n$ at $t = 0$, the tangent space of $\mathcal{OB}(n)$ at any oblique matrix Q is given by

$$
\begin{aligned}
\mathcal{T}_Q \mathcal{OB}(n) &= \{H \in \mathcal{M}(n)|\ \mathrm{diag}(H^\top Q + Q^\top H) = O_p\} \\
&= \{H \in \mathcal{M}(n)|\ Q^\top H = \mathrm{off}(S) + K\ \}\ ,
\end{aligned}
\tag{3.15}
$$

where $S \in \mathcal{S}(n)$, $K \in \mathcal{K}(n)$, and $\mathrm{off}(A) = A - \mathrm{diag}(A)$ for $A \in \mathcal{M}(n)$. Then any matrix normal to $\mathcal{OB}(n)$ at $Q \in \mathcal{OB}(n)$ with respect to the Frobenius inner product (2.17) is necessarily of the form $H_\perp = QD$, for some diagonal $n \times n$ matrix D. Thus any $A \in \mathcal{M}(n)$ at Q can be written as a direct sum of elements from the tangent and the normal space of $\mathcal{T}_Q\mathcal{OB}(n)$ at Q, i.e. $A = H + H_\perp$. A right-hand side multiplication of $A = H + H_\perp = H + QD$ by $Q^\top$ followed by taking the Diag operation gives

$$\mathrm{diag}(Q^\top H) = 0_p = \mathrm{diag}(Q^\top A) - \mathrm{diag}(Q^\top Q)D \ ,$$

and the unknown diagonal matrix D is identified. Then, we can define the following projection (Jennrich, 2002; Trendafilov, 1999):

Theorem 3.4.5. *Let* $A \in \mathcal{M}(n)$. *Then*

$$\pi_\mathcal{T}(A) := A - Q\mathrm{diag}\left(Q^\top A\right) \tag{3.16}$$

defines the projection of A *onto the tangent space* $\mathcal{T}_Q\mathcal{OB}(n)$.

3.4.6 $\mathcal{OB}(n,p)$

Similar considerations apply to the rectangular oblique matrices $\mathcal{OB}(n,p)$ (Absil and Gallivan, 2006):

Theorem 3.4.6. *Let* $A \in \mathcal{M}(n,p)$. *Then*

$$\pi_\mathcal{T}(A) := A - Q\mathrm{diag}\left(Q^\top A\right) \tag{3.17}$$

defines the projection of A *onto the tangent space* $\mathcal{T}_Q\mathcal{OB}(n,p)$.

3.4.7 $\mathcal{G}(n,p)$

The symmetric orthogonal matrices involved in the explicit Grassmannian (3.8) have the following rather simple spectral form $Q = VDV^\top$, where $V = [V_1 \ V_2] \in \mathcal{O}(n)$ is a block-matrix with $V_1 \in \mathcal{O}(n,p)$ and $V_2 \in \mathcal{O}(n, n{-}p)$, and $D = \mathrm{diag}_B(I_p, -I_{n-p}) = I_{p,n-p}$ (Lai et al., 2020, 1.1). The eigenvectors V_1 and V_2 are not unique, but the subspaces that they span are.

In this notation, the Grassmann manifold with its standard definition and its explicit definition in (3.8) are diffeomorphic. More precisely, let $\mathcal{U}$ be the span of the columns of U, i.e. $\mathcal{U} = \mathrm{span}(U)$. Then $\mathcal{U} \to Q$ is diffeomorphism, with $Q = UU^\top - U_\perp U_\perp^\top = 2UU^\top - I_n$, and $Q \to \mathrm{span}(V_1)$ is its inverse. To find its differential at $\mathcal{U} \in \mathcal{G}(n, p)$, a p-dimensional subspace in $\mathbb{R}^n$, we identify it with $U_0 = \begin{bmatrix} I_p \\ O \end{bmatrix} \in \mathcal{O}(n, p)$. As $\mathcal{G}(n, p)$ is $p(n - p)$-manifold, the general element of its tangent space has the form $\begin{bmatrix} O_p \\ X \end{bmatrix} \in \mathcal{M}(n, p)$ for some $X \neq O_{n-p,n}$. The differential of $2UU^\top - I_n$ at U_0 is $(dU)U_0^\top + U_0(dU)^\top = \begin{bmatrix} O_p \\ X \end{bmatrix}[I_p\ O] + \begin{bmatrix} I_p \\ O \end{bmatrix}[O_p\ X^\top] = \begin{bmatrix} O_p & X^\top \\ X & O_{n-p} \end{bmatrix}$, which is a non-singular matrix.

Making use of the defining formula (3.8), we can calculate the tangent space of $\mathcal{G}(n, p)$ at Q as follows. Assume that Q depends on the real parameter t, such that for all $t \geq 0$ the matrix $Q(t) \in \mathcal{G}(n, p)$, and for brevity, we simply write Q. The tangent matrix H at Q is the velocity of the smooth path $Q(t) \in \mathcal{G}(n, p)$ at $t = 0$. After differentiation $Q^\top Q = Q^\top Q = I_n$ at $t = 0$, we obtain that the tangent space at any $Q \in \mathcal{G}(n, p)$ is (Lai et al., 2020, 3.1)

$$\mathcal{T}_Q \mathcal{G}(n, p) = \{H \in \mathcal{S}(n) : HQ + QH = O_n, \ \mathrm{trace} H = 0\} \quad (3.18)$$
$$= \left\{ V \begin{bmatrix} O_p & B \\ B^\top & O_{n-p} \end{bmatrix} V^\top : B \in \mathcal{M}(p, n - p) \right\}.$$

Indeed, let $Q = VI_{p,n-p}V^\top = [V_1\ V_2]I_{p,n-p}[V_1\ V_2]^\top$. Note that as $HQ \in \mathcal{K}(n)$, so is $V^\top(HQ)V = (V^\top HV)I_{p,n-p}$. Now, the symmetric $V^\top HV$ can be expressed as a block-matrix, and we have

$$(V^\top HV)I_{p,n-p} = \begin{bmatrix} A & B \\ B^\top & C \end{bmatrix} I_{p,n-p} = \begin{bmatrix} A & -B \\ B^\top & -C \end{bmatrix} \in \mathcal{K}(n),$$

where $A \in \mathcal{S}(p)$ and $C \in \mathcal{S}(n - p)$, and thus they should be zero matrices. Thus, we finally obtain $H = V \begin{bmatrix} O & -B \\ B^\top & O \end{bmatrix} I_{p,n-p} V^\top$ as stated in (3.18).

To define the projection of a general $A \in \mathcal{M}(n)$ onto the tangent space $\mathcal{T}_Q \mathcal{G}(n, p)$, first we make use of Theorem 3.4.1 and project A on $\mathcal{O}(n)$ to find $Q \frac{Q^\top A - A^\top Q}{2} \in \mathcal{T}_Q \mathcal{O}(n)$. Then, we need to separate the part of it, specifically belonging to $\mathcal{T}_Q \mathcal{G}(n, p)$ from $\mathcal{T}_Q \mathcal{O}(n) = \mathcal{T}_Q \mathcal{G}(n, p) \oplus \mathcal{N}_Q \mathcal{G}(n, p)$.

Let the tangent vector $H_{\mathcal{O}(n)}$ at $\mathcal{T}_Q \mathcal{O}(n)$ be expressed as $H_{\mathcal{O}(n)} = H + H_\perp$, for $H \in \mathcal{T}_Q \mathcal{G}(n, p)$ and $H_\perp \in \mathcal{N}_Q \mathcal{G}(n, p)$. Note that H from (3.18) can be rewritten as

$$H = QK = QV \begin{bmatrix} O & -B \\ B^\top & O \end{bmatrix} V^\top = -KQ,$$

suggesting that the projection on $\mathcal{T}_Q\mathcal{G}(n,p)$ has the form $H = \frac{QK-KQ}{2}$ and, respectively, $H_\perp = \frac{QK+KQ}{2}$. Of course, $\text{trace}(H^\top H_\perp) = 0$.

Thus, the projection of a general $A \in \mathcal{M}(n)$ on $\mathcal{G}(n,p)$ is given by the H-part of $Q\frac{Q^\top A - A^\top Q}{2} \in \mathcal{T}_Q\mathcal{O}(n)$, which is

$$
\pi_\mathcal{T}(A) = \frac{Q\frac{QA-A^\top Q}{2} - \frac{QA-A^\top Q}{2}Q}{2} = \frac{(A+A^\top) - Q(A+A^\top)Q}{4}
$$
$$
= Q\frac{Q(A+A^\top) - (A+A^\top)Q}{4} . \tag{3.19}
$$

3.5 Optimization on matrix manifolds

The origins of the optimization methods on manifolds go back to at least the mid of the last century to the pioneering works of (Arrow et al., 1958) and (Rosen, 1960, 1961), and further developed in (Luenberger, 1972).

3.5.1 Dynamical systems

In the 1980s work had begun to appear that considered numerical linear algebra algorithms, such as EVD and SVD, as dynamical systems evolving on specific matrix manifolds (Ammar and Martin, 1986; Brockett, 1991; Deift et al., 1983; Flaschka, 1974; Watkins, 1983). In particular, the beginning was marked by the discovery that the (iterative) QR algorithm for computing EVD is a discretization of the (non-periodic) Toda flow (Deift et al., 1983; Flaschka, 1974). For a brief review, see (Chu, 1988), while more details are found in (Helmke and Moore, 1994) together with applications to signal processing and control theory.

The concluding remarks by Dean (1988) summarize quite well the benefits and drawbacks from such approaches which are still valid today: "Preliminary numerical solutions indicate that without special tailoring of the differential equation solution algorithms, conventional minimum seeking optimization algorithms are faster. However, special algorithmic end games may reduce the advantage of conventional algorithms substantially. On the other hand, the amount of geometrical understanding of the constraint manifolds provided by the geodesic flows is far superior. Finally, as opposed to conventional non-linear programming algorithmic approaches, the geodesic flow approach is in a mathematical form readily analysed by recent advances in global dynamic theory". Another very relevant and up-to-date reading is (Chu, 2008).

The idea of the dynamical system approach is to recast certain matrix (matching) optimization problem in the form of dynamical system governing a matrix flow on a (matrix) constrained manifold that starts at particular initial state/point and evolves until it reaches an equilibrium state/point, which is the solution of the original optimization problem. As mentioned above, the advantage is better theoretical, geometrical and global understanding of the problem.

The matrix optimization problem is given as a pair of objective/cost function, and constraint matrix manifold(s). The first step to approach the problem is to find the gradient of the objective/cost function. The standard gradient method, also known as the steepest descent method, is designed for optimization problems without constraints. It minimizes certain objective/-cost function, say, $f(X)$ by constructing a sequence of points $X_1, ..., X_t, ...$ or a flow $X(t)$ which follows the direction of its gradient $\nabla f(X)$ (which is the direction of the steepest descent) and eventually converges to the minimum of $f(X)$. The gradient method is realized by either an discrete dynamical system (iterative scheme) as

$$X_{t+1} = X_t - h_t \nabla f(X)|_{X=X_t} \text{ and } X_{t=0} = X_0, \qquad (3.20)$$

for some iteration step h_t, or as a continuous dynamical system:

$$\dot{X}(t) = -\nabla f(X(t)) \text{ and } X(0) = X_0, \qquad (3.21)$$

where $\dot{X}(t)$ denotes for short $\frac{dX}{dt}$ and X_0 is the initial value of the minimization. Such ordinary differential equation (ODE) where the independent variable t does not appear in the equation explicitly is usually called *autonomous*.

If X is restricted to move on certain constraint set $\mathcal{M}$, then the gradient $\nabla f(X)$ itself may move the sequence $X_1, ..., X_t, ...$ or the flow $X(t)$ out of $\mathcal{M}$. The projected gradient method extends the standard gradient method for constrained optimization problems. It keeps the steepest descent direction and moves on the constrained set $\mathcal{M}$ simultaneously. Let π project X onto $\mathcal{M}$, i.e. $\pi(X) \in \mathcal{T_M}$, where $\mathcal{T_M}$ denotes the tangent space of $\mathcal{M}$. Then, the projected gradient method modifies (3.20) and (3.21), such that the discrete dynamical system (iterative scheme) becomes

$$X_{t+1} = \pi(X_t - h_t \nabla f(X)|_{X=X_t}) \text{ and } X_{t=0} = X_0 \in \mathcal{M}, \qquad (3.22)$$

and, respectively, the continuous dynamical system (ODE) is

$$\dot{X}(t) = -\pi(\nabla f(X(t))) \text{ and } X(0) = X_0 \in \mathcal{M}, \qquad (3.23)$$

where X_0 is the initial value of the minimization.

Thus, the approach turns the optimization problems into *gradient dynamical systems* which enjoy rather decent behaviour, especially if defined for smooth cost functions on compact manifolds (Helmke and Moore, 1994; Hirsh and Smale, 1974; Stuart and Humphries, 1996). Particularly, if the cost function f is (real) *analytic*, i.e. smooth and locally $f(x) = \sum_{i=1}^{\infty} c_i(x-a)^i$ at every point a, then the gradient flow converges to a single limit point. Moreover, if the cost function is *Lyapunov stable*, then the limit point coincides with its minimum (Absil and Kurdyka, 2006; Absil et al., 2005). The results are based on the so-called *Lojasiewicz gradient inequality*: for open $\mathcal{U} \subset \mathbb{R}^n$ and real analytic $f : \mathcal{U} \to \mathbb{R}$, there exist constants $\alpha \in [0,1)$ and $c, \epsilon > 0$ such that

$$|f(x) - f(a)|^{\alpha} \leq c \|\nabla f(x)\| , \text{ for every } x \in \mathcal{B}(a, \epsilon) \subset \mathcal{U} .$$

Note that the polynomials are examples of real analytic functions, as well as the logarithmic and exponential functions. This implies that the MDA models (or at least most of them) and the related (least squares and likelihood) cost functions are analytical and the above conditions are fulfilled.

From numerical point of view, the limit points of the gradient flows can be found as solutions of initial (Cauchy) value problems for matrix ODE(s) of first order on some constraint manifold, considered embedded in high-dimensional Euclidean space. Thus, the standard ODE integrators can be used to solve such problems (Stuart and Humphries, 1996). In particular, one can rely on the MATLAB ODE numerical integrators (Shampine and Reichelt, 1997). Alternatively, there exists a number of other approaches for solving ODEs that preserve the specific structure of the matrix unknowns (e.g. orthogonality) such as *Runge–Kutta methods* on Lie groups (Del Buono and Lopez, 2001, 2002; Engø et al., 1999; Iserles et al., 2000; Owrem and Welfert, 2000) or *Cayley transformation* (Diele et al., 1998).

The dynamical system approach leads to matrix algorithms producing simultaneous solution for all matrix parameters and preserving the geometry of their specific matrix structures. Similar to the standard gradient method, these algorithms are *globally convergent*, i.e. convergence is reached independently of the starting (initial) point. However, the linear rate of convergence is also inherited. A review on how dynamical systems of the form (3.23) can be applied to a number of MDA problems is given in (Trendafilov, 2006).

The dynamical system approach is theoretically sound and leads to many new and interesting problems. However, its numerical implementation is very demanding and impractical, especially for large MDA applications.

3.5.2   Iterative schemes

Instead of solving ODEs defined by the dynamical system (3.23), one can approach the problem by considering iterative schemes. The standard methods solve (3.20) in some Euclidean space, while working with (3.22) needs their generalization to work on constraint matrix manifolds. The difficulty comes from the fact that we cannot operate with elements from different tangent spaces. Tools are needed to "transport" them from one tangent space to another.

The topological and differential properties of the manifolds considered in Section 3.1, Section 3.2 and Section 3.3 are not enough for this purpose. The manifolds need additional features to enable the standard algorithms as steepest descent, conjugate gradient, Newton method, etc. to work on manifolds utilizing their intrinsic geometry. In addition to the local topological equivalence to some Euclidean space, the manifolds also need further Euclidean-like metric features as length, angles, straight lines, parallel translations, etc. See, for example, one of the first globally convergent gradient descent methods with curvilinear search (along geodesic) proposed in (Gabay, 1982).

Let $\mathcal{M}$ be some matrix n-manifold. The first step in this direction is to define positive definite inner product $\langle,\rangle_X$ at every $X \in \mathcal{M}$, i.e. on every tangent space $\mathcal{T}_X\mathcal{M}$, such that $\langle,\rangle_X$ is smooth with respect to X (Sagle and Walde, 1973, A7). The collection of those inner products is called a *Riemannian metric* on $\mathcal{N}$. When a Riemannian metric is defined, then $\mathcal{M}$ is called a *Riemannian manifold*, and is denoted by $\{\mathcal{M}, \langle,\rangle_X\}$.

In practice, this usually means to define a PD $n \times n$ matrix A, such that $X \to A(X)$ is a diffeomorphism from $\mathcal{M}$ to $\mathcal{S}_+(n)$. Then, for $V_1, V_2 \in \mathcal{T}_X \mathcal{M}$, their inner product at X is defined as

$$\langle V_1, V_2 \rangle_X = V_1^\top A(X) V_2 \ ,$$

and the length of every $V \in \mathcal{T}_X \mathcal{M}$ is given by $\|V\| = \sqrt{\langle V, V \rangle_X}$. The *angle* between non-zero $V_1, V_2 \in \mathcal{T}_X \mathcal{M}$ is defined as the unique $\theta \in [0, \pi]$ such that

$$\cos \theta = \frac{\langle V_1, V_2 \rangle_X}{\|V_1\| \|V_2\|} \ ,$$

where $\|V\| = \sqrt{\langle V, V \rangle_X}$.

Riemannian metric exists on every smooth manifold in $\mathbb{R}^n$. For example, one can turn $\mathcal{O}(n, p)$ into a Riemannian manifold by defining an inner product inherited from $\mathbb{R}^{n \times p}$ and identical at every $Q \in \mathcal{O}(n, p)$. This is, of course, the standard Frobenius inner product (2.16):

$$\langle V_1, V_2 \rangle_Q := \langle V_1, V_2 \rangle = \text{trace}(V_1^\top V_2) \ .$$

We will see later that the choice of the inner product can lead to significant implications.

Another necessary tool for developing algorithms on manifolds is the analogue of straight line in $\mathbb{R}^n$, which brings us to the so-called *geodesic curves* on $\mathcal{M}$. A geodesic is (locally) the curve of shortest length between two points on $\mathcal{M}$. They are parametrized by their arc length. Let us remember of the matrix flow $X(t)$ on $\mathcal{M}$ defined by (3.23), where $\dot{X}(t)$ is interpreted as the *velocity* of the flow on $\mathcal{M}$. In a similar way we define *acceleration* of the flow by $\ddot{X}(t)$. We say that $X(t)$ follows geodesic direction on $\mathcal{M}$, if $\ddot{X}(t)$ is always orthogonal to $\dot{X}(t)$, i.e. to $\mathcal{T}_\mathcal{M}$ at $X(t)$.

For example, let $Q(t)$ be a curve/flow on $\mathcal{O}(n, p)$. For simplicity we drop t. Then, after taking two derivatives of $Q^\top Q = I_p$, one obtains

$$\ddot{Q}^\top Q + 2 \dot{Q}^\top \dot{Q} + Q^\top \ddot{Q} = O_p \ .$$

To be geodesic, $\ddot{Q}$ should be orthogonal to $\mathcal{T}_{\mathcal{O}(n, p)}$ at Q. This is possible if $\ddot{Q} = QS$ for some $S \in \mathcal{S}(p)$. Indeed, from Section 3.4.3 we know that every normal vector to $\mathcal{O}(n, p)$ at Q has the form QS with $S \in \mathcal{S}(p)$. Then, after substitution we identify $S = \dot{Q}^\top \dot{Q}$, and finally obtain the following geodesic ODE on $\mathcal{O}(n, p)$ (Edelman et al., 1998, 2.2.2.):

$$\ddot{Q} + Q[\dot{Q}^\top \dot{Q}] = O_{n \times p} \ . \tag{3.24}$$

The *parallel transport* or *translation* on manifold is essential to move tangent vectors along a geodesic from one to another point on the manifold, such that they remain tangent all the time, the same way this is realized along a straight line in a flat space. General formulas for parallel translation along the geodesic are derived in (Edelman et al., 1998, 2.2.3.). They are further used for conjugate gradient method to construct the conjugate directions.

Now, let us consider another inner product on $\mathcal{O}(n,p)$. First, we introduce $Q_\perp \in \mathcal{M}(n, n-p)$, such that $[Q\ Q_\perp] \in \mathcal{O}(n)$. Then, $[Q\ Q_\perp]$ forms a basis in $\mathbb{R}^n$ and every $A \in \mathcal{M}(n,p)$ can be written as a linear combination $A = [Q\ Q_\perp]C$ for some $C \in \mathcal{M}(n,p)$. It is convenient to split C into two blocks $C_1 \in \mathcal{M}(p,p)$ and $C_2 \in \mathcal{M}(n-p,p)$, such that $A = [Q\ Q_\perp]\begin{bmatrix} C_1 \\ C_2 \end{bmatrix} = QC_1 + Q_\perp C_2$. If $A \in \mathcal{T}_Q\mathcal{O}(n,p)$, then it follows from (3.11) that $C_1 + C_1^\top = 0_{p\times p}$, i.e. $C_1 \in \mathcal{K}(p)$. Note that C_2 does not have any particular structure here. Then, for $A = QC_1 + Q_\perp C_2$ consider:

$$\langle A, A \rangle_F = \text{trace}(A^\top A) = \text{trace}(C_1^\top C_1) + \text{trace}(C_2^\top C_2) \ ,$$

which shows that the independent $\frac{p(p-1)}{2}$ entries of C_1 are counted twice. In order to "balance" the contributions of C_1 and C_2, we observe that

$$\text{trace}\left[A^\top \left(I_n - \frac{QQ^\top}{2} \right) A \right] = \frac{1}{2}\text{trace}(C_1^\top C_1) + \text{trace}(C_2^\top C_2) \ .$$

With this in mind, one can introduce an alternative inner product:

$$\langle A_1, A_2 \rangle_c = \text{trace}\left[A_1^\top \left(I_n - \frac{QQ^\top}{2} \right) A_2 \right] \ , \tag{3.25}$$

which is known as the *canonical inner product* on $\mathcal{O}(n,p)$. This changes everything: the geodesic ODE (3.24), the parallel transport and the Stiefel gradient. For details, see (Edelman et al., 1998, 2.4).

For example, we find the Stiefel gradient $\nabla_Q f$ of the cost function f at Q with respect to the canonical inner product, from the following equation:

$$\langle \text{grad}_Q f, h \rangle_F = \langle \nabla_Q f, h \rangle_c = \text{trace}\left[\nabla_Q^\top f \left(I_n - \frac{QQ^\top}{2} \right) h \right] \ ,$$

making use of the notations in (2.23) of Section 2.5. Keeping in mind that $\nabla_Q^\top f \in \mathcal{K}(p)$, one obtains (Edelman et al., 1998, (2.53)):

$$\nabla_Q f = \text{grad}_Q f - Q(\text{grad}_Q f)^\top Q \ . \tag{3.26}$$

One can check that this Stiefel gradient $\nabla_Q f$ differs from the projections of $\text{grad}_Q f$ onto $\mathcal{T}_Q \mathcal{O}(n)$ and $\mathcal{T}_Q \mathcal{O}(n,p)$ proposed, respectively, in (3.10) and (3.12).

Unfortunately, the calculation of geodesics and parallel translations involves matrix exponential maps at every iteration step which is numerically expensive. A more practical alternative approach is to approximate the geodesics and the parallel translation with easier to compute *retractions* (Absil et al., 2008, 4.1). For every $Q \in \mathcal{M}$, the retraction maps $\mathcal{T}_Q \mathcal{M}$ onto $\mathcal{M}$, such that the zero element of $\mathcal{T}_Q \mathcal{M}$ goes to Q. Another important (pullback) feature of the retraction is that it transforms locally the cost function at $Q \in \mathcal{M}$ into a cost function on $\mathcal{T}_Q \mathcal{M}$.

3.5.3 Do it yourself!

A number of efficient retractions, vector translations and/or Riemannian inner products (metrics) are under investigation to construct more practical methods (Huang et al., 2017; Wen and Yin, 2013; Zhu, 2017).

For example, here we consider briefly a steepest descent method on the Stiefel manifold $\mathcal{O}(n,p)$ utilizing a very simple and efficient orthonormality preserving updating scheme (Wen and Yin, 2013). The retraction employed is the *Cayley transformation*. The provided formulas can be used by the reader to create their own codes for gradient descent on $\mathcal{O}(n,p)$.

For $Q \in \mathcal{O}(n,p)$, the classical steepest descent computes the next iterate as $P = Q - \alpha \nabla_f$, where α is the step size. However, in general, $P \notin \mathcal{O}(n,p)$. To overcome this problem, consider first the following two block matrices $U = [\text{grad}_Q f \;\; -Q]$ and $V = [Q \;\; \text{grad}_Q f]$. Then, the Stiefel gradient (3.26) becomes $\nabla_f = KQ$, where $K = UV^\top = \text{grad}_Q f Q^\top - Q\text{grad}_Q^\top f \in \mathcal{K}(n)$. In these notations, the next iterate is given by $P = Q - \alpha KQ$. Now, the main idea is to rewrite it in the form of an implicit (Crank–Nicolson-like) updating scheme:

$$P = Q - \alpha K \left(\frac{Q+P}{2} \right),$$

where P is considered a smooth function of the step size α, with $P(0) = Q$. By solving for P, one finds its expression as

$$P(\alpha) = W(\alpha)Q = \left(I_n + \frac{\alpha}{2}K \right)^{-1} \left(I_n - \frac{\alpha}{2}K \right) Q . \tag{3.27}$$

By direct calculation one can check that $P(\alpha)^\top P(\alpha) = Q^\top Q = I_p$ for $\alpha \geq 0$. This also follows from the fact that $W(\alpha)$ is the Cayley transformation of skew-symmetric K for every α, and thus orthogonal. This shows that $P(\alpha)$ is a curve in $\mathcal{O}(n, p)$, starting at $P(0) = Q$, and that it defines a retraction on it (Absil et al., 2008, p. 59).

To check that $P(\alpha)$ defines a descent direction for the (composite) cost function f at $\alpha = 0$, we need its derivative, which is given by

$$\dot{f}(0) = \frac{\partial f}{\partial P} \frac{dP}{d\alpha}\bigg|_0 = \text{trace}[\dot{f}(P(0))^\top \dot{P}(0)] \ .$$

Thus, we find

$$\dot{f}(P(0)) = \frac{\partial f}{\partial P}\bigg|_{P(0)} = \text{grad}_Q f \ ,$$

and

$$\dot{P}(\alpha) = -\left(I_n + \frac{\alpha}{2}K\right)^{-1} K \left(\frac{P(\alpha) + Q}{2}\right) \ ,$$

which gives $\dot{P}(0) = -KQ = -\nabla_f$. Then, the derivative of the cost function is

$$\dot{f}(0) = -\text{trace}[(\text{grad}_Q f)^\top (KQ)] = -\text{trace}[(Q\text{grad}_Q^\top f)K] = -\frac{1}{2} \|K\|^2 \ ,$$

which proves the assertion.

The updating scheme proposed in (3.27) requires inversion of a $n \times n$ matrix. This can be avoided by employing the *Sherman–Morrison–Woodbury (SMW)* formula, e.g. see (Golub and Van Loan, 2013, 2.1.4), which may simplify the calculation of the inverse matrix of a specific form:

$$(A + BC^\top)^{-1} = A^{-1} - A^{-1}B(I + C^\top A^{-1}B)^{-1}C^\top A^{-1} \ , \tag{3.28}$$

assuming that the involved inverses exist.

The correctness of the SMW formula (3.28) can be checked by direct manipulations. The bad news is that in practice the involved calculations are not stable. As a result one may need to check that $P(\alpha) \in \mathcal{O}(n, p)$, and correct it by QR decomposition (Gram–Schmidt orthogonalization).

In our particular situation with $K = UV^\top$, the SMW formula (3.28) gives

$$\left(I_n + \frac{\alpha}{2}UV^\top\right)^{-1} = I_n - \frac{\alpha}{2}U\left(I_p + \frac{\alpha}{2}V^\top U\right)^{-1} V^\top \ ,$$

which after substitution in $W(\alpha)$ leads to

$$
\begin{aligned}
W(\alpha) &= \left[I_n - \frac{\alpha}{2} U \left(I_p + \frac{\alpha}{2} V^\top U \right)^{-1} V^\top \right] \left(I_n - \frac{\alpha}{2} U V^\top \right) \\
&= \left(I_n - \frac{\alpha}{2} U V^\top \right) - \left[\frac{\alpha}{2} U \left(I_p + \frac{\alpha}{2} V^\top U \right)^{-1} V^\top \right] \left(I_n - \frac{\alpha}{2} U V^\top \right) \\
&= I_n - \frac{\alpha}{2} U \left[I_p + \left(I_p + \frac{\alpha}{2} V^\top U \right)^{-1} - \frac{\alpha}{2} \left(I_p + \frac{\alpha}{2} V^\top U \right)^{-1} V^\top U \right] V^\top \\
&= I_n - \frac{\alpha}{2} U \left[I_p + \left(I_p + \frac{\alpha}{2} V^\top U \right)^{-1} \left(I_p - \frac{\alpha}{2} V^\top U \right) \right] V^\top \\
&= I_n - \frac{\alpha}{2} U \left(I_p + \frac{\alpha}{2} V^\top U \right)^{-1} \left[\left(I_p + \frac{\alpha}{2} V^\top U \right) + \left(I_p - \frac{\alpha}{2} V^\top U \right) \right] V^\top \\
&= I_n - \alpha U \left(I_p + \frac{\alpha}{2} V^\top U \right)^{-1} V^\top .
\end{aligned}
$$

Then, the update formula (3.27) modifies to

$$
P(\alpha) = W(\alpha) Q = Q - \alpha U \left(I_p + \frac{\alpha}{2} V^\top U \right)^{-1} V^\top Q . \tag{3.29}
$$

Clearly, the new update formula (3.29) is to be preferred over (3.27) if $n \gg p$. Further details on the computation effectiveness related to the sizes n and p are provided in (Wen and Yin, 2013).

After these preparations, (Wen and Yin, 2013) propose a simple gradient descent algorithm with curvilinear search (in fact, linear search on $P(\alpha)$) and step size satisfying the *Armijo–Wolfe conditions* (Absil et al., 2008, 4.2). Further, they suggest to actually work with an accelerated algorithm (non-monotone line search) utilizing the *Barzilai–Borwein* step size (Barzilai and Borwein, 1988; Zhang and Hager, 2004,). Full details can be found in (Wen and Yin, 2013, Section 3), as well as possible ways to extend the approach to conjugate gradient and Newton-like methods/algorithms.

3.6   Optimization with Manopt

It was already mentioned that a number of more practical optimization methods on matrix manifolds, avoiding curve-linear search on geodesics, based on efficient retractions were developed and collected in (Absil et al., 2008). In addition, the authors and several of their Ph.D. and/or postdoc students (at that time) developed Manopt, a MATLAB-based software largely implementing the ideas and methods considered in (Absil et al., 2008; Boumal, 2020).

Manopt enjoys very clear online documentation/manual which can be found in the product website http://www.manopt.org/tutorial.html. Here we simply give a brief summary of the major facts one should have in mind before calling/using Manopt:

1) define your problem: state your objective/cost function and the constraint manifolds formed by the unknown matrices;

2) check that the involved constraint manifolds are among the supported by Manopt;

3) find the Euclidean gradient of the objective function (and its Hessian, if possible).

An important general note is that in Manopt, you can either supply Riemannian gradient of the cost function by `problem.grad`, or simply provide its Euclidean gradient by `problem.egrad`, from which Manopt finds its Riemannian counterpart. This is also explained in the documentation and illustrated with example.

Now, you are ready to apply Manopt for solving your problem. In the further sections of the book there will be given a number of Manopt codes for solving different problems. They are chosen such that to cover wide range of constraints. Most likely, the remaining task you need to take care of is to supply Manopt with the right choice of manifold, and your specific objective function and possibly its gradient (and Hessian).

3.6.1 Matrix manifolds in Manopt

The full list of the available matrix manifolds in Manopt can be found in the product manual http://www.manopt.org/tutorial.html. Here we list some of the most popular ones:

- Euclidean space (real and complex) $\mathbb{R}^{m \times n}$ and $\mathbb{C}^{m \times n}$.

- Symmetric matrices $\{X \in \mathcal{M}(n) : X = X^\top\}$.

- Skew-symmetric matrices $\{X \in \mathcal{M}(n) : X + X^\top = O_n\}$.

- Centred matrices (by rows) $\{X \in \mathcal{M}(n,p) : X1_p = 0_n^\top\}$.

- Sphere $\{X \in \mathcal{M}(n,p) : \|X\|_F = 1\}$.

- Oblique manifold $\{X \in \mathcal{M}(n,p) : \|X_{:1}\| = \cdots = \|X_{:p}\| = 1\}$.

- Stiefel manifold $\{X \in \mathcal{M}(n,p) : X^\top X = I_p\}$.

- Grassmann manifold $\{\text{span}(X) : X \in \mathcal{M}(n,p), X^\top X = I_p\}$.

- Rotation group $\{X \in \mathcal{M}(n) : X^\top X = I_n, \det(X) = 1\}$.

- Symmetric, positive definite matrices $\{X \in \mathcal{M}(n) : X = X^\top, X > O_n\}$.

- Symmetric positive semi-definite, fixed rank (complex) $\{X \in \mathcal{M}(n) : X = X^\top \geq O_n, \text{rank}(X) = k\}$.

- Symmetric positive semi-definite, fixed rank with unit diagonal $\{X \in \mathcal{M}(n) : X = X^\top \geq O_n, \text{rank}(X) = r, \text{diag}(X) = 1\}$.

- Symmetric positive semi-definite, fixed rank with unit trace $\{X \in \mathcal{M}(n) : X = X^\top \geq O_n, \text{rank}(X) = r, \text{trace}(X) = 1\}$.

3.6.2 Solvers

Manopt includes the following optimization methods/solvers, among which you are supposed to choose and/or try, depending on the particular problem:

- Trust regions (RTR).

- Adaptive regularization by cubics (ARC).

- Steepest descent.

- Conjugate gradient.

- Barzilai–Borwein.

- BFGS (Broyden–Fletcher–Goldfarb–Shanno algorithm).

- SGD (stochastic gradient descent).

- Particle swarm (PSO).

- Nelder–Mead.

The default option for Manopt is the (retraction) trust-region method (RTR). It is supposed to be the best option for optimizing smooth or analytic function on a (smooth) matrix manifold. This is especially true if the Hessian can be provided in analytical form, which can be fulfilled for many MDA problems. RTR approximates the cost function locally (in a sphere with trust-region radius) by a quadratic function and solves this modified problem at every step. Roughly speaking, RTR combines the strengths of the steepest descent and Newton's methods: exhibits global convergence features and at a higher rate than linear. However, it may not always perform best, because it is rather complex algorithmically. The steepest descent is attractive because it is simpler and is globally convergent, but its quick initial advance towards the optimum is followed by long and slow adjustment. Usually, the conjugate gradient is superior. Often, they both can be outperformed by the Barzilai–Borwein method which simplifies the step size search. The method of adaptive regularization by cubics (ARC) is relatively new and very promising approach to improve over the (search direction of) steepest descent and trust-region methods. The BFGS (Broyden–Fletcher–Goldfarb–Shanno) algorithm is a quasi-Newton method, in a sense that it does not need or compute the Hessian. The stochastic gradient descent on manifolds is in early stages of development, while the partial swarm and Nelder–Mead methods look unlikely options for MDA problems.

It is difficult to give a prescription which solver to apply in what situation. It was mentioned that the MDA problems considered in the book require optimization of smooth (even real analytic) functions on smooth constraint manifolds. The non-smooth penalties utilized to achieve sparseness are also approximated by smooth functions. In this respect, the MDA problems are not particularly "exotic" and full of weird stationary points. The main challenge is usually the size of the data involved and other related abnormalities, e.g. the finiteness of the data space dimension may start to fade away, which gradually affects its topological and metric properties (Giraud, 2015, Ch 1). The RTR and/or ARC methods should be tried first as the most advanced tools for non-convex optimization nowadays. However, simpler algorithms do still have their appeal when the main challenge is the size of the problem.

Using general-purpose optimization software for the MDA problems considered here, e.g. https://cran.r-project.org/web/packages/nloptr, is neither going to be easy nor advisable. They are not designed to preserve the structure of the MDA matrix constraints and are likely to produce unreliable results.

3.6.3 Portability

A weakness of the approach adopted in the book is that it assumes that everyone has access to MATLAB. Moreover, traditionally most statisticians work with R and they will be unsatisfied that the book enslaves them to use MATLAB. Luckily, this problem is pretty much resolved by the recently developed R package ManifoldOptim for optimization of real-valued functions on Riemannian manifolds (Martin et al., 2020). In its heart is the collection of C++ codes ROPTLIB (Huang et al., 2015). ROPTLIB does not cover the whole range of manifold scenarios available in Manopt, but provides great variety of solvers. In short, it gives excellent alternative for R users. ROPTLIB also has interfaces to MATLAB and Python.

We already mentioned that the Manopt developers recently proposed its Python and Julia versions, which will increase its popularity.

Originally, Manopt was compatible with the most popular MATLAB clone Octave. However, the further evolution of MATLAB (nested functions, object-oriented programming approach, etc.) damaged their compatibility. Depending on the Octave future evolution this may change and bring back the previous harmony.

3.7 Conclusion

The section is dedicated to several sets of matrices, $\mathcal{GL}(n), \mathcal{O}(n), \mathcal{O}(n, p)$, etc., which appear in different MDA problems. The corresponding MDA models contain unknown matrix parameters belonging to such sets of matrices. Their estimation turns into constraint optimization and these sets constitute the constraints of the problem.

Section 3.1 defines these sets and briefly gives their main topological properties. Section 3.2 provides basic facts from the theory of differentiable/smooth manifolds. In Section 3.3 it is shown that all of the above sets of matrices, frequently used in MDA, form smooth manifolds. Their main properties and dimensions are listed and their tangent spaces are described separately in Section 3.4.

After the introduction of a number of MDA constraint matrix manifolds, Section 3.5 revisits the two main approaches to (non-convex) optimization on matrix manifolds, adopting, respectively, dynamical systems and iterative schemes. The latter approach is a very active area of research in the last decade with increasing and diversifying number of applications.

Finally, Section 3.6 discusses the MDA utilization of the standard software Manopt for optimization on matrix manifolds. The most frequently used matrix manifolds implemented in Manopt are listed, as well as the available optimization methods/solvers.

3.8   Exercises

1. Let $A \in \mathcal{M}(n)$ such that $\|A\| < 1$.

   (a) Prove that $I_n + A \in \mathcal{GL}(n)$. Hint: Consider the following matrix geometric progression:

   $$(I_n + A)^{-1} = I_n - A + A^2 - A^3 + \ldots + (-1)^n A^n + \ldots \; ,$$

   and show that the series $I_n - A + A^2 - A^3 \ldots$ is convergent by proving that its partial sums form a Cauchy sequence.

   (b) Consider A as a linear operator in $\mathbb{R}^{n \times n}$, i.e. $A \in \mathcal{L}(\mathbb{R}^{n \times n})$. Then A is a linear bounded operator, and thus continuous. Then, show that the identity

   $$(I_n + A)(I_n - A + A^2 - A^3 \ldots) = (I_n + A)(I_n + A)^{-1} = I_n \; ,$$

   indicates that $I_n + A$ is bijection and $(I_n + A)^{-1}$ is its inverse bijection. As $I_n + A$ is linear and continuous, then $(I_n + A)^{-1}$ is also linear operator and continuous too.

   (c) Show that

   $$\|(I_n + A)^{-1}\| \leq \frac{1}{1 - \|A\|} \; .$$

(d) Prove that the derivative of the inverse matrix $d(X^{-1})$ is given by (2.26). Hint: $X + dX = X(I + X^{-1}dX)$ and take $A = X^{-1}dX$.

2. $\mathcal{GL}(n)$ is dense in $\mathcal{M}(n)$.

3. (a) Show that the matrix exponent (2.29) defines a diffeomorphism on some neighbourhood $\mathcal{U}$ of $O_n \in \mathcal{M}(n)$ to some neighbourhood $\mathcal{V}$ of $I_n \in \mathcal{LG}(n)$. The inverse map $\exp^{-1} : \mathcal{V} \to \mathcal{U}$ is called the logarithm for $\mathcal{LG}(n)$. Hint: Prove that $\exp : \mathcal{M}(n) \to \mathcal{M}(n)$ is smooth and its derivative at $X = O_n$ is non-singular, then use the inverse function theorem.

   (b) (Artin, 2011, (9.7.8)) In particular, the matrix exponent (2.29) defines a diffeomorphism from a neighbourhood of $O_n \in \mathcal{K}(n)$ to a neighbourhood of $I_n \in \mathcal{O}(n)$. Use this fact to prove that $\mathcal{O}(n)$ is a manifold of dimension $n(n-1)/2$. Hint: $\mathcal{K}(n)$ is a vector space, and thus a manifold of dimension $n(n-1)/2$. Then, the exponent makes $\mathcal{O}(n)$ a smooth manifold at the identity $I_n \in \mathcal{O}(n)$.

4. (Hall, 2015, Th 2.15) The matrix function $A : \mathbb{R} \to \mathcal{GL}(n)$ is called *one-parameter group*, if A is continuous, $A(0) = I_n$, and $A(\xi + \eta) = A(\xi)A(\eta)$ for all $\xi, \eta \in \mathbb{R}$. Prove that there exists a unique matrix $B \in \mathcal{M}(n)$, such that $A(\xi) = \exp(\xi B)$.

5. (Hall, 2015, Th 3.20) Let $\mathcal{G}$ be a matrix Lie group and $\mathfrak{g}$ be its Lie algebra. Prove that

   (a) if $A \in \mathcal{G}$ and $X \in \mathfrak{g}$, then $AXA^{-1} \in \mathfrak{g}$,

   (b) if $X, Y \in \mathfrak{g}$, their *bracket/commutator* $[X, Y] = XY - YX \in \mathfrak{g}$.

6. (Hiai and Petz, 2014, Th 4.5) If $A, B \in \mathcal{GL}(n)$ and $A \le B$, then $B^{-1} \le A^{-1}$. Hint: $B^{-1/2}AB^{-1/2} \le I_n$.

7. Check that (Borsuk, 1967, (3.10)), (Casini and Goebel, 2010)

   (a) $\{-1, 1\}$ is not a retract of $[-1, 1]$. Hint: Use features of connected sets.

   (b) (more generally) the unit sphere $\mathcal{S}^{n-1}$ is *not* a retract of the closed unit ball $\overline{\mathcal{B}(0, 1)} \subset \mathbb{R}^n$.

   (c) $\mathcal{S}^{n-1}$ is a retract of $\mathbb{R}^n - \{0\}$ and $\overline{\mathcal{B}(0, 1)}$ is a retract of $\mathbb{R}^n$.

   (d) The unit sphere $\mathcal{S}_{\ell^2} = \{(x_1, x_2, \ldots) \mid \sum_{i=1}^{\infty} x_i^2 = 1\} \subset \ell^2$ *is a* retract of $\overline{\mathcal{B}_{\ell^2}(0, 1)} = \{(x_1, x_2, \ldots) \mid \sum_{i=1}^{\infty} x_i^2 \le 1\} \subset \ell^2$.

8. A *homotopy* between two continuous functions $f, g : \mathcal{E} \to \mathcal{F}$ is defined to be a continuous function $h : \mathcal{E} \times [0, 1] \to \mathcal{F}$ such that $h(x, 0) = f(x)$ and $h(x, 1) = g(x)$ for all $x \in \mathcal{E}$. Then, f and g are called *homotopic*. Prove that for every convex subset $\mathcal{C} \subset \mathbb{R}^n$, the identity map $i : \mathcal{C} \to \mathcal{C}$ is homotopic to a constant map $f(\mathcal{C}) = \{c_0\}$ for some $c_0 \in \mathcal{C}$. Such a set is called *contractable*, i.e. the convex subsets of $\mathbb{R}^n$ are contractable.

9. A subspace $\mathcal{A} \subset \mathcal{E}$ is called a *deformation retract* of $\mathcal{E}$ if there is a homotopy $h : \mathcal{E} \times [0, 1] \to \mathcal{E}$ (called *deformation*) such that $h(x, 0) = x \in \mathcal{E}$ and $h(a, 1) = a \in \mathcal{A}$. Check that, $\mathcal{O}(n)$ is a deformation retract of $\mathcal{GL}(n)$. Hint: The cones $\mathcal{U}_+(n)$ and $\mathcal{S}_+(n)$ are convex, and thus contractable. Then, both QR and polar decompositions give deformation retraction from $\mathcal{GL}(n)$ to $\mathcal{O}(n)$. Indeed, if $X \in \mathcal{U}_+(n)/\mathcal{S}_+(n)$, then $h(X, t) = (1 - \alpha)X + \alpha I_n \in \mathcal{U}_+(n)/\mathcal{S}_+(n)$, for each $\alpha \in [0, 1]$. So, it is a homotopy between the identity on $\mathcal{U}_+(n)/\mathcal{S}_+(n)$ and the constant mapping to I_n.

10. Let $\mathcal{A} \subset \mathcal{E}$ be a retract of $\mathcal{E}$. Prove that $\mathcal{A}$ is closed in $\mathcal{E}$.

11. The set of all non-singular $n \times n$ real matrices with unit determinant is known as the *special linear group* and is denoted by $\mathcal{SL}(n)$. It is clear that $\mathcal{SL}(n) \subset \mathcal{GL}(n)$, and you can check that it is a subgroup of $\mathcal{GL}(n)$. Prove that $\mathcal{SL}(n)$ is a smooth manifold of dimension $n^2 - 1$. Hint: Show that the tangent space of $\mathcal{SL}(n)$ is the subspace of all matrices in $\mathcal{M}(n)$ with zero trace. You may consider a curve $A(t) = I_n + tX$ in $\mathcal{SL}(n)$ and make use of $\det(A(t)) = \det(I_n + tX) \approx 1 + t \times \operatorname{trace}(X)$ for small t. An alternative is to work with the derivative of $\det(A) - 1$, or show that 1 is a regular value of $\det : \mathcal{GL}(p) \to \mathbb{R}$, by checking that $d(\det) \neq 0$ on $\mathcal{GL}(p)$.

12. The orthogonal matrices $Q \in \mathcal{O}(n)$ correspond to the *isometric* linear functions, i.e. preserve length, in a sense that for any $x, y \in \mathbb{R}^n$

$$\langle Qx, Qy \rangle = \langle x, y \rangle . \tag{3.30}$$

Show that (3.30) implies the properties of the orthogonal matrices $\mathcal{O}(n)$ considered in Section 3.1.2.

13. Show that

(a) The function $f : \mathcal{S}^2 \to \mathbb{R}^2$ defined by:

$$v_1 = \left(\frac{u_1}{1 - u_3} , \frac{u_2}{1 - u_3} \right) , \quad v_2 = \left(\frac{u_1}{1 + u_3} , \frac{u_2}{1 + u_3} \right) ,$$

and $u_1^2 + u_2^2 + u_3^2 = 1$, gives stereographic projection of $\mathcal{S}^2$ on $\mathbb{R}^2$.

(b) The inverse $f^{-1} : \mathbb{R}^2 \to \mathcal{S}^2$ of the stereographic projection f is given by:

$$u_1 = \frac{2v_1}{v_1^2 + v_2^2 + 1} \ , \quad u_2 = \frac{2v_2}{v_1^2 + v_2^2 + 1} \ , \quad u_3 = \pm \frac{v_1^2 + v_2^2 - 1}{v_1^2 + v_2^2 + 1} \ ,$$

where $(v_1, v_2) \in \mathbb{R}^2$, and the above formulas give parametrization of the north and south (open) hemispheres of $\mathcal{S}^2$ with their poles $(0, 0, \pm 1)$ removed.

14. Prove that the set of all regular points of f is open. Hint: The determinant is non-zero in an open set containing the regular point.

15. Show that a set $\mathcal{U} \in \mathbb{R}^n$ with measure zero *cannot* contain a non-empty open set. Thus, prove that the complement of $\mathcal{U}$ ($\complement\,\mathcal{U}$) to $\mathbb{R}^n$ is *dense* in $\mathbb{R}^n$, i.e. that the closure of $\complement\,\mathcal{U}$ coincides with $\mathbb{R}^n$. Relate this to Sard's theorem to see that the set of regular values of f is dense (Section 3.2).

16. For a set $\mathcal{U} \subset \mathbb{R}^m$, the function $f : \mathcal{U} \to \mathbb{R}^n$ is said to satisfy the *Lipschitz condition* on $\mathcal{U}$, if there exists a constant λ such that

$$\|f(x) - f(y)\|_F \le \lambda \|x - y\|_F \ ,$$

for all $x, y \in \mathcal{U}$.

(a) Prove that any Lipschitz f on $\mathcal{U}$ is uniformly continuous on $\mathcal{U}$ (and thus, continuous on $\mathcal{U}$).

(b) Prove that any differentiable f, i.e. $f \in C^1$, is locally Lipschitz at each point. Hint: Use mean value theorem.

(c) Let $\mathcal{U} \subset \mathbb{R}^n$ be a set with measure 0 and $f : \mathcal{U} \to \mathbb{R}^n$ be Lipschitz on $\mathcal{U}$. Prove that $f(\mathcal{U})$ has measure 0.

(d) Let $\mathcal{U} \subset \mathbb{R}^m$ and the function $f : \mathcal{U} \to \mathbb{R}^n$ be Lipschitz on $\mathcal{U}$. If $m < n$, then $f(\mathcal{U})$ has measure 0. Check that this may not be true, if f is only continuous.

17. Show that the set $\mathcal{M}_r(n, p)$ of all matrices in $\mathcal{M}(n, p)$ with rank r is a smooth manifold of dimension $r(n + p - r)$. Hint: For every matrix $A \in \mathcal{M}_r(n, p)$, there exist $P \in \mathcal{GL}(n)$ and $Q \in \mathcal{GL}(p)$, such that

$$PAQ = \begin{bmatrix} A_{11} & A_{12} \\ A_{21} & A_{22} \end{bmatrix} \ ,$$

where $A_{11} \in \mathcal{GL}(r)$. For every $\epsilon > 0$, there exists $A_\epsilon \in \mathcal{GL}(r)$, such that $\|A_{11} - A_\epsilon\| < \epsilon$. The set $\mathcal{U}$ composed of such "close" matrices

$$\begin{bmatrix} A_\epsilon & A_{12} \\ A_{21} & A_{22} \end{bmatrix}$$

is open in $\mathcal{M}(n,p)$. Note that the identity

$$\begin{bmatrix} I_r & 0 \\ -A_{21}A_\epsilon^{-1} & I_{n-r} \end{bmatrix} \begin{bmatrix} A_\epsilon & A_{12} \\ A_{21} & A_{22} \end{bmatrix} = \begin{bmatrix} A_\epsilon & A_{12} \\ 0 & -A_{21}A_\epsilon^{-1}A_{12} + A_{22} \end{bmatrix}$$

implies that the matrices in $\mathcal{U}$ have rank r if and only if $A_{22} = A_{21}A_\epsilon^{-1}A_{12}$. Now, consider the set $\mathcal{V}$ of matrices $\begin{bmatrix} A_\epsilon & A_{12} \\ A_{21} & 0 \end{bmatrix}$, which is open in Euclidean space of dimension $r^2 + r(p - r) + (n - r)r = r(n + p - r)$. Then, the map defined by

$$\begin{bmatrix} A_\epsilon & A_{12} \\ A_{21} & 0 \end{bmatrix} \rightarrow \begin{bmatrix} A_\epsilon & A_{12} \\ 0 & -A_{21}A_\epsilon^{-1}A_{12} + A_{22} \end{bmatrix}$$

is a diffeomorphism from $\mathcal{V}$ onto $\mathcal{U} \cap \mathcal{M}(n,p)$ (Milnor, 1958).

18. Show that the set $\mathcal{M}_r(n,p)$ of all rank r matrices is not closed. Hint: Let A be such a matrix and consider the sequence $A/k, k = 1, 2, \ldots, \infty$.

19. Let $\mathcal{P}_r(n) \subset \mathcal{M}(n)$ be the subspace of all projection matrices of rank $r(\leq n)$, i.e. $\mathcal{P}_r(n) = \{P \in \mathcal{M}_r(n) : P^2 = P\}$. Prove that $\mathcal{P}_r(n)$ is a smooth $2r(n - r)$-submanifold of $\mathcal{M}(n)$. Hint: The spectral decomposition of $P \in \mathcal{P}_r(n)$ is $P = W\text{diag}_B(I_r, O_{n-r})W^{-1}$ for $W \in \mathcal{GL}(n)$.

20. Prove that the subset $\mathcal{M}_p(n,p)$ of all full rank matrices in $\mathcal{M}(n,p)$ is open (and thus it is a smooth manifold of dimension np). It is also known as the *non-compact Stiefel manifold*. Hint: Consider the map $\mathcal{M}_p(n,p) \rightarrow (0,\infty)$, relating every $A \in \mathcal{M}_p(n,p)$ to the sum of squares of the determinants of its $p \times p$ submatrices (Munkres, 1966, 3.10).

21. Prove that $\mathcal{G}(n,p)$ and $\mathcal{G}(n, n - p)$ are diffeomorphic.

22. (Lai et al., 2020, 2.4) Let Q be a point on the Grassmann manifold $\mathcal{G}(n,p)$ and according to (3.8) be written as $Q = VI_{p,n-p}V^\top = [V_1 \ V_2]I_{p,n-p}[V_1 \ V_2]^\top$. Prove that $\mathcal{G}(n,p)$ as defined in (3.8) is diffeomorphic to $\mathcal{O}(n,p)/\mathcal{O}(p)$. Hint: for $U \in \mathcal{O}(n,p)$ with span of its columns $\mathcal{U}$, consider $f(\mathcal{U}) = 2UU^\top - I_n$ and its inverse $f^{-1}(Q) = \mathcal{V}_1$, the span of $V_1 \in \mathcal{O}(n,p)$.

23. (Lai et al., 2020, 3.6) The normal element $H_\perp = QK \in T_Q \mathcal{O}(n)$ to $T_Q \mathcal{G}(n, p)$ is defined by $\text{trace}(H^\top H_\perp) = \text{trace}(H^\top QK) = 0$, for any $H \in T_Q \mathcal{G}(n, p)$ from (3.18). Check that the explicit formula for $H_\perp$ is given by: for some $K_1 \in \mathcal{K}(p)$ and $K_2 \in \mathcal{K}(n - p)$

$$H_\perp = QK = QV \begin{bmatrix} K_1 & O \\ O & K_2 \end{bmatrix} V^\top = KQ \ .$$

24. Let $\mathcal{A}$ and $\mathcal{B}$ be connected sets, say, in $\mathbb{R}^n$. Then:

    (a) their product $\mathcal{A} \times \mathcal{B}$ is connected;

    (b) if $\mathcal{A} \cap \mathcal{B} \neq \emptyset$, then $\mathcal{A} \cup \mathcal{B}$ is connected.

    Hint: Consider the opposite statement to reach contradiction.

25. Show that $\mathcal{M}(n) = \mathcal{S}(n) \oplus \mathcal{K}(n)$ is a direct sum of vector subspaces, i.e. every $n \times n$ matrix can be presented as a sum of symmetric matrix S and skew-symmetric matrix K, and that $\langle S, K \rangle = \text{trace}(S^\top K) = 0$.

26. The Cayley transformation and its inverse transformation.

    (a) For every $A \in \mathcal{M}(n)$, show that

    $$(I_n + A)^{-1}(I_n - A) = (I_n - A)(I_n + A)^{-1} \ ,$$

    provided the inverse exists. Hint: Use that $(I_n - A)$ and $(I_n + A)$ commute for every $A \in \mathcal{M}(n)$.

    (b) Show that if $K \in \mathcal{K}(n)$, then $I_n + K \in \mathcal{GL}(n)$ and

    $$Q = (I_n + K)^{-1}(I_n - K) \in \mathcal{O}(n) \ .$$

    Show that $\det(Q) = 1$ of such Q and that -1 cannot be its eigenvalue. Hint: Note that $x^\top K x = 0$, for every $x \in \mathbb{R}^n$.

    (c) Show that if $Q \in \mathcal{O}(n)$ and -1 is not among its eigenvalues, then $I_n + Q \in \mathcal{GL}(n)$ and

    $$K = (I_n - Q)(I_n + Q)^{-1} \in \mathcal{K}(n) \ .$$

    Hint: To find K, consider $(I_n + K)(I_n + Q)$ with Q from the previous part. Use $(I_n - Q)(I_n + Q)^{-1} = (I_n - Q)Q^\top Q(I_n + Q)^{-1}$, to prove that such $K \in \mathcal{K}(n)$.

27. Check that

    (a) if $Q \in \mathcal{O}(n)^+$ and n is odd, then $-Q \in \mathcal{O}(n)^-$;

    (b) if $Q \in \mathcal{O}(n)^-$, then $\det(I_n + Q) = 0$, implying that $\det(I_n - Q) = 0$ for $Q \in \mathcal{O}(n)^+$ and odd n.

28. Let $\mathcal{DS}(n)$ denote the set of all density matrices in $\mathcal{S}_+(n)$, i.e. $\mathcal{DS}(n) = \{A : A \in \mathcal{S}_+(n) \text{ and } \operatorname{trace}(A) = 1\}$.

    (a) Prove that $\mathcal{DS}(n)$ is an open set and subspace of $\mathcal{S}_+(n)$. Hint: $\mathcal{DS}(n)$ is intersection of an open set (convex cone) $\mathcal{S}_+(n)$ and a hyperplane in $\mathbb{R}^{n \times n}$.

    (b) Prove that $\mathcal{DS}(n)$ is a smooth manifold and $n(n+1)/2 - 1$ is its dimension. Identify its tangent space.

    (c) Consider $\mathcal{DS}(2) = \left\{ \begin{bmatrix} a & c \\ c & b \end{bmatrix} \right\}$. Show that $\mathcal{DS}(2)$ can be parametrized as $a^2 + c^2 < a$ and consider its geometric properties in $\mathbb{R}^2$.

    (d) The Cholesky decomposition $A = U^\top U$ is a standard parametrization for any $A \in \mathcal{S}_+(n)$, where $U \in \mathcal{U}_+(n)$, i.e U is an upper triangular matrix with positive diagonal entries. What constraints should be imposed on U to ensure that $U^\top U \in \mathcal{DS}(n)$? Let $\mathcal{U}_1(n)$ be the subset of such $U \in \mathcal{U}_+(n)$. Identify its tangent space $\mathcal{T}_U \mathcal{U}_1(n)$.

Chapter 4

Principal component analysis (PCA)

© Springer Nature Switzerland AG 2021 89
N. Trendafilov and M. Gallo, *Multivariate Data Analysis on Matrix Manifolds*,
Springer Series in the Data Sciences, https://doi.org/10.1007/978-3-030-76974-1_4

4.1   Introduction

Principal component analysis (PCA) was first defined in the form that is used nowadays by Pearson (1901). He found the best-fitting line in the least squares sense to the data points, which is known today as the first principal component. Hotelling (1933) showed that the loadings for the components are the eigenvectors of the sample covariance matrix. More details about the PCA history are provided in (Jolliffe, 2002, 1.2).

We should note that PCA as a scientific tool for data analysis appeared first in Psychometrika. This should not be very surprising if we recall that the first paper on singular value decomposition (SVD) was also published there (Eckart and Young, 1936). This simply reflects the reality that the quantitative measurement of the human intelligence and personality in the 30s of the last century should have been of the same challenging importance as the contemporary interest in big data analysis (gene and tissue engineering, signal processing, or analysing huge climate, financial or Internet data).

The chapter is organized as follows. Section 4.2 defines PCA and outlines briefly its main properties. Section 4.4 discusses the classical approach to PCA interpretation and its drawbacks. Section 4.5 considers the simple structure concept, which is the classical understanding for easily interpretable PCA solutions (loadings). We stress on the fact the simple structure in fact requires what is known today as sparse loadings. Then, the classical rotation approach to obtain "simple" component loadings is discussed and demonstrated using VARIMAX in Section 4.5.2. In a case study (Section 4.5.3) we illustrate the classical process of obtaining simple structure loadings, which is subjective and, in fact, finds sparse loadings. Before considering sparse PCA, we consider Section 4.6 a kind of an intermediate method which keeps some of the features of the rotation methods, but produces sparse component loadings. Section 4.7 is rather short and makes a brief characterization and classification of the huge number of methods for sparse PCA. Section 4.8 considers several different sparse PCA definitions which approach different features of the original PCA. Section 4.8.2 is of particular interest because the sparse PCA considered there is capable to produce very weakly correlated sparse components in which loadings are nearly orthogonal, and thus approximating the features of the original PCs. Finally, Section 4.9 introduces several new dimension reduction techniques aiming to replace PCA in case of very large data matrices.

4.2 Definition and main properties

Now, PCA is well-known and efficient technique for reducing the dimension of high-dimensional data (Jolliffe, 2002). Suppose that p random variables are observed on each of n individuals, and the measurements are collected in an $n \times p$ data matrix X. We assume that the data are standardized, i.e. $X^\top X$ is the sample correlation matrix. PCA forms p new "variables", called principal components (PCs), which are linear combinations of the original ones with the following *unique* properties: they are ordered according to their variances magnitudes, are uncorrelated and the vectors of their coefficients, called component loadings, are orthogonal.

Details on the PCA definition, and its main mathematical and statistical properties can be found in (Jolliffe, 2002, pp. 29–59). For our goals it is sufficient to mention only that formally, the ith PC ($i = 1, \ldots, p$) is the new variable $y_i = X a_i$ defined as a linear combination of the original variables (columns) of X, with maximum variance $a_i^\top X^\top X a_i$ subject to $a_i^\top a_i = 1$ and $a_i^\top a_j = 0$ for $i > j$. This corresponds exactly to the variational definition of the eigenvalue decomposition (EVD) of a symmetric matrix $X^\top X$ (Horn and Johnson, 1985, Ch 1, Ch 4). The vector a_i contains the component loadings of the ith PC and is the ith eigenvector corresponding to the ith largest eigenvalue of $X^\top X$. This most common PCA formulation can be expressed formally as follows:

$$\max_{\substack{\|a\|_2 = 1 \\ a \perp A_{i-1}}} a^\top X^\top X a , \qquad (4.1)$$

where the vector a of new PC loadings is orthogonal (denoted by $\perp$) to the already found vectors $a_1, \ldots, a_{i-1}$ of PC loadings collected in a $p \times (i-1)$ matrix $A_{i-1} = [a_1, \ldots, a_{i-1}]$. Alternatively, this constraint is written as $a^\top A_{i-1} = 0_{i-1}^\top$.

In practice, the first r ($\ll p$) PCs, accounting for the majority of the variation in the original variables, are used for further analysis which is known as *dimension reduction*. In this case, we usually speak for *truncated* EVD of $X^\top X \approx A_r D_r^2 A_r^\top$, where D_r^2 is $r \times r$ diagonal matrix containing the largest r eigenvalues of $X^\top X$ in descending order and $A_r \in \mathcal{O}(p, r)$ containing the corresponding eigenvectors. One can also express $A_r D_r^2 A_r^\top = \sum_{i=1}^r d_i^2 a_i a_i^\top$, which suggests another way of doing PCA sequentially: first, one solves

$$\max_{a^\top a = 1} a^\top X^\top X a ,$$

to find a_1 and $d_1^2 = a_1^\top X^\top X a_1$; next, to find a_2, one removes the contribution of a_1 to $X^\top X$ by considering $X^\top X - d_1^2 a_1 a_1^\top$, and solves:

$$\max_{a^\top a = 1} a^\top (X^\top X - d_1^2 a_1 a_1^\top) a ,$$

to find a_2, and then further "deflate" $X^\top X$ to $X^\top X - d_1^2 a_1 a_1^\top - d_2^2 a_2 a_2^\top$, etc. Such type of successive process for finding eigenvectors and eigenvalues is known as *deflation* method for PCA.

It is tempting to express (4.1) in matrix form as

$$\max_{A \in \mathcal{O}(p,r)} \text{trace}(A^\top X^\top X A) . \tag{4.2}$$

However, the problem (4.2) simply finds r vectors collected in A which maximize the total variance, while the vectors found in (4.1) are additionally ordered according to the magnitudes of the variance they explain. The problem is that any r orthonormal vectors spanning the same subspace as the eigenvectors corresponding to the r largest eigenvalues of $X^\top X$ give the same value of (4.2). In this sense, (4.2) depends only on the subspace generated by the columns of A. In other words, the solution of (4.1) is a solution of (4.2) too, but not every solution of (4.2) is solution of (4.1). In this sense, (4.2) is not really a PCA problem.

This is illustrated with a small example. Consider the following sample correlation matrix (Harman, 1976, p. 14):

$$X^\top X = \begin{bmatrix} 1.000 & .0098 & .9724 & .4389 & .0224 \\ .0098 & 1.000 & .1543 & .6914 & .8631 \\ .9724 & .1543 & 1.000 & .5147 & .1219 \\ .4389 & .6914 & .5147 & 1.000 & .7777 \\ .0224 & .8631 & .1219 & .7777 & 1.000 \end{bmatrix} , \tag{4.3}$$

and find two principal components by applying both of the above PCA definitions, i.e. solve (4.1) and (4.2) with $r = 2$. The results are given in Table 4.1.

We note that the variances of a_i obtained from (4.2) in Table 4.1 are not ordered according to their magnitudes. This is an obvious weakness if one is interested in dimension reduction: the next a_i may have bigger contribution to the total variance than the ones already found.

Vars	PCA (4.1)		PCA (4.2)	
	a_1	a_2	a_1	a_2
1	.3427	-.6016	.6905	-.0745
2	.4525	.4064	-.0571	.6052
3	.3967	-.5417	.6741	-.0002
4	.5501	.0778	.2494	.4897
5	.4667	.4164	-.0577	.6232
$a_i^\top X^\top X a_i$	2.8733	1.7967	2.1692	2.5005
Total	4.6700		4.6697	

Table 4.1: Solutions for two types of PCA problems.

In general, working with raw data X is preferable. The EVD of $X^\top X$ in the PCA definition given above can be replaced by the singular value decomposition (SVD) (Horn and Johnson, 1985, Ch 3) of the raw data matrix X, i.e. $X = FDA^\top$, where the diagonal elements of D are in decreasing order. For any r ($\leq p$), let $X_r = F_r D_r A_r^\top$ be the *truncated* SVD of X. As usual, F_r and A_r denote the first r columns of F and A, respectively, ($F_r^\top F_r = A_r^\top A_r = I_r$), and D_r is diagonal matrix with the first (largest) r singular values of X. The matrices $Y_r = F_r D_r$ and A_r contain the component scores and loadings, respectively, and D_r^2 contains the variances of the first r PCs. In PCA applications, we are usually interested in $r \ll p$.

A right multiplication of $X = FDA^\top$ by A_r gives

$$XA_r = FDA^\top A_r = FD \begin{bmatrix} I_r \\ O_{(p-r)\times r} \end{bmatrix} = F \begin{bmatrix} D_r \\ O_{(p-r)\times r} \end{bmatrix} = F_r D_r = Y_r \ , \quad (4.4)$$

i.e. the first r principal components Y_r are linear combination of the original variables X with coefficients A_r. Their variances are given by

$$Y_r^\top Y_r = D_r F_r^\top F_r D_r = D_r^2 = A_r^\top X_r^\top X_r A_r = A_r^\top X^\top X A_r \ , \quad (4.5)$$

which also shows that the principal components are uncorrelated, as $Y_r^\top Y_r$ is diagonal. Then, the total variance explained by the first r PCs is given by $\text{trace} D_r^2 = \text{trace}(A_r^\top X^\top X A_r) = \text{trace}(A_r^\top X_r^\top X_r A_r)$.

Further right multiplication of (4.4) by $A_r^\top$ gives

$$X_r = XA_r A_r^\top, \quad (4.6)$$

i.e. the best approximation X_r to X of rank r is given by the orthogonal projection of X onto the r-dimensional subspace in $\mathbb{R}^p$ spanned by the columns of A_r. That is, X_r is a low-dimensional representation of the original data X. In addition, Equation (4.6) helps to define PCA as a LS problem:

$$\min_{A \in \mathcal{O}(p,r)} \|X - XAA^\top\|_F , \qquad (4.7)$$

along with its definitions given above as maximization of a quadratic or bi-linear form (EVD or SVD).

Alternatively, the PCA formulation (4.7) can be expressed as a LS projection of the data onto the r-dimensional subspace in $\mathbb{R}^n$ spanned by the columns of F_r, the component scores (projections) of the data. Then, PCA of X can be rewritten as

$$\min_{F \in \mathcal{O}(n,r)} \|X - FF^\top X\|_F , \qquad (4.8)$$

and $X_r = F_r F_r^\top X$ is the best rank r (LS) approximation to X in $\mathbb{R}^n$.

Related to the truncated SVD of an $n \times p$ $(n \geq p)$ matrix X is its *"economy"* SVD. Let here r $(\leq p)$ be the rank of X. Then, the "economy" SVD of such X is written as

$$X = FDA^\top = \begin{bmatrix} F_1 & F_2 \end{bmatrix} \begin{bmatrix} D & O_{r \times (p-r)} \\ O_{(p-r) \times r} & O_{(p-r) \times (p-r)} \end{bmatrix} \begin{bmatrix} A_1^\top \\ A_2^\top \end{bmatrix} , \qquad (4.9)$$

where $F \in \mathcal{O}(n,p)$, $A \in \mathcal{O}(p)$ and $D \in \mathcal{D}(r)$. One can check easily that $F_1 \in \mathcal{O}(n,r)$, $F_2 \in \mathcal{O}(n,p-r)$, $A_1 \in \mathcal{O}(p,r)$ and $A_2 \in \mathcal{O}(p,p-r)$.

The following important orthogonal projections are associated with the SVD of X (Golub and Van Loan, 2013, 2.5.2):

- $F_1 F_1^\top$—projection onto the range of X;
- $A_1 A_1^\top$—projection onto the range of $X^\top$;
- $F_2 F_2^\top$—projection onto the null space of $X^\top$;
- $A_2 A_2^\top$—projection onto the null space of X.

Note that the orthogonal projectors $F_1 F_1^\top$ and $F_2 F_2^\top$ are not orthogonal matrices, but $F_1 F_1^\top - F_2 F_2^\top$ is.

Finally, it is worth mentioning another point of view on PCA, as a regression model (O'Hagan, 1984; Takeuchi et al., 1982). Instead of maximizing the PCs individual variances, it may look more natural from data analysis point of view to minimize the residual variance of the PCs. For this reason, consider the following alternative construction. Define r linear combinations (also called PCs) collected in the $n \times r$ matrix $Y_r = X A_r$. Note that no identification constraints are imposed on A_r. Then, one is interested to approximate the data matrix X by a linear combination of the PCs, say $X \approx Y_r W^\top$, where W is some $p \times r$ matrix of weights assumed full column rank. Then the problem is to

$$\min_{A_r, W} \|X - Y_r W^\top\|_F = \min_{A_r, W} \|X - X A_r W^\top\|_F , \tag{4.10}$$

which can be simplified by eliminating W. For this reason we fix A_r in (4.10), and find from the first-order optimality condition that the optimal W is given by

$$W = X^\top X A_r (A_r^\top X^\top X A_r)^{-1} . \tag{4.11}$$

Now, we impose the constraint $A_r^\top X^\top X A_r = I_r$ implying that $Y_r^\top Y_r = I_r$, i.e. that the "regression" PCs are also uncorrelated but they have *constant* variances equal to 1.

After substitution of (4.11) into (4.10), it transforms to

$$\min_{Y_r \in \mathcal{O}(n,r)} \|X - Y_r Y_r^\top X\|_F , \tag{4.12}$$

which is equivalent to

$$\max_{Y_r \in \mathcal{O}(n,r)} \operatorname{trace} Y_r^\top X X^\top Y_r . \tag{4.13}$$

We already know from (4.2) that among the solutions Y_r of (4.13) are the eigenvectors of $X X^\top$, corresponding to the r largest eigenvalues of $X^\top X$. Then, the "regression" PC loadings are found as $A_r = (X^\top X)^{-1} X^\top Y_r$, indicating that this approach requires non-singular $X^\top X$.

Let the r largest eigenvalues of $X X^\top$ (and $X^\top X$) be collected in $\Lambda \in \mathcal{D}(r)$. Then, the corresponding eigenvectors Y_r of $X X^\top$ fulfil $X X^\top Y_r = Y_r \Lambda$. In turn, this gives that $X^\top X A_r = A_r \Lambda$, meaning that the columns of A_r are the eigenvectors of $X^\top X$. Moreover, the identity $A_r^\top A_r \Lambda = I_r$ implies that $A_r^\top A_r = \Lambda^{-1}$, i.e. $A_r^\top A_r$ is a diagonal matrix, but not the identity.

A number of alternative PCA definitions/properties can be found in (Jolliffe, 2002, Ch2, Ch3). They are all equivalent in a sense that eventually one needs the eigenvectors of $X^\top X$ or their normalization involving the corresponding eigenvalues. The situation changes in sparse PCA where the different PCA definitions are utilized to impose sparseness on the eigenvectors differently.

4.3   Correspondence analysis (CA)

CA is a special case of PCA designed for analysis of categorical variables. Suppose we are interested in two *categorical variables* with n and p categories, respectively. Of course, any continuous variable can be turned into a categorical one. Let $f_{ij}, i = 1, \ldots, n; j = 1, \ldots, p$ be the (observed) frequency counts for all np possible outcomes of those categories. Suppose they are collected in a matrix $F \in \mathcal{M}(n, p)$, which is called (two-way) *contingency table*. Here is an example of a contingency table for two categorical variable having two categories each, i.e. $n = p = 2$:

	Airsickness	No airsickness
Male	24	31
Female	8	26

The usual question related to contingency tables is to check the (null) hypothesis H_0 that the two involved categorical variables are independent. In the above example, based on the given data, we are interested to check that the airsickness is independent from the gender of the passengers. This can be achieved with the χ^2-criterion

$$\chi_0^2 = \sum_{i=1}^{n} \sum_{j=1}^{p} \frac{(f_{ij} - e_{ij})^2}{e_{ij}} , \qquad (4.14)$$

where $e_{ij} = (\sum_{j=1}^{p} f_{ij})(\sum_{i=1}^{n} f_{ij})/(\sum_{i=1}^{n} \sum_{j=1}^{p} f_{ij})$ are known as the *expected* frequencies. If H_0 is true, then χ_0^2 has approximately χ^2 distribution with $(n-1)(p-1)$ degrees of freedom (Cramér, 1945, 30.5). Large values of χ_0^2 indicate significant deviation from H_0.

In order to specify the degree/level of association between the categorical variables, it is convenient to introduce a kind of "normalized" measure of association between them. For this reason, we introduce more compact notations also useful later for CA: $f_{i\cdot} = \sum_{j=1}^{p} f_{ij}, f_{\cdot j} = \sum_{i=1}^{n} f_{ij}$ and

$t = \sum_{i=1}^{n} \sum_{j=1}^{p} f_{ij}$. Then, $e_{ij} = f_{i\cdot}f_{\cdot j}/t$ and χ_0^2 can be rewritten as

$$\chi_0^2 = t \sum_{i=1}^{n} \sum_{j=1}^{p} \left(\frac{f_{ij}^2}{f_{i\cdot}f_{\cdot j}} - 1 \right), \tag{4.15}$$

which suggests the following measure (Cramér, 1945, 30.5):

$$\phi^2 = \frac{\chi_0^2}{t \times \min\{n-1, p-1\}},$$

with values between 0 and 1. Then, ϕ^2 is close to 1 in case of strong association between the categorical variables, and is close to 0 otherwise. Moreover, the upper bound 1 is attained if and only if every row (for $n \geq p$) or every column (for $p \geq n$) of F contains a single non-zero entry. The optimality of such a sparse contingency table is related in (Knüsel, 2008) to the general principle of PCA interpretability discussed in Section 4.5.1.

Originally, CA was designed to further enhance the analysis of contingency tables, but it was rapidly extended to the analysis of any data matrix with non-negative entries. The modern CA theory and its geometric interpretation is associated with the name of Jean-Paul Benzécri and it is now an emblem of the French school of data analysis (analyse des donées) in the 1960s. As a technique, it was often re-discovered under different names such as dual or optimal scaling, homogeneity analysis and reciprocal averaging. A brief introduction can be found in (Everitt and Dunn, 2001, 4) or (Izenman, 2008, 17.2). Standard sources for deeper study of CA are (Benzécri, 1992; Greenacre, 1984).

Consider again the χ^2-criterion (4.14), and arrange the terms of the sum in a $n \times p$ matrix X, with elements $x_{ij} = \frac{(f_{ij}-e_{ij})}{\sqrt{e_{ij}}}$. In brief, CA finds the SVD of $X = UDV^\top = \sum_{k=1}^{r} d_k u_k v_k^\top$, where r is the rank of X. Note that $r = \min\{n-1, p-1\}$. Then, the χ^2-criterion (4.14) can be expressed as

$$\chi^2 = \sum_{i=1}^{n} \sum_{j=1}^{p} x_{ij}^2 = \text{trace}(X^\top X) = \sum_{k=1}^{r} d_k^2 .$$

The values d_i/t are called *inertias*. CA works with the best two-dimensional approximation of X, i.e. $x_{ij} \approx d_1 u_{i1} v_{j1} + d_2 u_{i2} v_{j2}$. To simplify the CA interpretation, one simply looks at $x_{ij} \approx d_1 u_{i1} v_{j1}$, which indicates that

- if $u_{i1} v_{j1}$ is near zero, then the association between the ith and jth categories is weak, i.e. $f_{ij} \approx e_{ij}$;

- if u_{i1} and v_{j1} are both positive/negative and large, then the association between the ith and jth categories is positive and strong, i.e. $f_{ij} > e_{ij}$;

- if u_{i1} and v_{j1} are large, but with different signs, then the association between the ith and jth categories is negative, i.e. $f_{ij} < e_{ij}$.

4.4   PCA interpretation

The PCA is interpreted by considering the magnitudes of the component loadings, which indicate how strongly each of the original variables contribute to the PCs. However, PCs are really useful if they can be simply interpreted, which, unfortunately, is not often the case. Even if a reduced dimension of r PCs is considered for further analysis, each PC is still a linear combination of *all* original variables. This complicates the PCs interpretation, especially when p is large. The usual *ad hoc* practice in the PC interpretation is to ignore the variables with small absolute loadings or *set to zero* loadings smaller than some threshold value. Not surprisingly, such a practice is found to be misleading especially for PCs computed from a covariance matrix (Cadima and Jolliffe, 1995).

The oldest approach to solve the PCA interpretation problem is the so-called *simple structure rotation*. Initially, it is introduced in factor analysis (FA) (Thurstone, 1935) and later is adapted to PCA to make the components as interpretable as possible (Jolliffe, 2002, Chap. 11). It is based on the following simple identity:

$$X = FDA^\top = FQQ^{-1}DA^\top , \qquad (4.16)$$

where Q can be *any* non-singular transformation matrix. Two types of matrices Q are traditionally used in PCA (and FA): orthogonal and oblique, i.e. $Q \in \mathcal{O}(r)$ or $Q \in \mathcal{OB}(r)$.

Historically, the first methods to find a suitable Q are graphical and are performed by hand as a sequence of planar rotations (Mulaik, 1972, p. 219–224) or (Mulaik, 2010, Ch 10), which nowadays are called Jacobi or Givens rotations (Golub and Van Loan, 2013, Ch 5,8). This naturally explains why such methods are named "rotation". However, what looks less natural is that the first attempts to generalize the rotation approach to higher dimensions $r(\geq 3)$ insist on producing *pure* rotations Q, i.e. $Q \in \mathcal{SO}(r)$. This is clearly evident from the sentence: "The determinant of the third-order matrix must be $+1$, since the present problem concerns only rotation without reflection" (Thurstone, 1935, p. 214). Imposing such a restriction on Q sounds a bit strange, because (4.16) actually permits any $Q \in \mathcal{GL}(r)$. Probably for this reason, it is commonly accepted to define rotation problems/methods on the whole $\mathcal{O}(r)$ rather than on $\mathcal{SO}(r)$.

For a moment, let us concentrate our attention on $\mathcal{O}(r)$ alone. If we use rotation methods utilizing planar rotations, then the resulting Q belongs to $\mathcal{SO}(r)$ by construction. However, things become more complicated if matrix rotation algorithms are utilized which directly work with Q. Most (matrix) methods for manifold optimization find solution $Q \in \mathcal{SO}(r)$ by simply initiating the iterations from $\mathcal{SO}(r)$. More generally, if they start from certain connected component of $\mathcal{O}(r)$, then the iterations stay on that same component until convergence.

In this connection, the interesting question is how the particular rotation method is influenced when it starts from $Q \in \mathcal{SO}(r)$ or $Q \notin \mathcal{SO}(r)$. There are no indications that starting form $Q \in \mathcal{SO}(r)$ or $Q \notin \mathcal{SO}(r)$ influences systematically the optimal rotations. The likely explanation is that most rotation criteria (cost functions), e.g. VARIMAX (4.18) and (4.19), are polynomials containing only even degrees or are sums of absolute values, e.g. (4.22) or (4.23). Then, the signs of the elements of $Q \in \mathcal{O}(r)$ cannot affect the result. Nevertheless, some general results for the optima of the rotation criteria defined on $\mathcal{O}(r)$ and how they behave on its connected components would be useful. Related results for the trace of linear and quadratic functions of $Q \in \mathcal{O}(r)$ are considered in (Bozma et al., 2019).

Going back to the PCA interpretation, we stress that the traditional PCA employs (4.16) and refers to AD as a loadings matrix, which differs from the notations in Section 4.2. This is a consequence from the fact that the rotation methods were first introduced in FA and then adopted in PCA, which introduces a number of normalization problems (Jolliffe, 2002, p. 272). The traditional PC interpretation relies on either AD or its rotated version $ADQ^{-\top}$. For example, this will be the case in Section 4.5, where we illustrate an old approach for PCA interpretation. Many statistical packages also adopt this meaning of PC loadings, e.g. SPSS.

Alternatively, if one works in the notations of Section 4.2, where A is defined as the loadings matrix, then the identity (4.16) can be rewritten as

$$X = FDA^\top = FDQQ^{-1}A^\top , \tag{4.17}$$

and thus the interpretation is based on either A or its rotated version $AQ^{-\top}$. We will see in Section 4.7 that the modern PCA, and particularly sparse PCA, adopt A as a loadings matrix which is subject to interpretation (and sparsification).

The simple structure rotation approach proceeds as follows. First, the dimension reduction is performed, i.e. the appropriate r is chosen, and A and D are obtained. Then, a rotation Q is found by optimizing certain simple structure criterion, such that the resulting $ADQ^{-\top}$ (or $AQ^{-\top}$) has simple structure. The belief that the rotated component has its absolute loadings near 1 or 0 while avoiding intermediate values is usually false and makes the approach ambiguous. Its application is additionally complicated by the huge number of simplicity criteria to choose from. Another important shortcoming is that the rotated components lack the nice PCs properties to be uncorrelated and explain successively decreasing amount of variance.

The main types of rotation methods are briefly outlined in Section. Browne (2001) gives a comprehensive overview of the field. Some additional aspects are discussed in (Jennrich, 2007). Details can be found in the cited papers there as well as in the standard texts on FA (Harman, 1976; Mulaik, 1972, 2010). To save space, no reference to original papers on rotation methods is given.

4.5   Simple structure rotation in PCA (and FA)

4.5.1   Simple structure concept

It was mentioned in Section 4.4 that the goal of the simple structure approach to the PCA interpretation is to find orthogonal or oblique matrix Q such that the resulting loadings $ADQ^{-\top}$ have simple structure. To simplify the notations, we assume in this section that $A := AD$.

The original simple structure concept was introduced in FA as three rules (Thurstone, 1935, p. 156), which were later extended and elaborated in (Thurstone, 1947, p. 335). In more contemporary language, they look as follows (Harman, 1976, p. 98), where the term "factor matrix" used in FA context should be understood as component loadings matrix for the PCA case:

1. Each row of the factor matrix should have at least one *zero*.

2. If there are r common factors each column of the factor matrix should have at least r *zeros*.

3. For every pair of columns of the factor matrix there should be several variables whose entries *vanish* in one column but not in the other.

4. For every pair of columns of the factor matrix, a large proportion of the variables should have *vanishing* entries in both columns when there are four or more factors.

5. For every pair of columns of the factor matrix there should be only a small number of variables with *non-vanishing* entries in both columns.

The words in italic are made on purpose to stress that the original simple structure concept requires, in fact, *sparse* loadings. Broadly speaking $a \in \mathbb{R}^p$ is called sparse if $\|a\|_0 \ll p$, where $\|a\|_0$ is the number of the non-zero entries in a, also called its cardinality. Unfortunately, such clear-cut sparseness has never been achieved by the classical rotation methods.

There are situations when the first PC is a measure of "overall size" with non-trivial loadings for all variables. The same phenomenon occurs in the so-called bi-factor analysis (Harman, 1976, pp. 120–127). In such PCA/FA solutions, the five rules are applied to the remaining PCs/factors.

Let us take a closer look at how the simple structure concept is implemented in a number of rotation methods. It was mentioned that the oldest rotation methods were graphical. Their main conceptual drawback is that they are subjective. In order to make the interpretation process objective, analytic rotation methods were introduced. At the heart of these methods is the simplicity criterion, which is supposed to express Thurstone's five rules in a single mathematical formula. Unfortunately, it turned out that this is a very difficult task. As a result there were produced a huge number of criteria, because none of them is capable to produce satisfying solution for every problem.

4.5.2   Simple structure criteria and rotation methods

To get some flavour of what is this simple structure criterion, let us consider VARIMAX (**VAR**iance **MAX**imization) (Kaiser, 1958; Mulaik, 2010, p. 310). This is arguably the most popular (but not the best) criterion, because it produces satisfactory results in many applications and is based on a simple to explain concept. Let $B = AQ$ be the $p \times r$ matrix of orthogonally rotated loadings. The variance of the squared loadings of the jth rotated column b_j is

$$V_j = V(b_j) = \sum_{i=1}^{p} b_{ij}^4 - \frac{1}{p} \left(\sum_{i=1}^{p} b_{ij}^2 \right)^2 . \tag{4.18}$$

The variance V_j will be large if there are few large squared loadings and all the rest are near zero. The variance V_j will be small when all squared loadings have similar magnitudes. The VARIMAX rotation problem is to find an $r \times r$ orthogonal matrix Q such that the total variance $V = \sum_{j=1}^{r} V_j$ is maximized.

The original VARIMAX finds the optimal Q by making use of an algorithm based on successive planar rotations of all possible $r(r-1)/2$ pairs of factors. Alternatively, the VARIMAX criterion V can be defined in matrix form as follows (Magnus and Neudecker, 1988; Neudecker, 1981):

$$V(Q) = \text{trace} \, (AQ \odot AQ)^\top J_p(AQ \odot AQ) , \tag{4.19}$$

where $J_p = I_p - p^{-1}1_p 1_p^\top$ is the centring operator and $\odot$ denotes the Hadamard (element-wise) matrix product. If the expression under the trace in (4.19) is divided by $p - 1$ it presents the (sample) covariance matrix of the squared orthogonally transformed factor loadings $B(= AQ)$. Then the VARIMAX problem is to maximize the objective function (4.19) over all possible orthogonal rotations Q, i.e.

$$\max_{Q \in \mathcal{O}(r)} V(Q) . \tag{4.20}$$

The matrix VARIMAX formulation was further elaborated by ten Berge (1984). The resulting rotation problem requires solution of the following constrained optimization problem:

$$\max_{Q \in \mathcal{O}(r)} \text{trace} \left[\sum_{i=1}^{p} \left(\text{diag}(Q^T W_i Q) \right)^2 \right], \tag{4.21}$$

where $W_i = A^T A - p a_i a_i^T$, a_i^T denotes the i–th row of A. This reformulation of VARIMAX is interesting for two reasons. First, (4.21) is a problem for simultaneous diagonalization of several symmetric matrices (Chapter 9, Section 10.4.2). Second, it can be easily generalized to the one-parameter rotation family ORTHOMAX (Harman, 1976). ORTHOMAX is a class of rotation criteria/methods defined by $W_i(\lambda) = \lambda A^T A - p a_i a_i^T$, where $\lambda \in [0, 1]$. For example, $\lambda = 1$ yields VARIMAX, $\lambda = 0$ – QUARTIMAX (Harman, 1976), etc.

VARIMAX and ORTHOMAX can be solved on $\mathcal{O}(r)$ or $\mathcal{OB}(r)$ to produce orthogonally or obliquely rotated loadings. Both of these problems, as well as all other simple structure rotation problems are optimization problems subject to orthogonality or oblique constraints. Thus, they can be solved in a unified way by methods for optimization on matrix manifolds and utilizing Manopt.

Particularly, (4.21) was solved in (Chu and Trendafilov, 1998b) by using the dynamical system approach (Chu and Driessel, 1990; Helmke and Moore, 1994). Like Manopt, this approach also needs the gradient of the objective function of (4.21):

$$4 \sum_{i=1}^{n} W_i Q \mathrm{diag}(Q^T W_i Q)$$

and its projection onto the tangent space $\mathcal{T}_Q \mathcal{O}(r)$ defined in (3.4.1):

$$2Q \sum_{i=1}^{n} \left(Q^T W_i Q \mathrm{diag}(Q^T W_i Q) - \mathrm{diag}(Q^T W_i Q) \, Q^T W_i Q \right).$$

There are a number of alternative rotation methods which do not rely on any particular simplicity criterion. Instead, they search for well-defined configurations (of variables), e.g. hyperplanes. They can be divided into two categories: hyperplane fitting rotations and hyperplane counting methods.

The hyperplane fitting rotations "refine" the solution B from certain analytic rotation (Mulaik, 2010, p. 342–347). They construct a target matrix T reflecting the simple structure of the problem based on the already obtained B. Then the new solution BP_∞ is obtained by Procrustes fitting of B to the target T, where $P_\infty = \mathrm{argmin}_P \|T - BP\|_F$. It is worth noting that one of these methods, PROMAJ (**PRO**crustes **MAJ**orization)(Mulaik, 2010; Trendafilov, 1994), constructs a simple structure target matrix by a procedure identical to what is known now as *soft thresholding* (Donoho and Johnstone, 1994). The soft-thresholding operator is defined as $t_S(\alpha, \delta) = \text{sgn}(\alpha) \max\{|\alpha| - \delta, 0\}$, for any scalar α and threshold δ. In both cases, the expected result is the same: the soft-thresholding operator gives the best *sparse* linear regressor in LS sense subject to LASSO (**L**east **A**bsolute **S**hrinkage and **S**election **O**perator) constraint, while PROMAJ produces a *sparse* target matrix majorizing the relative contributions of each component (column) (Marshall et al., 2011, 5,B). Intuitively, the *vector majorization* is expressed by the following simple example with probability vectors in $\mathbb{R}^3$:

$$\left[\tfrac{1}{3}, \tfrac{1}{3}, \tfrac{1}{3}\right] \prec \left[0, \tfrac{1}{2}, \tfrac{1}{2}\right] \prec \left[0, 0, 1\right] \ ,$$

i.e. the "smallest" probability vector has equal entries (least diversity), while the "largest" one is the sparsest possible vector. One can use procedures generating majorization in order to achieve sparseness (Marshall et al., 2011, Ch 5). For example, let a_j be the j-th column of the component loadings matrix $A \in \mathcal{M}(p, r)$. Then, it follows from (Marshall and Olkin, 1979, B.2.) that

$$\left(\frac{a_{1j}^2}{\sum_{i=1}^{p} a_{ij}^2}, \frac{a_{2j}^2}{\sum_{i=1}^{p} a_{ij}^2}, \ldots, \frac{a_{pj}^2}{\sum_{i=1}^{p} a_{ij}^2}\right)^{\top} \prec$$

$$\left(\frac{\max(a_{1j}^2 - c_j^2, 0)}{\sum_{i=1}^{p} \max(a_{ij}^2 - c_j^2, 0)}, \ldots, \frac{\max(a_{pj}^2 - c_j^2, 0)}{\sum_{i=1}^{n} \max(a_{ij}^2 - c_j^2, 0)}\right)^{\top}$$

for any $c_j^2 < \max(a_{1j}^2, a_{2j}^2, \ldots, a_{nj}^2)$. After the proper c_js are specified for each column, one can define the target matrix $T = \{t_{ij}\}$ as follows:

$$t_{ij} = \text{sign}(a_{ij})\sqrt{\max(a_{ij}^2 - c_j^2, 0)} \ .$$

By taking, for example, $c_j^2 = \frac{1}{p}\sum_{i=1}^{p} a_{ij}^2$, one can achieve sparseness in A based on its loadings. Of course, the choice of c_js can be further elaborated and related to the fit of T to A.

The benefit from such an approach is that sparseness can be achieved without tuning parameters: the threshold is found by a certain majorization construction, instead of tuning parameters usually requiring multiple restarts and/or validation stage. Such a pattern construction can be further related to the fit and/or other desired features of the solution, e.g. classification error, etc.

Functions usually used to measure the "diversity" of the components of a vector are their variance (or simply their sum of squares) and their entropy (2.30) (Marshall et al., 2011, 13.F). For example, this is what VARIMAX does: it looks for an orthogonal Q that "diversifies" the loadings by making their squares to maximize their variance. As $Q \in \mathcal{O}(r)$, the initial and the rotated loadings, A and AQ, have the same sum of squares. However, the majorization $\text{vec}(A \odot A) \prec \text{vec}(AQ \odot AQ)$ cannot be guaranteed by VARIMAX (although it happens occasionally). Another potential weakness of VARIMAX is that it diversifies all loadings, as ordered in a long $pr \times 1$ vector, regardless the magnitudes of the other loadings in the particular column to which the element belongs. It might be worth constructing an alternative VARIMAX version that diversifies the loadings within every column separately, and/or achieving $\text{vec}(A \odot A) \prec \text{vec}(AQ \odot AQ)$, etc.

The hyperplane counting methods are also introduced by Thurstone (1947). As suggested by their name, they count the variables, close to certain hyperplane, and then try to maximize it. Recently the rationale behind these methods inspired the introduction of a class of rotation criteria called *component loss functions* (CLF) (Jennrich, 2004, 2006). The most intriguing of them is given by the ℓ_1 matrix norm of the rotated loadings, defined for any $p \times r$ matrix A as $\|A\|_{\ell_1} = \sum_i^p \sum_j^r |a_{ij}|$. The reason for referring to the CLF methods here is twofold: first, it seems they work well on a number of test classic data (Mulaik, 2010, p. 360-66), and second, its definition involves ℓ_1 norm, which is probably influenced by and corresponds to the recent considerable interest in sparse PCA penalized by LASSO (Jolliffe et al., 2003; Zou et al., 2006). The CLF method is defined for orthogonal rotations Q as (Jennrich, 2004)

$$\min_{Q \in \mathcal{O}(r)} \|AQ\|_{\ell_1} . \tag{4.22}$$

The orthogonal CLF simple structure rotation (4.22) does not always produce satisfying results. For this reason, the oblique CLF is usually applied instead (Jennrich, 2006). It is defined as

$$\min_{Q \in \mathcal{OB}(r)} \|AQ^{-\top}\|_{\ell_1} . \tag{4.23}$$

4.5.3   Case study: PCA interpretation via rotation methods

Traditionally, PCs are considered easily interpretable if there are plenty of small component loadings indicating the negligible importance of the corresponding variables. The rotation methods are incapable to produce vanishing (exactly zero) loadings. As a consequence, the "classic" way to apply the simple structure concept for interpretation is to ignore (effectively set to zero) component loadings whose absolute values fall below some threshold (Jolliffe, 2002, p. 269). Thus, the PCs simplicity and interpretability are implicitly associated with their *sparseness*. This will be illustrated by the following example, on a dataset, first used by Jolliffe et al. (2003), and then became a standard example in any work on sparse approximation of PCA.

Jeffers's Pitprop data example: The Pitprop data contain 13 variables measured for 180 pitprops cut from Corsican pine timber (Jeffers, 1967). Denote by $x_1, x_2, \ldots, x_{13}$ the variables in the order they appear in the cited paper. Unfortunately, the raw Pitprop data seem lost, and their correlation matrix is available only (Jeffers, 1967, Table 2). Following Jeffers (1967), six PCs are to be considered. The first six eigenvalues of the correlation matrix are: 4.2186, 2.3781, 1.8782, 1.1094, .9100 and .8154. Their sum is 11.3098, i.e. the total variance explained by the first six PCs is 86.9%. The component loadings are given in the first six columns of Table 4.2. The classical approach to their interpretation is reconstructed in (Trendafilov, 2014).

First, each column is normalized to have maximal magnitude of one. Then, the loadings greater than .7 are considered only (Jeffers, 1967, pp. 229-230). Effectively, the interpretation is based on the sparse loadings matrix given in the last six columns of Table 4.2. We can check that they fulfil the five rules of the simple structure. One may not be completely satisfied, because the first column has exactly six zeros. Indeed, this complies with the second rule for at least six zeros, but one feels tempted to improve it.

Can we obtain a clearer interpretation with the help of simple structure rotation? We apply VARIMAX from MATLAB (2019) to rotate the first six component loadings of Jeffers's Pitprop data. The rotated loadings are depicted in the first six columns of Table 4.3. Indeed, the VARIMAX loadings look better. This conclusion is mainly based on the observation that the last three columns have now clearly dominating loadings. Also, in the leading three columns the dominating and the vanishing loadings are more clearly highlighted. As Jeffers, to interpret the loadings, we normalize them first. Then we have to choose a threshold to discard the "non-contributing" variables. If we take .7 as in the previous example, we will end up with three empty (zero) rows, i.e. three variables do not contribute at all. Sup-

Table 4.2: Jeffers's Pitprop initial loadings and their interpretation (Trendafilov, 2014).

Vars	Component loadings (AD)						Jeffers's interpretation after normalization					
	1	2	3	4	5	6	1	2	3	4	5	6
x_1	.83	.34	-.28	-.10	.08	.11	1.0					
x_2	.83	.29	-.32	-.11	.11	.15	1.0					
x_3	.26	.83	.19	.08	-.33	-.25		1.0				
x_4	.36	.70	.48	.06	-.34	-.05		.84	.73			
x_5	.12	-.26	.66	.05	-.17	.56			1.0			1.0
x_6	.58	-.02	.65	-.07	.30	.05	.70		.99			
x_7	.82	-.29	.35	-.07	.21	.00	.99					
x_8	.60	-.29	-.33	.30	-.18	-.05	.72					
x_9	.73	.03	-.28	.10	.10	.03	.88					
x_{10}	.78	-.38	-.16	-.22	-.15	-.16	.93					
x_{11}	-.02	.32	-.10	.85	.33	.16				1.0		
x_{12}	-.24	.53	.13	-.32	.57	-.15					1.0	
x_{13}	-.23	.48	-.45	-.32	-.08	.57						1.0

pose, we want all original variables involved in the solution. This can be arranged by taking .59 as a threshold value. Then we find quite clear simple structure which is depicted in the last six columns of Table 4.3. Each variable contributes to only one component and the first column has now eight zeros. However, the weakness of both interpretations is that the choice of the threshold is completely subjective. The "sparse" matrices obtained after normalization and thresholding quite loosely reflect the fitting power of the loadings they are obtained from.

4.5.4 Rotation to independent components

In the beginning of Section 4.4 it was shown that the rotation methods are based on the identities (4.16) and (4.17):

$$ X = FDA^\top = FQQ^{-1}DA^\top = FDQQ^{-1}A^\top . $$

Depending on the choice of the loadings matrix, A or AD, ones optimizes certain simplicity criterion in order to obtain easily interpretable loadings. However, if the loadings are rotated, then the scores should be counter-rotated too. One cannot expect any particularly interesting findings for the rotated scores as the primary goal is to simplify the loadings.

In this section, on the contrary, we consider the rotation of the scores as a

Table 4.3: Jeffers's Pitprop rotated loadings and their interpretation (Trendafilov, 2014).

Vars	VARIMAX loadings						Normalized loadings greater than .59					
	1	2	3	4	5	6	1	2	3	4	5	6
x_1	.91	.26	-.01	.03	.01	.08	.97					
x_2	.94	.19	-.00	.03	.00	.10	1.0					
x_3	.13	.96	-.14	.08	.08	.04		1.0				
x_4	.13	.95	.24	.03	.06	-.03		.98				
x_5	-.14	.03	.90	-.03	-.18	-.03			1.0			
x_6	.36	.19	.61	-.03	.28	-.49			.68			
x_7	.62	-.02	.47	-.13	-.01	-.55	.66					
x_8	.54	-.10	-.10	.11	-.56	-.23					-.64	
x_9	.77	.03	-.03	.12	-.16	-.12	.82					
x_{10}	.68	-.10	.02	-.40	-.35	-.34	.73					
x_{11}	.03	.08	-.04	.97	.00	-.00				1.0		
x_{12}	-.06	.14	-.14	.04	.87	.09					1.0	
x_{13}	.10	.04	-.07	-.01	.15	.93						1.0

primary goal. To obtain reasonable results, we need to formulate somehow what the scores are supposed to obtain. A way to achieve this is to formulate the goal as a cost function, as the rotation criteria act for the loadings.

As an example we consider the independent component analysis (ICA) as a specific rotation method. This is inspired by the following suggestion (Hastie et al., 2009, p. 563): *ICA starts from essentially a factor analysis solution, and looks for rotations that lead to independent components. From this point of view, ICA is just another factor rotation method, along with the traditional "varimax" and "quartimax" methods used in psychometrics.*

A typical example for which ICA is useful is a party in a large room. We have records from several microphones scattered in the room. The purpose is to recover individual conversations from the microphone records. The blurred output signal is assumed a linear combination/mixture of independent signals. The ICA purpose is to find the weights of this linear combination, such that the unknown signals are as independent as possible.

First of all, ICA needs "whitened" data, i.e. X should be centred and than normalized to have unit length columns. Next, we need a cost function in which optimization will result in independent rotated scores. Unfortunately, it is well known that achieving numerically purely independent variables is an impossible task. For this reason, one of the central ICA problems is the reasonable approximation of independence (Hyvärinen et al., 2001).

If several variables are independent, their squares are also independent. Thus, the covariance matrix for these squared variables is diagonal. This simple fact is employed in (Jennrich and Trendafilov, 2005) to present ICA as a rotation method. They propose to diagonalize the covariance matrix of the *squared* rotated scores (compare to VARIMAX (4.19)):

$$S = \frac{1}{n-1}(FQ \odot FQ)^\top J_n(FQ \odot FQ) \ , \ J_n = I_n - n^{-1}E_n \ , \qquad (4.24)$$

i.e. to minimize the sum of the correlations among the squared component scores. Thus, the resulting ICA criterion is:

$$\min_{Q \in \mathcal{O}(r)} \ \mathrm{trace}[\mathrm{off}(S)]^2 = \min_{Q \in \mathcal{O}(r)} \ \mathrm{trace}[(S \odot S)(E_r - I_r)] \ , \qquad (4.25)$$

and resembles several other classical criteria for "independence" also based on fourth moments $E(x^4)$ or its normalized version kurtosis. This type of criteria are very simple, both computationally and theoretically, but in general not very robust.

The ICA problem (4.25) is another example of an optimization problem on matrix manifold, in this occasion $\mathcal{O}(r)$. Other more advanced ICA formulations also requiring optimization on matrix manifolds are listed in (Absil and Gallivan, 2006). They involve simultaneous diagonalization of several matrices and are related to common principal components (CPC) discussed in Section 9.

4.6 True simple structure: sparse loadings

Thurstone's simple structure concept mentioned in Section 4.5.1 clearly shows that the original idea is to achieve zero or vanishing loadings. In our modern jargon this is simply a call for some kind of sparse loadings.

We already discussed the main weakness of the rotation methods: they produce plenty of small non-zero loadings, part of which should be neglected subjectively. This problem is unavoidable with the rotation methods as they facilitate the interpretation while preserving the variance explained by the initial dimension reduction.

The common practice to interpret the (rotated) component loadings is to set all loadings with magnitudes less than some threshold to zero, e.g. 0.3. However, working with truncated loadings means working with sparse loadings which found in a subjective way. The rotation methods never tried to relate the thresholding process with the optimal features of the principal components when their loadings are replaced by truncated ones. Cadima and Jolliffe (1995) summarize this situation as follows: "Thresholding requires a degree of arbitrariness and affects the optimality and uncorrelatedness of the solutions with unforeseeable consequences". They also show that discarding small component loadings can be misleading, especially when working with covariance matrices. Applying significant test to decide which small loadings can be considered as zeros also has drawbacks. To apply certain test one needs to check a number of distributional assumptions, which frequently are not met. On top of that, making a decision for a single loading without considering the rest of them seems conceptually incorrect. Instead, it would make more sense to check the entire pattern of sparseness.

In the last decades, with the continuously increasing sizes of the data, it became clear that it would be desirable to put more emphasis on the interpretation simplicity than on the variance maximization. The attempts to achieve this goal led to the modern sparse PCA subject to additional sparseness inducing constraints (Trendafilov, 2014). It turns out that many original variables in large datasets are redundant and sparse solutions, involving only small proportion of them, can do the job reasonably well.

In the next section we consider a replacement of the rotation methods which produces true sparse loadings in an objective way.

4.6.1   SPARSIMAX: rotation-like sparse loadings

Trendafilov and Adachi (2015) assume that the dimension reduction is already performed and the appropriate number r of PCs to retain is chosen, and the corresponding A and D are obtained. The variance explained by the first r principal components and by the rotated once is $\text{trace}(D^2)$. We want to develop a new method producing component loadings some of which are exact zeros, called sparse, which can help for their unambiguous interpretation. Components with such sparse loadings are called sparse. However, the sparse components explain less variance than $\text{trace}(D^2)$. Thus, the new method should be designed to explain as much as possible variance, while producing sparse loadings, which will be called rotation-like sparse loadings. The new method is named SPARSIMAX to keep up with the tradition of the simple structure rotation methods.

With the PCA definition from Section 4.2 in mind, we consider first the following problem:

$$\min_{A \in \mathcal{O}(p,r)} [\text{trace}(A^\top X^\top X A) - \text{trace}(D^2)]^2 , \qquad (4.26)$$

which finds an orthonormal A such that the variance of XA equals the variance D^2 of the first r principal components of $X^\top X$. Clearly, the matrix A containing the first r eigenvectors of $X^\top X$ is a solution, and makes the objective function (4.26) zero. Note that this solution is not unique.

After having a way to find loadings A that explain the variance $\text{trace}(D^2)$, let us take care for their interpretability. We can follow the idea already employed by the CLF method (4.22) and require that the ℓ_1 matrix norm of A is minimized. This can be readily arranged by introducing the term $\|A\|_{\ell_1}$ in (4.26), which results in the following optimization problem:

$$\min_{A \in \mathcal{O}(p,r)} \|A\|_{\ell_1} + \tau[\text{trace}(A^\top X^\top X A) - \text{trace}(D^2)]^2 , \qquad (4.27)$$

where τ is a tuning parameter controlling the sparseness of A. In a sense, the new method is a compromise between sparse PCA and the classic rotation methods, as it relaxes the constraint $\text{trace}(A^\top X^\top X A) = \text{trace}(D^2)$, but is asking in reward for true sparse loadings A. The method is a particular case of the class of sparse PCA models introduced by Trendafilov (2014). They correspond to the Dantzig selector introduced by Candès and Tao (2007) for regression problems.

The similarity between (4.27) and (4.22) is that they both aim at the minimization of the sum of the absolute values of the loadings. Another similarity is that they both produce loadings which are *not* ordered according to decreasing variance. However, in contrast to the CLF problems (4.22), the solutions A of (4.27) only approximate the explained variance $\text{trace}(D^2)$ by the original dimension reduction.

It follows from the variational properties of the eigenvalues (Horn and Johnson, 1985) that the problem (4.27) can be simplified to the following SPARSIMAX problem:

$$\max_{A \in \mathcal{O}(p,r)} \|A\|_{\ell_1} - \tau\text{trace}(A^\top X^\top X A) . \qquad (4.28)$$

The clear benefit from working with (4.28) is that the original PCA solution is not required (for D^2). However, numerical experiments show that solving (4.28) is less stable than solving (4.27).

The solution of (4.27) requires minimization on the Stiefel manifold $\mathcal{O}(p, r)$, which is numerically more demanding task than solving (4.22). For small datasets, the problem (4.27) can be readily solved by the projected gradient method (Jennrich, 2001), as well as the CLF problem (4.22). For such problems it is straightforward to use optimization methods on matrix manifolds (Absil et al., 2008), and the relevant algorithms with much better convergence properties (Boumal et al., 2014).

The sparse components XA are correlated for sparse loadings A, even for orthonormal A. Then $\text{trace}(A^\top X^\top XA)$ is not any more a suitable measure for the total variance explained by the sparse components. To take into account the correlations among the sparse components, Zou et al. (2006) proposed the following. Assuming that XA has rank r, its QR-decomposition $XA = QU$ finds a basis of r orthonormal vectors collected in the orthonormal $p \times r$ matrix Q, such that they span the same subspace in $\mathbb{R}^p$ as XA. The upper triangular $r \times r$ matrix U necessarily has non-zero diagonal entries. Then, the adjusted variance is defined as: $\text{AdjVar}(XA) = \text{trace}[\text{Diag}(U)^2] \leq \text{trace}(U^\top U) = \text{trace}(A^\top X^\top XA)$. This adjusted variance is not completely satisfying, e.g. works only with orthonormal sparse loadings A. Other authors proposed alternative methods to measure the variance explained by the sparse components (Mackey, 2009; Shen and Huang, 2008), which also need further improvement. In the sequel, we calculate the adjusted variance (Zou et al., 2006) of the SPARSIMAX sparse components, because the solutions A of (4.27) are orthonormal.

In large applications, the tuning parameter τ is usually found by cross-validation. Another popular option is employing information criteria. For small applications, as those considered in the paper, the optimal tuning parameter τ can be easily located by solving the problem for several values of τ and compromising between sparseness and fit. Another option is to solve (4.27) for a range of values of τ and choose the most appropriate of them based on some index of sparseness. Here we use the following one:

$$\text{IS} = \frac{V_a V_s}{V_o^2} \times \frac{\#_0}{pr} , \qquad (4.29)$$

which is introduced in (Trendafilov, 2014). V_a, V_s and V_o are the adjusted, unadjusted and ordinary total variances for the problem, and $\#_0$ is the

Table 4.4: Loadings obtained by solving (4.27) with $\tau = 0$ and $\tau = 2000$, starting from (random) initial loadings A_0 (Trendafilov and Adachi, 2015).

Vars	A_0		$A_{\tau=0}$		$A_{\tau=2000}$	
	I	II	I	II	I	II
1	-.094	-.516	-.000	.000	0.050	-0.695
2	-.201	-.600	-.000	-1.00	-0.603	0.063
3	.819	-.423	1.00	-.000	0.000	-0.675
4	-.166	.168	.000	.000	-0.496	-0.229
5	.503	.408	-.000	.000	-0.623	0.065

number of zeros among all pr loadings of A. IS increases with the goodness-of-fit (V_s/V_o), the higher adjusted variance (V_a/V_o) and the sparseness. For equal or similar values of IS, the one with the largest adjusted variance is preferred.

4.6.2   Know-how for applying SPARSIMAX

Trendafilov and Adachi (2015) illustrate how SPARSIMAX works on the small well-known dataset of Five Socio-Economic Variables (Harman, 1976, Table 2.1, p. 14) containing $n = 12$ observations and $p = 5$ variables. This also helps to demonstrate the behaviour of its objective function (4.27).

The variance explained by the first two $(r = 2)$ principal components is $2.873 + 1.797 = 4.67$, which is 93.4% of the total variance. When $\tau = 0$, the second term in (4.27) switches off, and the solution does not depend on X. Each column of the solution has only one non-zero entry equal to ± 1, in which location corresponds to the entry with the largest magnitude in the same column of the initial A_0 used to start the algorithm. This is illustrated in Table 4.4 with an (random) initial matrix A_0 for solving (4.27) depicted in its first two columns. The SPARSIMAX solution with $\tau = 0$ is reproduced in the middle two columns of Table 4.4. Respectively, when τ has large enough value, then the second term in (4.27) completely wipes out the first term. Indeed, the solution with $\tau = 2000$ is depicted in the last two columns of Table 4.4. The same A is obtained if one solves (4.26). The variances of the corresponding components are 2.4972 and 2.1726 which is 93.4% of the total variance.

Table 4.5: Simple structure (VARIMAX) solution, Harman's sparse pattern and two SPARSIMAX solutions (sparse loadings) of (4.27) with $\tau = 19$ and $\tau = 9$ (Trendafilov and Adachi, 2015).

Vars	A_V		A_P		$A_{\tau=19}$		$A_{\tau=9}$	
	I	II	I	II	I	II	I	II
1	.016	.994		.994		-.707		.707
2	.941	-.009	.941		.579		.582	
3	.137	.980		.980		-.707		.707
4	.825	.447	.825	.447	.546	-.001	.523	
5	.968	-.006	.968		.606		.623	

Now, let see how SPARSIMAX is actually applied to get interpretable loadings and how they are related to rotated loadings. The first two columns of Table 4.5 reproduce the VARIMAX rotated loadings A_V (Harman, 1976, p. 295). The next two columns give the pattern A_P suggested in (Harman, 1976, p. 295) to interpret the loadings: "...In this sense the factor matrix exhibits one zero in each row, with the exception of variable 4; two zeros in each column; and several variables whose entries vanish in one column but not in the other. Even the exception of variable 4 is not contradictory to the set of simple structure criteria (see number 5)..."

Now, we solve (4.27) for a range of values of τ from 1 to 100. The IS values and the adjusted variances of these solutions are plotted in Figure 4.1. The IS graph on the upper plot has a ladder-like shape. Each "step" corresponds to certain number of zero loadings, starting with five and going down to two zero loadings. If one needs a solution with six zero loadings then (4.27) should be solved with $\tau < 1$. The solutions obtained by tuning parameter roughly between 5 and 15 have the highest IS values. These are the τ values, for which the SPARSIMAX solutions have five or four zero loadings. First, we are looking for SPARSIMAX solutions with four zeros as suggested by Harman's simple structure pattern from Table 4.5. They correspond to the τ values from 16 to 24 in Figure 4.1. The solution with the highest IS value (4.29) among all other solutions with four zeros is obtained with $\tau = 19$. It is denoted as $A_{\tau=19}$ in Table 4.5. Its IS value is 0.3683 and its adjusted variance is 88.66% of the total variance. Clearly, this is almost a solution with five zeros and thus differs considerably from Harman's simple structure pattern. Then, we look for SPARSIMAX solutions with five zeros. They correspond to the τ values from 1 to 15 in Figure 4.1. The solution with the highest IS value among them is obtained with $\tau = 9$. It is denoted as $A_{\tau=9}$ in Table 4.5. Its IS value is 0.4602 and its adjusted variance is 88.69% of the total variance.

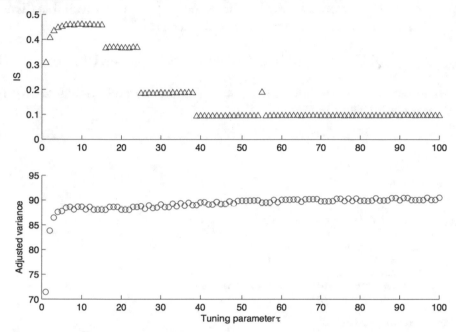

Figure 4.1: IS values (4.29) and adjusted variances for SPARSIMAX solutions with $\tau \in [1, 100]$ (Trendafilov and Adachi, 2015).

In short, SPARSIMAX does not find the classical simple structure A_P optimal. Instead, SPARSIMAX finds that the fourth loading of the second component is redundant, while the classical approach intuitively considers it significantly non-zero.

The readers are encouraged to repeat these experiments using the codes provided in Section 4.6.3. The results should be the same, though they might be obtained for different values of τ.

4.6.3 Manopt code for SPARSIMAX

To make/keep the optimization problem (4.27) smooth, $\|A\|_{\ell_1}$ is replaced by the following approximation:

$$\|A\|_{\ell_1} \approx \sum_{ij} \left(\sqrt{A_{ij}^2 + \epsilon^2} - \epsilon \right) , \tag{4.30}$$

where ϵ is a small positive number. The smaller ϵ, the better the approximation of $\|A\|_{\ell_1}$. Usually ϵ is fixed to 10^{-6} or 10^{-7}.

A number of other smooth approximations for $\|A\|_{\ell_1}$ are also available (Trendafi 2014). For example, one can employ the following:

$$\|A\|_{\ell_1} = \text{trace}[A^\top \text{sign}(A)] \approx \text{trace}[A^\top \tanh(\gamma A)] , \qquad (4.31)$$

for some large $\gamma > 0$ (Jolliffe et al., 2003; Trendafilov and Jolliffe, 2006). Its application is utilized in Section 11.2.6.

```
1   function [A,more] = sparsimax(X,r,mu,varargin)
2
3   close all;
4
5   eps = .0000001;
6
7   %% Verify the specified solver
8   solver_list = { 'TR' 'SD' 'CG' 'PS' 'NM'};
9
10  if isempty(varargin)
11      solver = 'TR';
12  elseif length(varargin) == 1
13      if any(ismember(solver_list ,varargin{1}))
14          solver = varargin{1};
15      else
16          error('Unknown solver')
17      end
18  else
19      error('Incorrect number of input arguments')
20  end
21
22  %% initialization with original PCA
23  [n p] = size(X);
24  if n==p
25      R = X;
26  else
27      R = corrcoef(X);
28  end
29  [A,D] = eigs(R,r);
30
31  % Create the problem structure.
32  manifold = stiefelfactory(p, r);
33  problem.M = manifold;
34
35  %options.maxiter = 5000;
36  %options.tolgradnorm = .01;
37  options.verbosity = 1; % or 0 for no output
38
39  % Define the problem cost function and its gradient.
40  problem.costgrad = @(A) mycostgrad(A);
41
42  % Numerically check gradient consistency.
43  % warning('off', 'manopt:stiefel:exp');
```

```
warning('off', 'manopt:getHessian:approx');
checkgradient(problem);

% Solve
if solver == 'TR'
    [A, xcost, info, options] = trustregions(problem,[],options);
elseif solver == 'SD'
    [A, xcost, info, options] = steepestdescent(problem,[],options);
elseif solver == 'CG'
    [A, xcost, info, options] = conjugategradient(problem,[],options);
elseif solver == 'PS'
    [A, xcost, info, options] = pso(problem,[]);
elseif solver == 'NM'
    [A, xcost, info, options] = neldermead(problem,[]);
end

% Display some statistics.
    plot_ok = 'Y' % 'N' %
    if plot_ok == 'Y'
        figure(2);
        %semilogy([info.iter], [info.gradnorm], 'k.-');
        plot([info.iter], [info.gradnorm], 'k.-');
        hold on;
        %semilogy([info.iter], [info.cost], 'ro-');
        plot([info.iter], [info.cost], 'ro-');
        xlabel('Iteration #'); ylabel('Gradient norm/Cost (red)');
    end

%Subroutine for the objective function f and its gradient
function [f g] = mycostgrad(A)
    W = R*A;
    w = sum(sum(A.*W)) - sum(diag(D));
    W1 = sqrt(A.^2 + eps^2);

    % objective function
    f = sum(sum(W1) - eps) + mu*w^2;

    % gradient
    if nargout == 2
        % Euclidean gradient
        g = A./W1 + mu*4*W*w;

        % projected gradient
        W2 = A'*g;
        %g = .5*A*(W2 - W2') + (eye(p) - A*A')*g;
        g = g - .5*A*(W2 + W2');
    end
end

more.time = info(end).time;
more.iter = info(end).iter;
more.cost = info(end).cost;
%options

disp('Number of zeros per component'), nz = sum(abs(A)<.00005)
disp('Total number of zeros'), sum(nz)
```

```
100
101  disp('Compare with identity matrix!'), A'*A
102  RS = A'*R*A; TV = trace(R);
103  disp('Variances'), diag(RS)'/TV, sum(diag(RS))*100/TV
104  disp('Adjusted Variances'), AdV = (diag(chol(RS)).^2)/TV, sum(AdV)*100
105  disp('Correlations amomng sparse comonents'), corrcov(RS)
106
107  end
```

4.7 Sparse PCA

In the previous sections we discussed that the classical approach to PCA interpretation—the simple structure rotation—implicitly relies on the sparse representation of the component loadings.

There is already a great number of highly efficient methods directly producing components with sparse loadings which significantly facilitates the PCA interpretation. The most influential are outlined in (Trendafilov, 2014).

In this section, we list briefly the most popular methods for sparse PCA. Then, they are classified according to the form of the objective function and constraints involved in their definitions. It will be seen that there is one unexplored yet way to define sparse PCA, called the *function-constrained form*, which corresponds to the Dantzig selector introduced for regression problems (Candès and Tao, 2007). The remaining part of the section is dedicated to follow this way and define several alternatives for sparse PCA.

4.7.1 Sparse components: genesis, history and present times

It was shown in the previous sections that the need of sparse components is realized long time ago and is related to simplifying their interpretation (Thurstone, 1935). The first approach to serve this purpose was the rotation approach. The rotated components explain the same amount of variance as the principal components do. However, the rotated components are not orthogonal any more to each other, and do not explain decreasing amount of variance reflecting their order. Additional severe weakness of the rotation methods is that the component loading of the rotated components are usually still difficult to interpret. They require a subjective choice of a threshold value in order to decide which component loadings can be regarded as zeros.

An alternative mode of attack is to look for a number of components with genuine zero loadings and sacrifice part of the variance explained by the same number of principal components. The first such method was proposed by Hausman (1982) who modified PCA to explicitly produce simple PCs in a sense that the values of the component loadings are restricted. The method finds PC loadings from a prescribed subset of values, say $S = \{-1, 0, 1\}$. Later this idea was extended to arbitrary integers (Vines, 2000).

Jolliffe and Uddin (2000) are the first to propose a penalized version (SCoT) of PCA which produces PCs additionally satisfying the VARIMAX simple structure criterion. Unfortunately, the resulting loadings contain plenty of small non-zero values, i.e. they are still not sparse. Then, Jolliffe et al. (2003) modify the original PCs to additionally satisfy the LASSO constraint (Tibshirani, 1996), which drives indeed many loadings to *exact* zeros. This PCA modification is known now under the abbreviation SCoTLASS (Jolliffe et al., 2003; Trendafilov and Jolliffe, 2006). Since then, there appeared a great number of methods producing sparse component loadings, e.g. d'Aspremont et al. (2007, 2008); Journée et al. (2010); Moghaddam et al. (2006); Qi et al. (2013); Sriperumbudur et al. (2011); Witten et al. (2009); Zou et al. (2006). Sparse PCA can also be attacked through Bayesian approach by replacing PCA with its probabilistic version and the LASSO (or ℓ_1) regularization term of the loadings—by independent Laplace (double-exponential) priors for each of the loadings, e.g. Ding et al. (2011); Guan and Dy (2009).

The main conceptual difficulty related to the methods employing LASSO/-cardinality constraint is the right choice of the LASSO threshold/number of zero loadings per component that compromise between sparseness and explained variance. This necessarily requires some kind of cross-validation, e.g. (Zou et al., 2006), or employing information criteria as AIC, BIC, etc. (Guo et al., 2010; Qi et al., 2013; Zou et al., 2007). However, they can be time-consuming for large data applications. Moreover, there is still plenty of freedom for subjective choice of threshold/number of zero loadings per component. That is why, there exists another large group of methods which totally avoid employing any LASSO/cardinality constraint penalty, e.g. Chipman and Gu (2005); Johnstone and Lu (2009); Rousson and Gasser (2004).

A weakness of most of the listed methods is that they produce sparse loadings that are not completely orthonormal and the corresponding components are correlated. To our knowledge, only SCoTLASS (Jolliffe et al., 2003; Trendafilov and Jolliffe, 2006) and the method proposed by Qi et al. (2013) are capable to produce either orthonormal loadings or uncorrelated sparse components. Lu and Zhang (2012) consider a novel type of sparse PCA explicitly controlling the orthonormality of the loadings and the correlations among components.

In contrast to the existing methods, the function-constraint approach makes it possible to achieve simultaneously nearly uncorrelated sparse components with nearly orthonormal loadings. The presumption is that such sparse PCA preserves better the optimal features of PCA. Further research is needed to make clear the benefits from such solutions, e.g. with respect to consistency in high dimensions, etc. The performance of the new method was tested and illustrated on artificial and real large datasets (Genicot et al., 2015). It is demonstrated there that the new method outperforms some of the best available methods for sparse PCA in many occasions.

4.7.2   Taxonomy of PCA subject to ℓ_1 constraint (LASSO)

Wright (2011) proposed the following taxonomy of problems seeking for sparse minimizers a of $f(a)$ through the ℓ_1 norm:

- Weighted form: $\min f(a) + \tau\|a\|_1$, for some $\tau > 0$.

- ℓ_1-constrained form (variable selection): $\min f(a)$ subject to $\|a\|_1 \leq \tau$.

- Function-constrained form: $\min \|a\|_1$ subject to $f(a) \leq \bar{f}$.

All three options are explored for regression type of problems when $f(a) = Xa + b$ and X is a given data matrix (Candès and Tao, 2007; Tibshirani, 1996). This taxonomy can be restated accordingly for sparse PCA. For a given $p \times p$ correlation matrix R, let $f(a) = a^\top Ra$ and consider finding a vector of loadings a, $(\|a\|_2 = 1)$, by solving one of the following:

- Weighted form: $\max a^\top Ra + \tau\|a\|_1$, for some $\tau > 0$.

- ℓ_1-constrained form (variable selection): $\max a^\top Ra$ subject to $\|a\|_1 \leq \tau$, $\tau \in [1, \sqrt{p}]$.

- Function-constrained form: $\min \|a\|_1$ subject to $a^\top Ra \leq \lambda$ for any sparse a, where λ is eigenvalue of R.

The first two forms are explored in a number of papers. For example, SCoT-LASS is in the ℓ_1-constrained form (variable selection), while SPCA (Zou et al., 2006) utilizes the weighted form to define the sparsification problem. It is interesting that the function-constrained form has never been used to attack the PCA sparsification. In Section 4.8, sparse component loadings will be constructed by considering the function-constrained form of PCA. Several possible definitions of sparse PCA will be listed aiming to approximate different features of the original PCs. They were illustrated in (Trendafilov, 2014) on the well-known Jeffers's Pitprop data (Jeffers, 1967).

4.8 Function-constrained sparse components

4.8.1 Orthonormal sparse component loadings

First we define problems involving orthonormal sparse component loadings A, i.e. $A^\top A = I_r$, which necessarily produce correlated sparse components. They involve optimization on the Stiefel manifold $\mathcal{O}(p, r)$.

Weakly correlated sparse components

The following version of sparse PCA

$$\min_{A \in \mathcal{O}(p,r)} \|A\|_{\ell_1} + \tau \|A^\top R A - D^2\|_F^2 \tag{4.32}$$

seeks for sparse loadings A which additionally diagonalize R, i.e. they are supposed to produce sparse components which are as weakly correlated as possible. D^2 is diagonal matrix of the original PCs' variances. The loadings obtained by solving (4.32) are typically not very sparse, but pretty uncorrelated.

Sparse components approximating the PCs variances

The next version of sparse PCA is

$$\min_{A \in \mathcal{O}(p,r)} \|A\|_{\ell_1} + \tau \|\mathrm{diag}(A^\top R A) - D^2\|_F^2 \, , \tag{4.33}$$

in which the variances of the sparse components should fit better the initial variances D^2, without paying attention to the off-diagonal elements of R. As a result $A^\top RA$ is expected to be less similar to a diagonal matrix than in (4.32) and the resulting sparse components—more correlated. The sparse loadings obtained by solving (4.33) are comparable to the well-known results obtained by other methods, e.g. Shen and Huang (2008); Zou et al. (2006).

Sparse components approximating the total variance

Instead of fitting the variances of individual sparse components to the initial ones as in (4.33), one can consider sparse components in which total variance fits the total variance of the first r PCs. This leads to the following problem:

$$\min_{A \in \mathcal{O}(p,r)} \|A\|_{\ell_1} + \tau [\text{trace}(A^\top RA) - \text{trace}(D^2)]^2 ,$$

which was considered in detail in Section 4.6.1. The obvious drawback of this sparse PCA formulation is that the resulting sparse components will not be ordered according to the magnitudes of their variances. As the total variance is fitted only, the explained variance will be higher than with the previous formulations (4.32) and (4.33). However, the adjusted variance of such components considerably departs from the total explained variance.

Sequential sparse components approximating the PCs variance

Finally, problem (4.33) can be rewritten in the following vectorial form:

$$\min_{\substack{a^\top a = 1 \\ a \perp A_{i-1}}} \|a\|_1 + \tau (a^\top Ra - d_i^2)^2 , \tag{4.34}$$

where $A_0 := 0$ and $A_{i-1} = [a_1, a_2, ..., a_{i-1}]$. The sparse loadings obtained by solving (4.34) are comparable to the results obtained by (4.33). The vector algorithm is faster, than the matrix one.

4.8.2 Uncorrelated sparse components

Sequential sparse components approximating the PCs variances

The problem (4.34) can be modified for obtaining uncorrelated sparse components (however, loosing $A^\top A = I_r$) as follows:

$$\min_{\substack{a^\top a = 1 \\ Ra \perp A_{i-1}}} \|a\|_1 + \tau(a^\top Ra - d_i^2)^2 , \qquad (4.35)$$

where $A_0 := 0$ and $A_{i-1} = [a_1, a_2, ..., a_{i-1}]$.

Weakly correlated sparse components with oblique loadings

Along with problem (4.35), one can consider solving the following matrix optimization problem:

$$\min_{A \in \mathcal{OB}(p,r)} \|A\|_{\ell_1} + \tau\|A^\top RA - D^2\|_F^2 . \qquad (4.36)$$

Clearly, the components obtained from (4.36) can be uncorrelated only approximately. However, this sparse PCA formulation is very interesting, because the resulting loadings A stay nearly orthonormal and $A^\top RA$ is nearly diagonal, as in (4.32). This is explained by the first-order optimality condition for minimizing the penalty term in (4.36) subject to $A \in \mathcal{OB}(p,r)$. It is given by

$$A^\top RA(A^\top RA - D^2) = (A^\top A)\mathrm{diag}[A^\top RA(A^\top RA - D^2)] , \qquad (4.37)$$

and implies that when the minimum of (4.36) is approached and $A^\top RA$ approaches D^2, then both sides of (4.37) become diagonal, enforcing $A^\top A$ to approach I_r.

As we know from Section 3, the set of all oblique matrices is *open*, which may cause convergence problems and makes the utilization of (4.36) questionable for large data. In reality, this is not a problem because $\mathcal{OB}(p,r)$ is the interior of $\prod_r \mathcal{S}^{p-1}$ (Proposition 3.1.5) and is replaced for computational purposes. Thus, instead of (4.36), Manopt actually solves

$$\min_{A \in \prod_r \mathcal{S}^{p-1}} \|A\|_{\ell_1} + \tau\|A^\top RA - D^2\|_F^2 .$$

As a conclusion of Section 4.8, we can mention a general weakness of the considered approach to sparse PCA: all problems contain D^2, i.e. one needs to find first the largest r eigenvalues from the original data.

4.8.3   Numerical Example

Problem (4.36) was introduced and illustrated on a small dataset in (Trendafilov, 2014). Genicot et al. (2015) performed tests on real DNA methylation datasets using Manopt. As a preprocessing step, 2000 genes are randomly selected and standardized. Three measures are of interest to evaluate the performances of the method: the variance explained the orthogonality of the loading factors and the correlations among the resulting components. Each of these measures is analysed in regard to the sparsity achieved. The many runs carried out showed very stable performance and results. They are compared to GPower (Journée et al., 2010), which is the state-of-the-art method for sparse PCA. A weakness of the method is that it produces many small, but not exact zeros loadings. For this reason, a threshold τ is used to cut off to 0 loadings with magnitudes below it.

As stated above, a large value of τ increases the importance of the second term in (4.36) and thus decreases the sparsity. Genicot et al. (2015) show that the solutions of (4.36) are able to explain quite reasonable part of the variance even for a high degree of sparsity. It is very important that they preserve simultaneously high level of orthogonality of the loadings while the correlations among the resulting components are very weak. Thus, the solutions of (4.36) preserve very well the optimal PCA features outlined in Section 4.2. Here, we use the same approximation (4.30) for $\|A\|_{\ell_1}$.

4.8.4   Manopt code for weakly correlated sparse components with nearly orthonormal loadings

The following is a slight modification of the Manopt code used in (Genicot et al., 2015) to carry out the numerical experiments.

```
1   function [A, more] = weakSPCA(X, r, mu, varargin)
2
3   close all;
4
5   eps = .0000001;
6
```

```
%% Verify the specified solver
solver_list = { 'TR' 'SD' 'CG' 'PS' 'NM'};

if isempty(varargin)
     solver = 'TR';
elseif length(varargin) == 1
     if any(ismember(solver_list ,varargin{1}))
          solver = varargin{1};
     else
          error('Unknown solver')
     end
else
     error('Incorrect number of input arguments')
end

%% initialization with original PCA
[n p] = size(X);
if n==p
     R = X;
else
     R = corrcoef(X);
end
[A,D] = eigs(R,r);

% Create the problem structure.
manifold = obliquefactory(p, r);
problem.M = manifold;

%options.maxiter = 5000;
%options.tolgradnorm = .01;
options.verbosity = 1; % or 0 for no output

% Define the problem cost function and its gradient.
problem.costgrad = @(A) mycostgrad(A);

% Numerically check gradient consistency.
% warning('off', 'manopt:stiefel:exp');
warning('off', 'manopt:getHessian:approx');
checkgradient(problem);

% Solve
if solver == 'TR'
     [A, xcost, info, options] = trustregions(problem,[],options);
elseif solver == 'SD'
     [A, xcost, info, options] = steepestdescent(problem,[],options);
elseif solver == 'CG'
     [A, xcost, info, options] = conjugategradient(problem,[],options);
elseif solver == 'PS'
     [A, xcost, info, options] = pso(problem,[]);
elseif solver == 'NM'
     [A, xcost, info, options] = neldermead(problem,[]);
end

% Display some statistics.
     plot_ok = 'Y' % 'N' %
     if plot_ok == 'Y'
```

```
63            figure(2);
64            %semilogy([info.iter], [info.gradnorm], 'k.-');
65            plot([info.iter], [info.gradnorm], 'k.-');
66            hold on;
67            %semilogy([info.iter], [info.cost], 'ro-');
68            plot([info.iter], [info.cost], 'ro-');
69            xlabel('Iteration #'); ylabel('Gradient norm/Cost (red)');
70        end
71
72  %Subroutine for the objective function f and its gradient
73  function [f g] = mycostgrad(A)
74      W = R*A;
75      WW = A'*W - D;
76      W1 = sqrt(A.^2 + eps^2);
77
78      % objective function
79      f = sum(sum(W1 - eps)) + mu*norm(WW,'fro')^2;
80
81      % gradient
82      if nargout == 2
83          % Euclidean gradient
84          g = A./W1 + mu*4*W*WW;
85          % projected gradient
86          g = g - A.*repmat(sum(g.*A),p,1);
87      end
88  end
89
90  more.time = info(end).time;
91  more.iter = info(end).iter;
92  more.cost = info(end).cost;
93  %options
94
95  disp('Number of zeros per component'), sum(abs(A)<.00005)
96  disp('Total number of zeros'), sum(ans)
97
98  disp('Compare with identity matrix!'), A'*A
99  RS = A'*R*A;
100 disp('Variances'), diag(RS)'/p, sum(diag(RS))*100/p
101 disp('Adjusted Variances'), AdV = diag(chol(RS)).^2; AdV', sum(AdV)
         *100/p
102 disp('Correlations amomng sparse comonents'), corrcov(RS)
103 end
104
105 Q = Qout; wD =Q'*R*Q;
106 [diag(wD) d0;sum(diag(wD)) sum(d0)], corrcov(wD)
107 disp('Variance:' ), diag(wD)'/p, cumsum(diag(wD)/p)'
108 AdjVar1 = diag(chol(wD)).^2;
109 display('Total adjusted explained variance'), cumsum(AdjVar1)'/p
110 display('Rotation-like loadings'),
111 Qout = Q*diag(sqrt(diag(wD)))
112 disp('Number of zeros per component:' ), nzc = sum(abs(Q)<.0005)
113 disp('Total number of zeros:' ), sum(nzc)
```

4.9 New generation dimension reduction

In this section we briefly consider several new approaches for dimension reduction which are supposed to be faster and/or cheaper than SVD. We first consider the centroid method which is a very old procedure but recently revived. The rest of the considered methods comprise a very active research area shaping the modern large data analysis. An extended survey of such methods is given in (Kumar and Schneider, 2017).

4.9.1 Centroid method

The *centroid method* is probably the first method for approximate decomposition/factoring of a covariance/correlation matrix, i.e. a symmetric PSD matrix. According to (Harman, 1976, 8.8.), the centroid method was first employed by Cyril Burt in 1917, but elaborated further by Thurstone and presented as a developed method in (Thurstone, 1935, Ch III). The persistent problem with the centroid method is that its definition relies on geometric reasoning only. The lack of formal mathematical definition even triggers complaints "...that it is unclear what is accomplished by the centroid method" (Choulakian, 2003).

The centroid idea/method is more than 100 years old, but it is still attractive because of its simplicity: being designed for pocket calculator! For this same reason, the centroid method is recently revived to possibly serve for analysis of very large matrices together with SVD (Chu and Funderlic, 2001). To achieve this, a generalized centroid algorithm is proposed that works directly with the data matrix. This improves the classical algorithm requiring correlation matrix, which is expensive in modern data with huge number of variables.

The first algorithm-like presentation of the centroid method is given in (Horst, 1965, Ch 5), where the method is also recognized as a specific dimension reduction process, reducing the rank of the input correlation matrix by one at every step. Another very important contribution is that it is clearly stated that the involved weight vectors in the dimension reduction are sign vectors, i.e. vectors having components 1 or -1 only. A contemporary description of the classical centroid method/algorithm is given in (Chu and Funderlic, 2001), as well as its generalization for large data.

Here we prefer to stress on another look at the centroid method (Choulakian, 2003, 2005). Let X be given standardized $n \times p$ $(n > p)$ data matrix and $R = X^\top X$ be its (sample) correlation matrix. Then, the vector of the first centroid component loadings is given by

$$c_1 = \frac{R1_p}{\sqrt{1_p^\top R1_p}} \ ,$$

which is frequently suggested as an initial value for power iterations in PCA/EVD. One would expect to *deflate* R to $R_1 = R - c_1 c_1^\top$, and repeat the process by calculating

$$c_2 = \frac{R_1 1_p}{\sqrt{1_p^\top R_1 1_p}} \ ,$$

etc. However, one can see that

$$R_1 1_p = R1_p - \frac{R1_p 1_p^\top R}{1_p^\top R1_p}1_p = 0 \ ,$$

which shows that the second vector of centroid loadings has the form $\frac{0}{0}$, and is not well defined. To continue this process and find a proper c_2, the geometric reasoning suggests to replace 1_p by some p-dimensional sign vector s and produce centroid loadings of the form:

$$c = \frac{Rs}{\sqrt{s^\top Rs}} \ .$$

Then, the optimal s is found by maximizing the quadratic form $s^\top Rs$, over all p-dimensional sign vectors s. In other words, the centroid method requires the solution of

$$\max_{s \in \{-1,1\}^p} s^\top Rs \ . \tag{4.38}$$

Suppose that we do not know how to do SVD of X, and cannot do PCA exactly. Instead, we try to obtain some kind of approximate solution. For this reason, we follow the standard dimension reduction scenario and present X as a sum of rank 1 matrices $vc^\top$, where $v \in \mathbb{R}^n$ and $c \in \mathbb{R}^p$. We assume that v is normalized $\|v\|_2 = 1$. As the total variance of the data is expressed by $\mathrm{trace}(X^\top X) = \mathrm{trace}(cc^\top) = c^\top c$, it is desirable to find the loadings c with the maximal length $\|c\|_2$. However, we do not know how to do quadratic optimization and look for an alternative way to express this. A reasonable option is to look for $c \in \mathbb{R}^p$, such that the sum of the absolute values of its components is maximal, i.e.

$$\max \|c\|_1 = \max(|c_1| + |c_2| + \ldots + |c_p|) = c^\top \text{sign}(c) .$$

To do this, express the loadings as $c = X^\top v$. Then, the sum of the absolute values of the components of c is the maximum of $s^\top X^\top v$ over all sign vectors $s \in \{-1, 1\}^p$. It follows from the Cauchy–Schwartz inequality (2.4) that

$$s^\top X^\top v = (Xs)^\top v \leq \sqrt{s^\top X^\top X s} \sqrt{v^\top v} = \sqrt{s^\top R s} . \qquad (4.39)$$

This estimate shows the upper bound for certain sign vector s. Of course, one can improve/maximize the upper bound (4.39) by solving (4.38) for the optimal sign vector s. The maximum is reached in case of equality in (4.39), which can happen only if the two involved multipliers Xs and v are proportional, i.e. $v = \alpha Xs$. From the normalization condition it follows that $\alpha = \frac{1}{\sqrt{s^\top R s}}$. Thus, for any particular sign vector s, the optimal scores are given by $v = \frac{Xs}{\sqrt{s^\top R s}}$. Then, the corresponding loadings are $c = \frac{Rs}{\sqrt{s^\top R s}}$. They are called centroid components/scores and centroid loadings, respectively.

This approach is important because it reveals the connection between the centroid method and the so-called ℓ_1-PCA (Choulakian, 2005). For example, the standard PCA defined in (4.1) as

$$\max_{\|a\|_2=1} \|Xa\|_2 \quad \text{or} \quad \max_{\|b\|_2=1} \|X^\top b\|_2 ,$$

can be "robustified" by solving:

$$\max_{\|a\|_2=1} \|Xa\|_1 \quad \text{or} \quad \max_{\|b\|_2=1} \|X^\top b\|_1 ,$$

depending on the dimensions of X. In these notations the centroid method (4.38) can be written as

$$\max_{s \in \{-1,1\}^p} \|Xs\|_2 \quad \text{or} \quad \max_{\|v\|_2=1} \|X^\top v\|_1 .$$

Of course, an alternative option for ℓ_1-PCA would be to robustify (4.7), by solving:

$$\max_{\|a\|_2=1} \|X(I_p - aa^\top)\|_1 .$$

4.9.2   Randomized SVD

The purpose of these kind of methods is to become cheaper surrogates for SVD of large data matrix $X \in \mathcal{M}(n,p)$. The idea is, for a specified/desired rank r, to generate a random and approximate basis $Q \in \mathcal{O}(n,r)$ for the range of X, such that $X \approx QQ^\top X$ (Halko et al., 2011; Mahoney, 2011). Then, one can find $Y = Q^\top X \in \mathcal{M}(r,p)$ with reduced dimensions and cheaper SVD as $Y = UDV^\top$, which finally gives low-rank approximation as $X \approx QUDV^\top$.

In practice, the generation of a random and approximate basis Q is as follows. Let $\Omega \in \mathcal{M}(p,r)$ be a random matrix. Then, orthogonalize $X\Omega \in \mathcal{M}(n,r)$ and keep the result in $Q \in \mathcal{O}(p,r)$. Instead of r, it is suggested to use $r+k$, where k is called *oversampling parameter* and is usually taken five or ten. The goal of the algorithm is to produce $Q \in \mathcal{O}(p,r+k)$ that achieves $\|(I_p - QQ^\top)X\| \leq \epsilon$, for some pre-specified tolerance ϵ.

It turns out that the major computational burden for large data is the matrix-to-matrix product $X\Omega$. For this reason, Ω with special structure, e.g. Gaussian, are used that facilitate the computation of $X\Omega$. A standard *Gaussian matrix* is a random matrix whose entries are independent normally distributed random variables with zero mean and unit standard deviation. For Gaussian $\Omega \in \mathcal{M}(p,r+k)$, it is proven under mild assumptions that the probability of the following estimation of the spectral norm

$$\|X - QQ^\top X\| \leq \left[1 + 9\sqrt{r+k}\sqrt{\min\{n,p\}}\right]\sigma_{r+1} \,,$$

is at least $1 - 3k^{-k}$, which justifies the choice of $k = 5$ (Halko et al., 2011, 10.3.). The following probabilistic bound for the average Frobenius error is also valid for Gaussian $\Omega \in \mathcal{M}(p,r+k)$ (Halko et al., 2011, Th 10.5):

$$E\|(I_n - QQ^\top)X\|_F \leq \sqrt{1 + \frac{r}{k-1}}\|D_2\|_F \,,$$

where $E()$ is the expectation operator and $X = [U_1\, U_2]\mathrm{Diag}(D_1, D_2)[V_1\, V_2]^\top$ is the SVD of X. Here, $\mathrm{Diag}(D_1, D_2)$ is a block-diagonal matrix where D_1 contains the largest r singular values of X.

Important class of randomized algorithms are the so-called Monte Carlo algorithms designed to analyse large data from areas as latent semantic indexing, DNA microarray analysis, facial and object recognition, and web search models (Kannan and Vempala, 2017). They sample rows/columns from the large data matrix X using certain sampling probabilities. The approach makes it possible to obtain/prove error bounds.

For example, calculate the following $n \times 1$ and $p \times 1$ vectors of row and column sample probabilities:

$$p_{I:} = \frac{(X \odot X)1_p}{1_n^\top(X \odot X)1_p} \quad \text{and} \quad p_{:J} = \frac{1_n^\top(X \odot X)}{1_n^\top(X \odot X)1_p} ,$$

and sample rows/columns of X, such that ith row and jth column is chosen with probability $p_{i:}$ and $p_{:j}$, respectively. Usually, the sampling process is performed sequentially, e.g. first choose rows and collect them in $X_{I:}$, and then chose the columns from $X_{I:}$, to find their intersection X_{IJ}. Further development in this direction is obtained through adaptive and volume sampling, which helps to reduce the low-rank approximation error produced by the sampled rows/columns (Deshpande et al., 2006). The clustered low-rank matrix approximation is designed to take into account and preserve structural information of the data (Savas and Dhillon, 2016).

4.9.3 CUR approximations

This class of methods are very closely related to the ones considered in Section 4.9.2, at least in terms of sampling subsets of rows and/or columns. They are considered separately simply because they are developed around another matrix decomposition than SVD.

The research on CUR matrix approximation/decomposition was probably initiated by the introduction of the so-called *pseudo-skeleton matrix decomposition* (Goreinov et al., 1997). The idea comes from the following result. Let $X \in \mathcal{M}(n,p)$ has rank r, i.e. X has non-singular submatrix $X_{IJ} \in \mathcal{GL}(r)$. Suppose that X_{IJ} is composed of r rows and columns of X, in which indexes are collected in I and J, respectively. Form $C = X(:,J) \in \mathcal{M}(n,r)$ and $R = X(I,:) \in \mathcal{M}(r,p)$. Then, $X = CX_{IJ}^{-1}R$, and is called the *skeleton decomposition* of X (Friedland, 2015, 4.14).

In reality, we do not know the rank of X or we know it only approximately. Also, we do not know C and R, and X_{IJ} is most likely ill-conditioned. Then, it makes more sense to replace X_{IJ} with some "better" non-singular matrix, such that $X \approx CUR$. This was initially called the pseudo-skeleton decomposition of X. Intuitively, among the submatrices X_{IJ} with rank r, one is interested to choose the one with the largest determinant/volume. Thus, such decomposition is related to the problem of finding the maximum determinant (volume) submatrix form $\mathcal{GL}(r)$, or a good approximation of it. On the other hand, it is well known that the determinant measures the *generalized variance* of the involved variables, e.g. the determinant of their correlation matrix (Anderson, 1984, 7.5.). In this sense, this decomposition generalizes the classical PCA.

It is well known that for given C and R, the solution of

$$\min_{U} \|X - CUR\|_F \tag{4.40}$$

is given by $U = C^{\dagger} X R^{\dagger}$, where † denotes the *Moore–Penrose (pseudo)inverse* of its argument (Golub and Van Loan, 2013, p. 290). Note that this result extends to rectangular U.

The algorithms for solving (4.40) use block Gaussian elimination with a suitable block-matrix to determine a rank r approximation. They have complexity $O(r^2(n+p))$ once an appropriate block is found. This is considerably better than the Arnoldi or Lanczos methods to find the r-truncated SVD, requiring $O(npr))$ operations (Golub and Van Loan, 2013, Ch 10).

Nowadays, such type of decompositions are simply referred to as CUR. In the last two decades, this is a very active area of research (Mahoney and Drineas, 2009). There exists a great number of approaches how to choose C and/or R for (4.40), called sampling techniques, e.g. adaptive, volume, as well as how to solve (4.40) approximately. There are also two different ways to look at the CUR decomposition: either as an exact matrix decomposition/factorization of X, or as a low-rank approximation to it.

4.9.4 The Nyström Method

In many machine learning and data analysis applications, one is interested in symmetric PSD matrices, e.g. large covariance/correlation matrices, kernel matrices, etc. One common column-sampling-based approach to low-rank approximation of such matrices is the so-called Nyström method (Drineas and Mahoney, 2005). It is closely related to CUR, and directly benefits from the progress in CUR techniques. An interesting illustration of the Nyström method is made by demonstrating that three modern algorithms for *multi-dimensional scaling* are in fact its realizations (Platt, 2005).

Roughly, the method selects columns from the original data matrix uniformly at random and then uses them to construct a low-rank PSD approximation. Let $R \in \mathcal{S}(n)$ be PSD and $C \in \mathcal{M}(n, r)$ contain the selected columns from R, where $r \ll n$. After a suitable permutation one can assume that the selected columns are the first r columns of R. In other words, we have the following partition:

$$C = \begin{bmatrix} W \\ R_{21} \end{bmatrix} \quad \text{and} \quad R = \begin{bmatrix} W & R_{21}^\top \\ R_{21} & R_{22} \end{bmatrix} .$$

Then, $CW^\dagger C^\top$ gives a low-rank approximation to R of at most r. Another approximation of R is given by $CC^\dagger R (CC^\dagger)^\top$, where C is obtained by adaptive sampling (Wang and Zhang, 2013). Again, the stress is on establishing error bounds for the low-rank approximation resulting from certain sampling scheme. Many tools are adopted from the CUR methods.

4.10 Conclusion

Section 4.2 provides several alternative PCA definitions useful in different scenarios. In general, it is advisable to analyse raw data, because the potential noise/errors in X multiply in the correlation matrix $X^\top X$. The sequential/deflation definitions are preferable for PCA with unspecified/unknown dimension of the reduced space. On the contrary, problems concerning maximum explained (or minimum residual) variance may benefit from simultaneous solutions. The regression approach to PCA (4.7) turns it to a projection method. The particular case of PCA applied to count/frequency data, known as correspondence analysis, is briefly outlined in Section 4.3.

The PCA interpretation is a central issue because the same principles apply to the interpretation of any MDA technique. Section 4.4 describes two types of PCA linear combinations serving this purpose. Section 4.5 discusses the meaning of the PCA (and MDA) interpretation and shows that the idea of "sparse" solutions (linear combinations) is implicitly used for nearly 100 years in the form of the "simple structure" concept. We briefly review the classical rotation methods to achieve solutions with simple structure. In a real data example we illustrate how the classical PCA interpretation process constructs a hypothetical sparse approximation of the solution which actually gives the final interpretation. An important feature of the classical rotation methods is that they preserve the variance captured by the initial PCA solution. This is relaxed in Section 4.6 in order to construct sparse PC loadings preserving as much as possible of the initially captured variance.

Section 4.7 gives a brief review of the existing methods for sparse PCA. Section 4.8 provides a general approach to define a spectre of sparse PCA problems which approximate different aspects of the original PCA and may suit various scenarios.

Finally, Section 4.9 introduces several modern methods for dimension reduction of very large data.

4.11   Exercises

1. Consider the following vector version of "regression" PCA (4.13):

$$\max_{y^\top y=1} y^\top XX^\top y , \qquad (4.41)$$

   and show that this is a standard eigenvalue problem. Hint: Introduce $y = Xa$, and rewrite (4.41) with Lagrange multipliers.

2. (Lai et al., 2020, 8.1) Reconsider the PCA problem (4.2)

$$\max_{A \in \mathcal{O}(p,r)} \text{trace}(A^\top XX^\top A) ,$$

   by solving it on the Grassmann manifold $\mathcal{G}(p,r)$, i.e.

$$\max_{Q \in \mathcal{G}(p,r)} \text{trace}(XX^\top Q) .$$

   Hint: Change the variables $Q = 2AA^\top - I_p$.

3. Let A and B be $n \times n$ symmetric matrices, and A is PSD. Prove that

$$\lambda_{min}(B)\text{trace}(A) \leq \text{trace}(AB) \leq \lambda_{max}(B)\text{trace}(A) ,$$

   where $\lambda_{min}(B)$ and $\lambda_{max}(B)$ denote the smallest and largest eigenvalues of B. Hint: As a PSD matrix, A has non-negative elements on its main diagonal.

4. Let $F \in \mathcal{M}(n, p)$ be contingency table, and $r = F1_p$ and $c = F^\top 1_n$ be vectors containing the row and column sums of F, and the total is $t = 1_n^\top F 1_p$. Check that

(a) the χ_0^2 statistic (4.14) is given in matrix form by

$$\chi_0^2 = t \times \text{trace} \left[D_c^{-1} (F - E)^\top D_r^{-1} (F - E) \right] ,$$

where $E = \frac{rc^\top}{t}$, and $D_v = \text{diag}(v)(= I_p \odot v1_p^\top)$ for any $v \in \mathbb{R}^p$;

(b) CA is SVD/PCA of $X = \sqrt{t} \times D_r^{-1/2}(F - E)D_c^{-1/2}$;

(c) $\sqrt{r}$ and $\sqrt{c}$ are left and right singular values of $Y = D_r^{-1/2}FD_c^{-1/2}$, and 1 is the corresponding singular value; thus, show that

$$\frac{X}{\sqrt{t}} = Y - \frac{D_r^{1/2} 1_n 1_p^\top D_c^{1/2}}{t} = \frac{1}{\sqrt{t}} \sum_2^r d_i u_i v_i^\top ,$$

where u_i and v_i are the ith columns of U and V, and d_i is the ith entry of D from the SVD of $X = UDV^\top$;

(d) the row and column *principal coordinate matrices* in CA, defined as $R = D_r^{-1/2}UD$ and $C = D_c^{-1/2}VD$, are related through the following transition formulas:

$$R = D_r^{-1}FCD^{-1} \quad \text{and} \quad C = D_c^{-1}F^\top RD^{-1} .$$

Hint: $R = D_r^{-1/2}UD^{-1} = D_r^{-1/2}YV = D_r^{-1/2}YD_C^{1/2}CD^{-1}$.

5. Check that in Section 4.3:

(a) $\chi_0^2 \leq t \times \min\{n - 1, p - 1\}$;

(b) the rank of X is at most $r = \min\{n - 1, p - 1\}$. Hint: See (b) and (c) in the previous exercise.

6. Let $A \in \mathcal{M}(p, r)$ be an initial PCA (EFA) loadings matrix and denote $B = AQ$ and $C = B \odot B$. Write Manopt codes for the following χ^2 rotation criterion (Knüsel, 2008):

$$\max_Q \text{trace}[C^\top \text{diag}^{-1}(C1_r)C\text{diag}^{-1}(1_p^\top C)] ,$$

with $Q \in \mathcal{O}(r)$ and $Q \in \mathcal{OB}(r)$. Compare with other rotation criteria and their rotated loadings.

7. Check, that

   (a) For $B = AQ$, the VARIMAX function (4.19) can be written as

   $$V(Q) = \text{trace}[B^\top(B \odot B \odot B)] - \frac{1}{p}\text{trace}\{B^\top B[I_p \odot (B^\top B)]\} \ .$$

   (b) For any $A \in \mathcal{M}(p, r)$, the following identity holds:

   $$A^\top(A \odot A \odot A) - \frac{1}{p}(A^\top A)[I_p \odot (A^\top A)] = \sum_{i=1}^{p} W_i[I_p \odot W_i] \ ,$$

   where $W_i = A^T A - p a_i a_i^T$, and a_i^T denotes here the i-th row of A. Hint: $A^\top A = \sum_i^p a_i a_i^T$ (ten Berge, 1984).

   (c) Consider the VARIMAX version for PCA with $A \in \mathcal{O}(p, r)$, simplifying the above expressions, and possibly produce (non-rotationa[l] algorithm for) sparse A having single non-zero entry in every row $a_i^T, i = 1, \ldots, p$, and maximizing V among all loadings A with such pattern.

8. (Mackey, 2009) The deflation methods solve

   $$\max_{a^\top a=1} a^\top R_{i-1} a \ \text{ for } i = 1, \ldots, r \ ,$$

   where R_0 is a given $p \times p$ correlation matrix, and the next R_i is calculated from the previous a and R_{i-1}, in several ways, as follows:

   (a) Hotelling's deflation: $R_i = R_{i-1} - a a^\top R_{i-1} a a^\top$.

   (b) Projection deflation: $R_i = (I_p - a a^\top) R_{i-1} (I_p - a a^\top)$.

   (c) Schur complement deflation: $R_i = R_{i-1} - \frac{R_{i-1} a a^\top R_{i-1}}{a^\top R_{i-1} a}$.

   - Check that (b) and (c) reduce to (a), when a is eigenvector of R.
   - In sparse PCA, a is *not* eigenvector of R, and is different for (a), (b) and (c). Check for which of them, the sparse a
     - is orthogonal to the deflated matrix, i.e. $R_i a_i = 0_p$;
     - renders PSD deflated matrix, i.e. if R_{i-1} is PSD, so is R_i.

9. Consider the following absolute value approximation for LASSO-related problems. Note that $|x| = \max\{x, -x\}$, and employ the corresponding smoothing $\max\{a, b\} \approx u \log(e^{a/u} + e^{b/u})$, where $u > 0$ is a smoothing parameter (Chen and Mangasarian, 1995).

10. Plot the unit "circle" with respect to the l_4 norm, defined as $\{x \in \mathbb{R}^2 : \|x\|_4 = \sqrt[4]{x_1^4 + x_2^4} = 1\}$. For sparse results, explore the maximization of $\|A\|_{\ell_4} = \sqrt[4]{\text{trace}(A \odot A)^\top (A \odot A)} = \sqrt{\|A \odot A\|_{\ell_2}}$ to replace the minimization of $\|A\|_{\ell_1}$ (Zhai et al., 2020).

11. Let R be $p \times p$ block-diagonal correlation matrix:

$$R = \begin{bmatrix} R_{q_1} & O_{p_1 \times p_2} & \cdots & O_{p_1 \times p_k} \\ O_{p_2 \times p_1} & R_{p_2} & \cdots & O_{p_2 \times p_k} \\ \vdots & \vdots & \ddots & \vdots \\ O_{p_k \times p_1} & O_{p_k \times p_2} & \cdots & R_{p_k} \end{bmatrix},$$

where each block R_{p_i} is a $p_i \times p_i$ correlation matrix and $\sum_{i=1}^k p_i = p$ (Horn and Johnson, 2013, 0.9.2). Then, the eigenvalues of R can be found by solving k smaller eigenvalue problems for $R_{p_1}, ..., R_{q_k}$ (Horn and Johnson, 2013, 1.1.P4). Consider the EVD of $R_{q_i} = V_{p_i} D_{p_i}^2 V_{p_i}^\top$, and check that the eigenvectors of R are given by

$$V = \begin{bmatrix} V_{p_1} & O_{p_1 \times 1} & \cdots & O_{p_1 \times 1} \\ O_{p_2 \times 1} & V_{p_2} & \cdots & O_{p_2 \times 1} \\ \vdots & \vdots & \ddots & \vdots \\ O_{p_k \times 1} & O_{p_k \times 1} & \cdots & V_{p_k} \end{bmatrix}.$$

Thus, PCA of a block-diagonal correlation matrix R results in a sparse loadings matrix V. In practice, this feature can be utilized for sparse PCA, by preliminary

   (a) clustering of the variables (Enki and Trendafilov, 2012; Enki et al., 2013; Rousson and Gasser, 2004);

   (b) thresholding the entries of R (Deshpande and Montanari, 2016; Johnstone and Lu, 2009).

12. To avoid the dependence on D^2 in (4.36), consider

$$\min_{A \in \mathcal{OB}(p,r)} \|A\|_{\ell_1} - \tau \text{trace}(A^\top R A).$$

Show that the first-order optimality condition for maximizing its penalty term implies that $A^\top A$ becomes close to identity at its maximum. Adapt the codes from Section 4.8.4 to solve this problem, and compare the solutions with those of (4.36).

13. Check that the centroid method for PCA of $X^\top X$ (Chu and Funderlic, 2001) is equivalent to the ℓ_1-version of PCA defined as (Kwak, 2008):

$$\max_{a^\top a=1} \|Xa\|_1 \ .$$

14. Consider a matrix version of the centroid method, by utilizing the matrix version of the Cauchy–Schwartz inequality (2.28). Generalize (4.39) to obtain matrices of centroid loadings and scores, as well as sign matrices S, maximizing trace$(S^\top RS)$.

15. Let X be a standardized $n \times p$ data matrix. Consider sparse centroid method by solving

$$\max_{\substack{\|b\|_2 = 1 \\ a \in \{-1,0,1\}^p}} b^\top Xa \ .$$

The solution will produce sparse loadings a and "dense" scores b. See for comparison (Kolda and O'Leary, 1998).

16. Prove the skeleton decomposition of X (see Section 4.9.3), i.e. that $X = CX_{IJ}^{-1}R$. Hint: As rank$(C) = $ rank$(R) = $ rank$(X) = r$, there exists $W \in \mathcal{M}(r,n)$ such that $X = CW$. Then, consider the matrix of zeros and ones $P_I \in \mathcal{M}(r,n)$, $(P_I P_I^\top = I_r)$, which picks up the rows of X according to the index set I. Thus, we have $P_I X = R$ and, respectively, $P_I C = X_{IJ}$.

17. Probabilistic approach to *non-linear* dimension reduction (Hinton and Roweis, 2002; van der Maaten and Hinton, 2008): Let $X \in \mathcal{M}(n,p)$ be given data matrix with rows $x_1, \ldots, x_n$, and $Y \in \mathcal{M}(n,r)$ be its unknown r-dimensional projection with rows $y_1, \ldots, y_n$. The (conditional) probability that x_i chooses x_j as a neighbour is:

$$p_{i \to j} = \frac{\exp(-\|y_i - y_j\|^2 / 2\sigma^2)}{\sum_{k \neq i} \exp(-\|y_i - y_k\|^2 / 2\sigma^2)} \ .$$

The probability of y_i choosing y_j and vice versa in the low-dimensional space is symmetric and is taken as either:

$$q_{ij} = \frac{\exp(-\|y_i - y_j\|^2)}{\sum_{k \neq l} \exp(-\|y_k - y_l\|^2)} \text{ or } q_{ij} = \frac{(1 + \|y_i - y_j\|^2)^{-1}}{\sum_{k \neq l}(1 + \|y_k - y_l\|^2)^{-1}} \ .$$

Let $P = \{(p_{i \to j} + p_{j \to i})/2n\}$ and $Q = \{q_{ij}\}$, assuming $p_{ii} = q_{ii} = 0$. The non-linear projection Y of X is found by minimizing the Kullback–Leibler divergence (2.32) between the joint distributions p and q:

$$s(p|q) = \text{trace}[P(\ln P - \ln Q)] \,,$$

with element-wise logarithm. Check that the gradients of s for the two choices of Q are:

$$\nabla_{y_i} = 4 \sum_{i=j}^{n} (y_i - y_j)(p_{ij} - q_{ij})$$

and, respectively,

$$\nabla_{y_i} = 4 \sum_{j=1}^{n} (y_i - y_j)(p_{ij} - q_{ij})(1 + \|y_i - y_j\|^2)^{-1} \,.$$

Hint: Introduce new variables, e.g. $d_{ij} = \|y_i - y_j\|$, $z = \sum_{k \neq l} f(d_{kl})$ with $f(t) = e^{-t^2}$ and $f(t) = (1 + t^2)^{-1}$, and note that $q_{ij}z = f(d_{ij})$ (van der Maaten and Hinton, 2008, Appendix A).

Chapter 5

Factor analysis (FA)

© Springer Nature Switzerland AG 2021
N. Trendafilov and M. Gallo, *Multivariate Data Analysis on Matrix Manifolds*,
Springer Series in the Data Sciences, https://doi.org/10.1007/978-3-030-76974-1_5

5.1   Introduction

Factor analysis (FA) originated in psychometrics as a theoretical underpinning and rationale to the measurement of individual differences in human ability. In 1904, Charles Spearman published his seminal paper (Spearman, 1904), which is considered the beginning of FA. The purpose was to measure the human intelligence, thought at that time to be a single unobservable entity, called general intelligence g. In his paper, Spearman tried to quantify the influence of g on the examinee's test scores on several topics: Pitch, Light, Weight, Classics, French, English and Mathematics. He used regression-like equations but with the difference that g is not available (observable) as in the regression problems. Soon, the theory of intelligence evolved and Spearman's (one factor) model with a single latent construct became less popular and replaced by the so-called multiple FA (Thurstone, 1935, 1947). The early history and the early detection of problems with the FA model are briefly discussed in (Steiger, 1979).

Now, FA is widely used in the social and behavioural sciences (Bartholomew et al., 2011). Most of the relatively recent innovations in FA has tended either towards models of increasing complexity, such as dynamic factor models (Browne and Zhang, 2007), non-linear FA (Wall and Amemiya, 2007), or sparse factor (regression) models incorporated from a Bayesian perspective (Rockova and George, 2016; West, 2003). Whereas dynamic factor models have become popular in econometrics, non-linear FA has found applications in signal processing and pattern recognition. Sparse factor (regression) models are designed to analyse high-dimensional data, e.g. gene expression data. The surge in interest and methodological work has been motivated by scientific and practical needs and has been driven by advancement in computing. It has broadened the scope and the applicability of FA as a whole.

Here, we are concerned solely with *exploratory* FA (EFA), used as a (linear) technique to investigate the relationships between manifest and latent variables without making any prior assumptions about which variables are related to which factors (e.g. Mulaik, 2010). The FA branch that takes such information into account is known as *confirmatory* FA.

5.2 Fundamental equations of EFA

5.2.1 Population EFA definition

Let x be a *random* vector of p standardized manifest variables, i.e. $x \in \mathbb{R}^p$. The classical (linear) EFA model states that x can be written as a linear combination of two types of unobservable (latent) random variables called *common factors* $f \in \mathbb{R}^r$ and *unique* factors $u \in \mathbb{R}^p$. Formally, this is expressed by the fundamental EFA equation as follows (Mulaik, 2010):

$$x = \Lambda f + \Psi u , \tag{5.1}$$

where Λ and Ψ (diagonal) are parametric matrices with sizes $p \times r$ and $p \times p$, respectively. Λ is known as the matrix of *factor loadings*, while Ψ^2 contains the variances of the unique factors u. It is important to stress and remember that the classic EFA assumes that Ψ^2 is PD, i.e. $\Psi^2 > 0$. The choice of r is subject to some limitations, which will not be commented here (Harman, 1976; Mulaik, 2010). The r-factor model (5.1) assumes that $E(x) = 0$, $E(f) = 0$, $E(u) = 0$ and $E(uu^\top) = I_p$, where $E()$ denotes the expectation operator. It is assumed also that $E(xx^\top) = R_{xx}$ and $E(ff^\top) = R_{ff}$ are correlation matrices, i.e. symmetric positive definite (pd) matrices with ones as diagonal entries. Additionally, the common and unique factors are assumed uncorrelated, i.e. $E(fu^\top) = E(uf^\top) = 0$. Following the r-model defined above and the assumptions made, it can be found that

$$
\begin{aligned}
R_{xx} &= E(xx^\top) = E[(\Lambda f + \Psi u)(\Lambda f + \Psi u)^\top] = \\
&\quad \Lambda R_{ff}\Lambda^\top + \Lambda R_{fu}\Psi^\top + \Psi R_{uf}\Lambda^\top + \Psi^2 = \\
&\quad \Lambda R_{ff}\Lambda^\top + \Psi^2 .
\end{aligned}
\tag{5.2}
$$

Usually the common factors are assumed uncorrelated ($R_{ff} = I_r$) and they are called *orthogonal*. Then (5.2) simply reduces to

$$R_{xx} = \Lambda\Lambda^\top + \Psi^2 . \tag{5.3}$$

If $R_{ff} = \Phi \neq I_r$, then we have an EFA model with correlated factors:

$$R_{xx} = \Lambda\Phi\Lambda^\top + \Psi^2 , \tag{5.4}$$

which is rarely used in practice. We will see later that the new EFA parametrization proposed in Section 5.3.2 helps to absorb both EFA models (5.3) and (5.4) into a single more general EFA model.

The EFA definition (5.1) involves two types of unknowns: random variables f, u and fixed parameters Λ, Ψ. The classical EFA is mainly focused on the estimation of Λ, Ψ, based on the derived model (5.3) for the correlation structure of the data. Dealing with f, u is much more complicated issue in this classical setting (Mulaik, 2010, Ch.13).

5.2.2 Sample EFA definition

Most of the EFA population difficulties can be easily avoided by simply employing the sample EFA definition. Let X be a given $n \times p(n > p)$ data matrix of p observed standardized variables on n cases, i.e. $X^\top X$ be their sample correlation matrix R. The classical (linear) EFA models the data as follows:

$$X = F\Lambda^\top + U\Psi = [F\ U][\Lambda\ \Psi]^\top, \tag{5.5}$$

where, as before, Λ and Ψ (diagonal) are matrices with sizes $p \times r$ and $p \times p$, respectively. The columns of the $n \times r$ matrix F are called *common* factors and those of the $n \times p$ matrix U – *unique* factors.

Similarly to the population EFA model (5.1), in the r-factor model (5.5) we have that $1_n^\top X = 0_p, 1_n^\top F = 0_r, 1_n^\top U = 0_p$ and $U^\top U = I_p$. Additionally, the sample EFA assumes that the common and unique factors are orthogonal to each other, i.e. $F^\top U = O_{r\times p}$, and that $F^\top F$ is the correlation matrix of the common factors. Following the EFA r-model defined above and the assumptions made, it can be found that the corresponding model for the correlation matrix of these data is

$$
\begin{aligned}
R_{XX} &= (F\Lambda^\top + U\Psi)^\top (F\Lambda^\top + U\Psi) = \\
&\quad \Lambda F^\top F\Lambda^\top + \Lambda F^\top U\Psi^\top + \Psi U^\top F\Lambda^\top + \Psi U^\top U\Psi = \\
&\quad \Lambda F^\top F\Lambda^\top + \Psi^2 .
\end{aligned} \tag{5.6}
$$

Usually the common factors are assumed orthogonal $F^\top F = I_r$, and (5.6) simply reduces to

$$R_{XX} = \Lambda\Lambda^\top + \Psi^2 , \tag{5.7}$$

which coincides with the EFA population result (5.3) for the model correlation structure of the data. The classical EFA problem is to find Λ and Ψ such that R_{XX} fits R best in some sense.

The big difference with the population model from Section 5.2.1 is that here EFA can be based on (5.5): one can estimate *all* involved parameters F, U, Λ, Ψ by fitting directly the EFA model (5.5) to X (Trendafilov and Unkel, 2011; Unkel and Trendafilov, 2013). Moreover, this approach leads to a very fast EFA algorithms suitable for analysis of very large data. Some details are briefly outlined in Section 6.7.

In this chapter we reconsider the classical EFA and thus, we work with the defining models (5.3) and (5.7). The methods use correlations only, i.e. they involve R_{XX} and R only, without reference to the raw data X.

5.3 EFA parameters estimation

The estimation of the EFA parameters is a problem for finding the pair $\{\Lambda, \Psi\}$ which gives the best fit (for certain r) to the (sample) correlation matrix R of the data. If the data are assumed normally distributed the maximum likelihood principle can be applied (Magnus and Neudecker, 1988; Mardia et al., 1979). Then the factor extraction problem can be formulated as optimization of a certain log-likelihood function, which is equivalent to the following fitting problem (Lawley and Maxwell, 1971; Magnus and Neudecker, 1988; Mardia et al., 1979):

$$\min_{\Lambda, \Psi} \; \log[\det(\Lambda\Lambda^\top + \Psi^2)] + \operatorname{trace}[(\Lambda\Lambda^\top + \Psi^2)^{-1}R] \, , \qquad (5.8)$$

where $\det()$ denotes the determinant of the argument. For short problem (5.8) is called ML-EFA.

If nothing is assumed about the distribution of the data the log-likelihood function (5.8) can still be used as a measure of the discrepancy between the model and sample correlation matrices, R_{xx} and R. This is based on the proof that equivalent estimates to the ML-EFA ones can be obtained without normality assumptions, by only assuming finite second-order moments (Howe, 1955). Working with the partial correlations between observed and latent variables, it is shown that

$$\max_{\Lambda, \Psi} \; \frac{\det(R - \Lambda\Lambda^\top)}{\det(\Psi^2)} \qquad (5.9)$$

leads to the same ML-EFA equations.

In fact, the EFA problem (5.9) follows from a general algebraic result, Hadamar's inequality (Horn and Johnson, 2013, p. 505).

Proposition 5.3.1. *For any square matrix A, $det(A) \leq det(diag(A))$. When A is PD, then the equality holds if and only if A is diagonal.*

Thus, the EFA formulation (5.9), as well as the ML-EFA (5.8), can be used without *any* assumptions on the data.

There are a number of other discrepancy measures which are used in place of (5.8). A natural choice is the least squares approach for fitting the factor analysis model (5.3) to the data. It can be formulated as a special case of the following general class of weighted least squares problems (Bartholomew et al., 2011):

$$\min_{\Lambda, \Psi} \text{ trace} \left((R - \Lambda\Lambda^\top - \Psi^2)W \right)^2, \tag{5.10}$$

where W is a matrix of weights. In this chapter we consider in detail two special cases which are most popular in practice. The first one $W = I_p$ is known as (unweighted) least squares EFA, for short LS (Harman, 1976; Jöreskog, 1977; Mulaik, 2010). The second special case $W = R^{-1}$ is known as the generalized LS minimization problem (Anderson, 1984; Jöreskog, 1977). For short this problem is called GLS-EFA.

5.3.1 Classical EFA estimation

The minimization problems ML, LS and GLS listed above are not *unconstrained*. The unknowns Λ and Ψ are sought subject to the following constraints:

$$\Lambda^\top \Psi^{-2} \Lambda \text{ to be diagonal} \tag{5.11}$$

in ML and GLS, and

$$\Lambda^\top \Lambda \text{ to be diagonal} \tag{5.12}$$

in LS (Jöreskog, 1977; Mardia et al., 1979). It is common to say that these constraints eliminate the indeterminacy in (5.3). Indeed, let Λ_1 and Λ_2 be two solutions for which $\Lambda_1\Lambda_1^\top = \Lambda_2\Lambda_2^\top$, and $\Lambda_1^\top \Psi^{-2}\Lambda_1$ and $\Lambda_2^\top \Psi^{-2}\Lambda_2$ be diagonal. Then Λ_1 and Λ_2 have identical columns up to sign changes in whole columns. Thus the number $pq + p$ of the unknowns in P and Ψ is reduced by $r(r-1)/2$—the number of the zeros required in (5.11) and (5.12).

This indeterminacy-elimination feature is not always helpful for EFA, because such solutions are usually difficult for interpretation (Mardia et al., 1979, p. 268). Instead, the parameter estimation is followed by some kind of "simple structure" rotation (Section 4.5), which, in turn, gives solutions violating (5.11) and (5.12). In fact, the constraints (5.11) and (5.12) facilitate the algorithms for numerical solution of the ML, LS and GLS factor analysis (5.8) and (5.10), respectively, see for details (Jöreskog, 1977; Lawley and Maxwell, 1971). The existence of the constraint (5.11) can be statistically motivated—when it holds the factors are independent given the observed variables (Bartholomew et al., 2011) and thus, may also facilitate the interpretation. This implies that "simple structure" rotation may not be necessary for such problems.

In this relation, one can choose alternatively Λ to be an $p \times r$ lower triangular matrix (Anderson and Rubin, 1956). Such a choice serves both goals: eliminates the rotational indeterminacy and simplifies the factor loadings by making $r(r-1)/2$ of them zero.

The standard numerical solutions of the ML, LS and GLS factor analysis problems (5.8) and (5.10) are iterative, usually based on gradient or Newton-Raphson procedure, (Harman, 1976; Jöreskog, 1977; Lawley and Maxwell, 1971; Mulaik, 2010). The corresponding penalty function is minimized alternatively over one of the unknowns of the problem, say Λ, and keeping the other one fixed. An imaginary column(s) may appear in Λ, but the *major* difficulty in factor analysis is to maintain Ψ^2 to be a PD matrix. This is also crucial for the performance of the ML and GLS minimization (estimation) procedures, because (5.11) is not defined, e.g. Lawley and Maxwell (1971); Mardia et al. (1979). Such singular solutions are commonly referred to as "Heywood cases" (Harman, 1976; Jöreskog, 1977; Mardia et al., 1979). The case with negative entries in Ψ^2 is considered separately by some authors (e.g. SAS/STAT User's Guide, 1990) and referred to as ultra-"Heywood case". There we read (SAS/STAT User's Guide, 1990): *"An ultra-Heywood case renders a factor solution invalid. Factor analysts disagree about whether or not a factor solution with a Heywood case (zero entries in Ψ^2) can be considered legitimate."* Indeed, in such situations some authors say that the Heywood-case-variable is explained entirely by the common factors (Anderson, 1984; Bartholomew et al., 2011), while many others find it unrealistic. In general, the Heywood cases are considered statistical phenomenon (Bartholomew et al., 2011), which origin may range from simple sampling errors to wrong number r of factors. There are strategies to avoid or deal with such singular solutions (Bartholomew et al., 2011; Mardia et al., 1979). Numerically, the problem with maintaining $\Psi^2 > 0$ can be solved by enforcing its components to stay above some small positive number $\epsilon (= .005)$. This strategy is considered rather clumsy and likely to cause additional complications (Lawley and Maxwell, 1971, p. 32). Some modern aspects of this discussion are considered in Section 5.7, where EFA is treated as a correlation matrix *completion* problem.

In the next Section 5.3.2 we will see how the optimization on matrix manifolds can help to avoid the estimation problems in the classical EFA, including the notorious Heywood cases. An alternative way to approach these difficulties is by reconsidering EFA as a *semi-definite programming* problem which is briefly outlined in Section 5.7.

5.3.2   EFA estimation on manifolds

Here we consider a new formulation of the factor estimation problems (5.8) and (5.10) proposed in (Trendafilov, 2003). The constraints (5.11) and (5.12) will not be needed any more. As it has been discussed above that (5.11) and (5.12) are not really intrinsic for the minimization problems ML, LS and GLS, but are rather invoked by the needs of specific numerical procedures. The only *natural* constraints inferred from the r-factor analysis model (5.3) are that

1. an $p \times r$ matrix Λ of full column rank, and

2. a PD $p \times p$ diagonal matrix Ψ^2

should be sought.

There are a number of ways to fulfil the first constraint. Probably the simplest possible way is if Λ is sought to be an $p \times r$ lower triangular matrix. Of course, to maintain the requirement for full rank Λ, one should make sure that the main diagonal of such Λ has non-zero entries only. Moreover, as already mentioned, such a choice would facilitate its interpretation. Such an EFA parametrisation can be viewed as a kind of "truncated" Cholesky factorization of R. In contrast to PCA, the EFA model aims to improve the fit to R by adding Ψ^2.

Nevertheless, lower triangular loadings Λ are not very popular in EFA. Here, we concentrate on the central EFA goal to produce a general matrix of loadings with prescribed rank r. This can be achieved by making use of the following parametrization of the classical EFA model (5.3). Consider the EVD of the PSD product $\Lambda\Lambda^\top$ of rank at most r in (5.3), i.e. let $\Lambda\Lambda^\top = QD^2Q^\top$, where D^2 be an $r \times r$ diagonal matrix composed of the largest (non-negative) r eigenvalues of $\Lambda\Lambda^\top$ arranged in descending order and Q be an $p \times r$ orthonormal matrix containing the corresponding eigenvectors. Note that for this re-parameterization $\Lambda^\top\Lambda$ is diagonal, i.e. the condition (5.12) is fulfilled automatically. Then (5.7) can be rewritten as

$$R_{XX} = QD^2Q^\top + \Psi^2 \ . \tag{5.13}$$

Thus, instead of the pair $\{\Lambda, \Psi\}$, a triple $\{Q, D, \Psi\}$ will be sought. Formally, it means that we look for the best PSD symmetric approximant of rank r of the PSD sample correlation matrix R of rank p, plus a PD diagonal adjustment of rank p. Note, that the model (5.13) does not permit rotations, only permutations are possible. In order to maintain the factor analysis constraints, the triple $\{Q, D, \Psi\}$ should be sought such that Q be an $p \times r$ orthonormal matrix of full column rank, $D \in \mathcal{D}_{\neq}(r)$ and $\Psi \in \mathcal{D}_{\neq}(p)$, where the set of all $r \times r$ non-singular diagonal matrices is denoted by $\mathcal{D}_{\neq}(r)$. One can easily check that the number of the unknowns in both of the factor analysis formulations (5.7) and (5.13) is the same. Indeed, the number of the unknowns in (5.13) is $pr + r + p$ minus the number $r(r + 1)/2$ of the orthonormality equations for Q.

Thus, we define the following constraint set:

$$\mathcal{C}_{\neq} = \mathcal{O}(p, r) \times \mathcal{D}_{\neq}(r) \times \mathcal{D}_{\neq}(p),$$

where $\mathcal{O}(p, r)$ is the Stiefel manifold. However, working with $\mathcal{D}_{\neq}$ to achieve strictly PD D^2 and/or Ψ^2 is rather tricky and complicates the problem considerably. Indeed, for $r = 1$, we need to work with $\mathbb{R}\backslash\{0\}$, which has two (non overlapping) components $(-\infty, 0)$ and $(0, +\infty)$. Similarly, for $r = 2$, we work on $\mathbb{R} \times \mathbb{R}$ with removed coordinate axes, which has four components, etc. The problem is that we do not have tools to keep the computations on a single component without slipping on its boundary.

A possible way to avoid such difficulties is to employ a variable Frobenius inner product for the calculation of the partial gradients with respect to D and/or Ψ: for any diagonal A, B, define

$$\langle A, B \rangle_{\Delta,1} := \text{tr}(A \Delta^{-1} B) \text{ or } \langle A, B \rangle_{\Delta,2} := \text{tr}(\Delta^{-1} A \Delta^{-1} B) , \qquad (5.14)$$

for some appropriate PD diagonal matrix Δ. In the EFA situation, the natural choice is $\Delta^{-1} \equiv D^{-2}$ and/or Ψ^{-2}. Such a norm keeps the unknown parameter away from 0 (Helmke and Moore, 1994). Unfortunately, this approach cannot be directly utilized through Manopt programming.

The alternative way to deal with the problem is to consider D^2 and/or Ψ^2 in (5.13) directly as unknowns D and/or Ψ on the set of all PD diagonal matrices. Then, the EFA correlation model (5.13) can be redefined as

$$R_{XX} = QDQ^\top + \Psi , \qquad (5.15)$$

where D and Ψ are PD diagonal matrices. Then, the constrained set is modified to

$$\mathcal{C}_+ = \mathcal{O}(p,r) \times \mathcal{D}_+(r) \times \mathcal{D}_+(p) , \qquad (5.16)$$

where $\mathcal{D}_+(r)$ denotes the set of all $r \times r$ PD diagonal matrices which forms an open convex cone, i.e. if $D_1, D_2 \in \mathcal{D}_+(r)$, then $D_1 + \alpha D_2 \in \mathcal{D}_+(r)$ for any $\alpha > 0$. The tangent space of $\mathcal{D}_+(r)$ at Δ is furnished with the inner product $\langle A, B \rangle_{\Delta,2}$ from (5.14) (Bhatia, 1997, Ch 6). Solving the EFA problems (ML, GLS and LS) by utilizing (5.15)–(5.16) will guarantee that such $QDQ^\top$ has rank r and that such Ψ is PD, i.e. such EFA solutions are Heywood case–free! The Manopt codes are provided in Section 5.6.2. The drawback is that working with (5.15)–(5.16), is computationally more demanding, because the gradients with respect to D and/or Ψ are more complicated in the $\mathcal{D}_+()$ geometry. One would face the same effect, if the variable metric strategy (5.14) would have been directly employable.

All these complications can be easily avoided by simply accepting that Ψ^2 in the EFA correlation model (5.13) can be PSD, i.e. *can admit* zero entries. It turns out, that in most practical situations factor solutions with PD D^2 and Ψ^2 are obtained by simply solving the EFA problems (ML, GLS and LS) with correlation model (5.13) and requiring the less restrictive condition:

$$\mathcal{C} = \mathcal{O}(p,r) \times \mathcal{D}(r) \times \mathcal{D}(p), \qquad (5.17)$$

where the subspace of *all* $r \times r$ diagonal matrices is denoted by $\mathcal{D}(r)$. Although the positive definiteness of Ψ^2 is traditionally in the centre of discussions, the positive definiteness of D^2 is in fact the more crucial one because it shapes the EFA model. From now on, we consider EFA estimation under the correlation model (5.13) on the constraint set $\mathcal{C}$ from (5.17).

The new formulation of the EFA estimation problems is straightforward. Indeed, for a given sample correlation matrix R, the ML-EFA is reformulated as follows:

$$\min_{\mathcal{C}} \log(\det(QD^2Q^\top + \Psi^2)) + \text{trace}((QD^2Q^\top + \Psi^2)^{-1}R) . \qquad (5.18)$$

For short, problem (5.18) is referred to as ML-EFA.

In these terms, the LS and GLS estimation problems are concerned with the following constrained optimization problems:

$$\min_{\mathcal{C}} \text{trace}\left((R - QD^2Q^\top - \Psi^2)W\right)^2 . \qquad (5.19)$$

Hereafter, for short, problem (5.19) with $W = I_p$ is referred to as LS-EFA, and for the case $W = R^{-1}$—as GLS-EFA.

5.4 ML exploratory factor analysis

5.4.1 Gradients

The ML-EFA problem (5.18) is equivalent to the following constrained optimization problem: for a given $p \times p$ correlation matrix R

$$\min_{\mathcal{C}} \frac{1}{2}[\log\left(\det(R_{XX})/\det(R)\right) + \text{trace}((R_{XX}^{-1} - R^{-1})R)] , \qquad (5.20)$$

where $R_{XX} = QD^2Q^T + \Psi^2$ as defined in (5.13).

Let $E_{ML}(Q, D, \Psi)$ denotes the ML-EFA objective function in (5.20). By the chain rule and the product rule it is easy to obtain the Euclidean gradient $\nabla E_{ML}(Q, D, \Psi)$ with respect to the induced Frobenius norm of the function

(5.20) to be minimized:

$$\nabla E_{ML}(Q, D, \Psi) = (\nabla_Q E_{ML}, \nabla_D E_{ML}, \nabla_\Psi E_{ML}) =$$
$$(-YQD^2, -Q^T YQ \odot D, -Y \odot \Psi) , \quad (5.21)$$

where $Y = R_{XX}^{-1}(R - R_{XX})R_{XX}^{-1}$.

Then the projection $g_Q(Q, D, \Psi)$ of $\nabla_Q E_{ML}(Q, D, \Psi)$ onto the tangent space $\mathcal{T}_Q \mathcal{O}(p, q)$ is given by

$$g_Q(Q, D, \Psi) = -\frac{Q}{2}\left[Q^T YQ, D^2\right] - (I_n - QQ^T)YQD^2$$
$$= -\frac{Q}{2}\left[Q^T R_{XX}^{-1}(R - R_{XX})R_{XX}^{-1}Q, D^2\right] -$$
$$(I_n - QQ^T)R_{XX}^{-1}(R - R_{XX})R_{XX}^{-1}QD^2 , \quad (5.22)$$

where the Lie bracket notation $[Z_1, Z_2] = Z_1 Z_2 - Z_2 Z_1$ is adopted.

The projection $g_D(Q, D, \Psi)$ of $\nabla_D E_{ML}(Q, D, \Psi)$ onto the tangent space of all $r \times r$ diagonal matrices has simply the form:

$$g_D(Q, D, \Psi) = -[Q^T R_{XX}^{-1}(R - R_{XX})R_{XX}^{-1}Q] \odot D . \quad (5.23)$$

Similarly, the projection $g_\Psi(Q, D, \Psi)$ of $\nabla_\Psi E_{ML}(Q, D, \Psi)$ onto the tangent space of all $p \times p$ diagonal matrices is

$$g_\Psi(Q, D, \Psi) = -[R_{XX}^{-1}(R - R_{XX})R_{XX}^{-1}] \odot \Psi . \quad (5.24)$$

5.4.2   Optimality conditions

Now we can derive first-order derivative necessary conditions for stationary point identification by making all gradients equal to zero (matrices).

Theorem 5.4.1. *A necessary condition for $(Q, D, \Psi) \in \mathcal{O}(p, r) \times \mathcal{D}(r) \times \mathcal{D}(p)$ to be a stationary point of the ML-EFA is that the following equations should hold simultaneously:*

- D^2 and $Q^T R_{XX}^{-1}(R - R_{XX})R_{XX}^{-1}Q$ must commute ;
- $(I_p - QQ^T)R_{XX}^{-1}(R - R_{XX})R_{XX}^{-1}QD^2 = O_{p \times r}$;

- $[Q^T R_{XX}^{-1}(R - R_{XX})R_{XX}^{-1})Q] \odot D = O_r$;
- $[R_{XX}^{-1}(R - R_{XX})R_{XX}^{-1})] \odot \Psi = O_p$.

If D has distinct diagonal entries then $Q^T R_{XX}^{-1}(R - R_{XX})R_{XX}^{-1}Q$ must be a diagonal matrix. Moreover $Q^T R_{XX}^{-1}(R - R_{XX})R_{XX}^{-1}Q = O_r$, if its entries are all non-zero. In case Ψ has non-zero diagonal entries, then the fourth condition is equivalent to $\mathrm{diag}[R_{XX}^{-1}(R - R_{XX})R_{XX}^{-1}] = O_p$.

Corollary 5.4.2. *If D has distinct diagonal entries, a necessary condition for $(Q, D, \Psi) \in \mathcal{O}(p, r) \times \mathcal{D}_+(r) \times \mathcal{D}_+(p)$ to be a stationary point of the ML factor analysis problem is that*

- $Q^T R_{XX}^{-1}(R - R_{XX})R_{XX}^{-1}Q = O_r$ or $RR_{XX}^{-1}Q = Q$;
- $\mathrm{diag}R_{XX}^{-1} = \mathrm{diag}(R_{XX}^{-1}RR_{XX}^{-1})$

hold simultaneously.

Corollary 5.4.2 is identical to the first-order necessary conditions well known from the classical ML-EFA (Mardia et al., 1979, Th 9.4.2), where Ψ^2 is assumed PD.

5.5 LS and GLS exploratory factor analysis

5.5.1 Gradients

The LS and GLS estimation problems (5.19) is: for a given $p \times p$ correlation matrix R

$$\min_{\mathcal{C}} \frac{1}{4}\mathrm{trace}\left((R - QD^2Q^T - \Psi^2)W\right)^2 , \qquad (5.25)$$

where W is assumed a constant symmetric matrix. By the chain rule and the product rule it is straightforward to obtain the Euclidean gradient of the objective function $\nabla E_{GLS}(Q, D, \Psi)$ in (5.25) with respect to the induced Frobenius norm:

$$\nabla E_{GLS}(Q, D, \Psi) = (\nabla_Q E_{GLS}, \nabla_D E_{GLS}, \nabla_\Psi E_{GLS}) =$$
$$(-YQD^2, -Q^TYQ \odot D, -Y \odot \Psi) , \qquad (5.26)$$

where $Y = W(R - R_{XX})W$. The projections of the gradients on the corresponding tangent spaces are

$$g_Q = \frac{Q}{2} \left[Q^T W(R - R_{XX})WQ, D^2 \right] + (I_n - QQ^T)W(R - R_{XX})WQD^2 , \tag{5.27}$$

$$g_D = Q^T W(R - R_{XX})WQ \odot D , \tag{5.28}$$

and

$$g_\Psi = W(R - R_{XX})W \odot \Psi . \tag{5.29}$$

5.5.2   Optimality conditions

Further on we consider the LS and GLS cases separately. First let $W = I_p$. Then the following first-order optimality conditions hold:

Theorem 5.5.1. *A necessary condition for $(Q, D, \Psi) \in \mathcal{O}(p, r) \times \mathcal{D}(r) \times \mathcal{D}(p)$ to be a stationary point of the LS-EFA problem is that*

- D^2 *and* $Q^T(R - \Psi^2)Q$ *must commute* ;
- $(I_p - QQ^T)(R - \Psi^2)QD^2 = O_{p \times r}$;
- $[Q^T(R - \Psi^2)Q - D^2] \odot D = O_r$;
- $(R - \Psi^2 - QD^2Q^T) \odot \Psi = O_p$

hold simultaneously.

The first condition simply says that if D has distinct diagonal entries then $Q^T(R - \Psi^2)Q$ must be a diagonal matrix. Moreover if all diagonal entries of D are non-zero then the third condition implies $Q^T(R - \Psi^2)Q = D^2$. If the diagonal entries of Ψ are non-zero the fourth condition necessarily leads to $\operatorname{diag}(R) = \operatorname{diag}(\Psi^2 + QD^2Q^T)$. The fourth condition of Theorem 5.5.1 explains the existence of the Heywood solutions in LS-EFA analysis. Indeed, if some of the diagonal entries of Ψ approach zero, the corresponding diagonal entry in $R - \Psi^2 - QD^2Q^T$ may depart away from 0 and a Heywood solution will be exhibited.

Corollary 5.5.2. *A necessary condition for $(Q, D, \Psi) \in \mathcal{O}(p, r) \times \mathcal{D}_+(r) \times \mathcal{D}_+(p)$ to be a stationary point of the LS-EFA (in this case the EFA model is given in 5.15), assuming D has distinct entries, is that*

- $Q^T(R - \Psi)Q = D$;

- $diag(R - \Psi) = diag(QDQ^T)$

hold simultaneously, or simply

- $diag(R - \Psi) = diag[QQ^T(R - \Psi)QQ^T]$.

Finally, we consider the GLS-EFA, i.e. let $W = R^{-1}$ in (5.27)–(5.29). Then we have the following first-order optimality characterizations:

Theorem 5.5.3. *A necessary condition for* $(Q, D, \Psi) \in \mathcal{O}(p,r) \times \mathcal{D}(r) \times \mathcal{D}(p)$ *to be a stationary point of the GLS-EFA is that*

- D^2 *and* $Q^T R^{-1}(R - R_{XX})R^{-1}Q$ *must commute* ;

- $(I_p - Q^T)R^{-1}(R - R_{XX})R^{-1}QD^2 = O_{p \times r}$;

- $[Q^T R^{-1}(R - R_{XX})R^{-1}Q] \odot D = O_r$;

- $[R^{-1}(R - R_{XX})R^{-1}] \odot \Psi = O_p$

hold simultaneously.

If D has distinct diagonal entries then $Q^T R^{-1}(R - R_{XX})R^{-1}Q$ must be a diagonal matrix. Moreover if its (D) entries are all non-zero then $Q^T R^{-1}(R - R_{XX})R^{-1}Q = O_q$. In case the diagonal entries of Ψ are non-zero then the fourth condition of Theorem 5.5.3 is equivalent to $diag[R^{-1}(R - R_{XX})R^{-1}] = O_p$.

Corollary 5.5.4. *If* D *has distinct entries, a necessary condition for* $(Q, D, \Psi) \in \mathcal{O}(p,r) \times \mathcal{D}_+(r) \times \mathcal{D}_+(p)$ *to be a stationary point of the GLS-EFA is that*

- $Q^T R^{-1}(R_{XX} - R)R^{-1}Q = O_r$ *or* $R_{XX}^{-1}RQ = Q$;

- $diag\, R^{-1} = diag(R^{-1}R_{XX}R^{-1})$

hold simultaneously .

This result looks quite similar to the optimality conditions for ML-EFA listed in Corollary 5.4.2. Nevertheless, the nice ML feature $\mathrm{diag}R = \mathrm{diag}R_{XX}$ (e.g. Morrison, 1976, p. 310) does *not* hold for the GLS case. Instead, we have

$$\mathrm{diag}(R - R_{XX}) = 2\mathrm{diag}[QD^2Q^T R_{XX}^{-1}(R - R_{XX})] .$$

5.6   Manopt codes for classical ML-, LS-, and GLS-EFA

5.6.1   Standard case: $\Psi^2 \geq O_p$

The following codes solve the ML-, LS-, and GLS-EFA problems utilizing the new EFA parameterization introduced in Section 5.4.1 and Section 5.5.1. These EFA problems are defined on a product of three manifolds. Thus, the following codes also demonstrate how Manopt can be applied to problems with more than one unknown.

```
1   function [f S] = EFA(R,r,method)
2   % solves ML- ,LS-, and GLS–EFA using new parameterization
3   % R – input covariance/correlation matrix
4   % r – number of factors to retain
5   % method = 'LS' % 'ML' % 'GLS' %
6   %
7
8   p = size(R,1);
9
10  %========random_start============================
11
12   [A,Rw] = qr(rand(p,r) −.5,0);
13   idx = find(diag(Rw)<=0); A(:,idx) = −A(:,idx);
14   S.A = A; S.D = rand(r,1) − .5;
15  %Psi = rand(p,1);
16   S.Psi = diag(R − A*diag(S.D.^2)*A');
17  %=================================================
18
19  % Create the problem structure.
20   elements = struct();
21   elements.A = stiefelfactory(p,r);
22   elements.D = euclideanfactory(r,1);
23   elements.Psi = euclideanfactory(p,1);
24   M = productmanifold(elements);
25
26   problem.M = M;
27
28  %M.name()
29
```

```
% Define the problem cost function and its gradient.
problem.costgrad = @(S) mycostgrad(S.A,S.D,S.Psi,method);

% Numerically check gradient consistency.
warning('off', 'manopt:getHessian:approx');
checkgradient(problem);

    % Solve.
    tic
    options.verbosity=1; % or 0 for no output
%    [S f info] = conjugategradient(problem,[],options);
    [S f info] = trustregions(problem,[],options);
    toc

%    Out = ; Psi = S.Psi;
    [S.A*diag(S.D) S.Psi.^2]

    % Display some statistics.
    figure;
    if strcmp(method, 'ML')
        plot([info.iter], [info.gradnorm], '.-');
        hold on;
        plot([info.iter], [info.cost], 'ro-');
    else
        semilogy([info.iter], [info.gradnorm], '.-');
        hold on;
        semilogy([info.iter], [info.cost], 'ro-');
    end
    xlabel('Iteration #');
    ylabel('Gradient norm');


%Subroutine for the objective function f and its gradient
function [f G] = mycostgrad(A,D,Psi,method)
    [p,r] = size(A);
    D2 = D.^2;
    Psi2 = Psi.^2;
    AD = A.*repmat(D',p,1);
    RM = AD*AD' + diag(Psi2);

    Y = R - RM;
    % objective function
    switch method
        case {'ML'}
            Y = (RM\Y)/RM;
            f = (log(det(RM)) + trace(RM\R))/2;
        case{'GLS'}
            W = Y/R;
            Y = R\W;
            f = trace(W*W)/4;
        otherwise
            f = trace(Y'*Y)/4;
    end

    % gradient
    if nargout == 2
```

```
86              YA = Y*A;
87              gradA = YA.*repmat(D2',p,1);
88
89              ww = A'*gradA;
90              %G.A = .5*A*(ww' - ww) + (A*A' - eye(p))*gradA;
91              G.A = - gradA + .5*A*(ww + ww');
92              G.D = - diag(A'*YA).*D;
93              G.Psi = - diag(Y).*Psi;
94          end
95      end
96      end
```

For illustration, we reexamine the following example taken from Spearman (1904) and considered in (Mardia et al., 1979, Example 9.1.1). We are interested in a rank-one EFA approximation of the following (pd) correlation matrix R_S:

$$R_S = \begin{bmatrix} 1.00 & 0.83 & 0.78 \\ 0.83 & 1.00 & 0.67 \\ 0.78 & 0.67 & 1.00 \end{bmatrix}.$$

The exact (unique) one-factor solution can be found by the formulas given in (Mardia et al., 1979, p. 260):

$$R_S = \begin{bmatrix} 0.983 \\ 0.844 \\ 0.794 \end{bmatrix} \begin{bmatrix} 0.983 & 0.844 & 0.794 \end{bmatrix} + \begin{bmatrix} 0.034 & 0 & 0 \\ 0 & 0.287 & 0 \\ 0 & 0 & 0.370 \end{bmatrix}.$$

The LS- and ML-EFA solutions found by **EFA** reproduce perfectly this exact solution, and both D^2 and Ψ^2 are automatically PD without taking special care for this.

5.6.2   Avoiding Heywood cases: $\Psi^2 > O_p$

In the previous Section 5.6.1, the construction problem.costgrad expects Riemannian gradient. Here, another construction is used where the cost and its gradient are separated. In this case, one simply supplies the Euclidean gradient, and Manopt calculates the Rimannian one. This is organized by problem.egrad, which is explicitly reserved for the Euclidean gradient. More details can be found in the Manopt tutorial, together with example how to avoid redundant computations by exploiting the caching structure.

The following code is also created by Bamdev Mishra. It solves the ML-, LS-, and GLS-EFA problems utilizing the new EFA parameterization introduced in Section 5.3.2. Here, D^2 and/or Ψ^2 are directly treated as unknowns (D and Ψ, respectively) on the set of all PD diagonal matrices $\mathcal{D}_+$ of appropriate dimension. Then, the EFA problems are solved utilizing the correlation model (5.15) on the modified constrained set (5.16).

```
function [f S] = EFA_BM(R, r, method)
    % solves ML- ,LS-, and GLS-EFA using new parameterization
    % R - input covariance/correlation matrix
    % r - number of factors to retain
    % method = 'LS' % 'ML' % 'GLS' %
    %
    % FA model here is: ADA' + Psi,
    % where both D and Psi diagonal positive definite
    % codes provided by Bamdev Mishra

    p = size(R,1);

    %========random_start==================

    % [A,Rw] = qr(rand(p,r) -.5,0);
    % idx = find(diag(Rw)<=0); A(:,idx) = -A(:,idx);
    % S.A = A; S.D = rand(r,1) - .5;
    % %Psi = rand(p,1);
    % S.Psi = diag(R - A*diag(S.D.^2)*A');
    %==================================

    % Create the problem structure.
    elements = struct();
    elements.A = stiefelfactory(p,r);
    elements.D = positivefactory(r,1);
    elements.Psi = positivefactory(p,1);
    M = productmanifold(elements);

    problem.M = M;

    % Define the problem cost function and its gradient.
    problem.cost = @cost;
    function [f, store] = cost(S, store)
        if ~isfield(store, 'Res')
            A = S.A;
            D = S.D;
            Psi = S.Psi;

            RM = A*diag(D)*A' + diag(Psi); % BM: note that we don't
                need to square D because D is already positive.

            Y = R - RM;

            % objective function
            switch method
                case {'ML'}
```

```
46                    f = (log(det(RM)) + trace(RM\R))/2;
47                    Res = (RM\Y)/RM;
48                case{'GLS'}
49                    W = Y/R;
50                    f = trace(W*W)/4;
51                    Res = R\W;
52                otherwise
53                    f = trace(Y'*Y)/4;
54                    Res = Y;
55            end

57            store.Res = Res;
58            store.f = f;
59        end

61        f = store.f;

63    end

65    problem.egrad = @egrad; % Euclidean gradient
66    function [G, store] = egrad(S, store)
67        if ~isfield(store, 'Res')
68            [~, store] = cost(S, store);
69        end
70        Res = store.Res;
71        A = S.A;
72        D = S.D;
73        ResA = Res*A;

75        G.A = - ResA*diag(D);
76        G.D = - 0.5*diag(A'*ResA);
77        G.Psi = -0.5*diag(Res);
78    end


81    % Numerically check gradient consistency.
82    warning('off', 'manopt:getHessian:approx');
83    checkgradient(problem);
84    %pause;

86    % Solve.
87    tic
88    options.verbosity=1; % or 0 for no output
89    [S f info] = conjugategradient(problem,[],options);
90    %   [S f info] = trustregions(problem,[],options);
91    toc

93    %   Out = ; Psi = S.Psi;
94    [S.A*diag(sqrt(S.D)) S.Psi] % BM: note

96    S.D


99    % Display some statistics.
100   figure;
101   if strcmp(method, 'ML')
```

```
        plot([info.iter], [info.gradnorm], '.-');
        hold on;
        plot([info.iter], [info.cost], 'ro-');
    else
        semilogy([info.iter], [info.gradnorm], '.-');
        hold on;
        semilogy([info.iter], [info.cost], 'ro-');
    end
    xlabel('Iteration #');
    ylabel('Gradient norm');
end
```

For completeness, **positivefactory** is also provided. The code is created by Bamdev Mishra.

```
function M = positivefactory(m,n)
% Manifold of m-by-n matrices with element-wise positive entries, each
    with
% the bi-invariant geometry.
%
% function M = positivefactory(m,n)
%
% A point X on the manifold is represented as matrix X (mxn).
% Tangent vectors are  matrices of the same size
% (but not necessarily element-wise positive).
%
% The Riemannian metric is the bi-invariant metric for scalars,
    derived
% positive definite matrices mentioned in
% Chapter 6 of the 2007 book "Positive definite matrices"
% by Rajendra Bhatia, Princeton University Press.


% This file is part of Manopt: www.manopt.org.
% Original author: Bamdev Mishra, Dec 03, 2017.

    if ~exist('n', 'var') || isempty(n)
        n = 1;
    end

    M.name = @() sprintf('Element-wise positive %dx%d matrices', m, n)
        ;

    M.dim = @() m*n;

    % Choice of the metric on the orthonormal space is motivated by
    % the positive definite cone is its natural bi-invariant metric.
    M.inner = @myinner;
    function innerproduct = myinner(X, eta, zeta)
        innerproduct = (eta(:)./X(:))'*(zeta(:)./X(:));
    end

    M.norm = @(X, eta) sqrt(myinner(X, eta, eta));

    M.dist = @(X, Y) M.norm(X, log(Y./X), log(Y./X));
```

```
39      M. typicaldist  = @()  m*n; % To do
40
41      M. egrad2rgrad  =  @egrad2rgrad;
42      function eta  =  egrad2rgrad (X,  eta)
43           eta = X.*( eta ).*X;
44      end
45
46      M. ehess2rhess  =  @ehess2rhess ;
47      function Hess  =  ehess2rhess (X,  egrad ,  ehess ,  eta)
48           % Directional  derivatives  of  the  Riemannian  gradient
49           Hess = X.*( ehess ).*X + 2*( eta.*( egrad ).*X);
50
51           % Correction  factor  for  the  non-constant  metric
52           Hess = Hess - ( eta.*( egrad ).*X);
53      end
54
55
56      M. proj  = @(X,  eta)  eta ;
57
58      M. tangent  = M. proj ;
59      M. tangent2ambient  = @(X,  eta)  eta ;
60
61      M. retr  = @exponential ;
62
63      M. exp  = @exponential ;
64      function Y  = exponential (X,  eta ,  t)
65           if  nargin < 3
66                t = 1.0;
67           end
68           % The symm() and real () calls are mathematically not necessary
                    but
69           % are  numerically  necessary .
70           Y = (X.*( exp ((t*eta )./X)));
71      end
72
73      M. log  = @logarithm ;
74      function H = logarithm (X,  Y)
75           % Same remark  regarding  the  calls  to  symm() and real ().
76           H = (X.*( log (Y./X)));
77      end
78
79      M. hash  = @(X)  [ 'z ' hashmd5 (X(:))];
80
81      % Generate  a  random  element-wise  positive  matrix  following  a
82      % certain  distribution . The  particular  choice  of  a  distribution  is
                    of
83      % course  arbitrary , and  specific  applications  might  require
                    different
84      % ones .
85      M. rand  = @random ;
86      function X = random ()
87           X = rand (m,n)  ;
88      end
89
90      % Generate  a  uniformly  random  unit-norm  tangent  vector  at  X.
91      M. randvec  = @randomvec ;
```

```
function  eta  =  randomvec (X)
    eta  =  randn (m, n);
    nrm  =  M. norm (X,  eta);
    eta  =  eta  /  nrm;
end

M. lincomb  =  @matrixlincomb;

M. zerovec  =  @(X)  zeros (m, n);


M. transp  =  @(X1,  X2,  eta)  eta;

% For  reference ,  a  proper  vector  transport  is  given  here ,
    following
% work  by  Sra  and  Hosseini:  "Conic  geometric  optimisation  on  the
% manifold  of  positive  definite  matrices",  to  appear  in  SIAM  J.
    Optim .
% in  2015;  also  available  here:  http:// arxiv . org/abs/1312.1039
% This  will  not  be  used  by  default .  To  force  the  use  of  this
    transport ,
% execute  "M. transp  = M. paralleltransp;"  on  your  M  returned  by  the
% present  factory .
M. paralleltransp  =  @parallel_transport;
function  zeta  =  parallel_transport (X,  Y,  eta)
    E =  sqrt ((Y./X));
    zeta  = E. * eta . * E;
end

% vec  and  mat  are  not  isometries ,  because  of  the  unusual  inner
    metric .
M. vec  =  @(X,  U)  U(:);
M. mat  =  @(X,  u)  reshape (u,  m,  n);
M. vecmatareisometries  =  @()  true;

end
```

Here, we consider the following artificial example taken from (Mardia et al., 1979, Exercise 9.2.6), we are interested in rank-one EFA approximation of the following (pd) correlation matrix R:

$$R = \begin{bmatrix} 1 & 1/3 & 1/3 \\ 1/3 & 1 & 1/10 \\ 1/3 & 1/10 & 1 \end{bmatrix} .$$

The exact (unique) one-factor solution can be found by the formulas given in (Mardia et al., 1979, p. 260). However, this solution is not acceptable because it is a Heywood case: Ψ^2 has one negative entry (-0.1111). In this situation, the exact EFA solution achieving the exact fit to R is not very valuable because it does not take into account the additional info that Ψ^2 contains variances, and thus, cannot have negative entry.

The EFA numerical solutions are more useful as they necessarily produce $\Psi^2 \geq 0$. The results for LS and ML found by EFA from Section 5.6.1 are summarized in the first four columns of the following Table 5.1. The results for LS and ML found by EFA_BM (provided above) are given in the last four columns of Table 5.1.

VARS	EFA				EFA_BM			
	LS		ML		LS		ML	
	QD	Ψ^2	QD	Ψ^2	QD	Ψ^2	QD	Ψ^2
1	1.0010	0.0000	1.0000	0.0000	1.0006	0.0009	0.9997	0.0005
2	0.3301	0.8911	0.3333	0.8889	0.3302	0.8910	0.3334	0.8888
3	0.3301	0.8911	0.3333	0.8889	0.3302	0.8910	0.3334	0.8888

Table 5.1: LS– and ML–EFA solutions for an artificial correlation matrix R (Mardia et al., 1979, Exercise 9.2.6).

The reconstructed correlation matrices from the LS and ML solutions obtained by EFA are as follows (Trendafilov, 2003):

$$R_{LS} = \begin{bmatrix} 1.0019 & 0.3304 & 0.3304 \\ 0.3304 & 1.0000 & 0.1089 \\ 0.3304 & 0.1089 & 1.0000 \end{bmatrix} \text{ and } R_{ML} = \begin{bmatrix} 1.0000 & 0.3333 & 0.3333 \\ 0.3333 & 1.0000 & 0.1111 \\ 0.3333 & 0.1111 & 1.0000 \end{bmatrix} ,$$

and, respectively, obtained by EFA_BM, as

$$R_{LS} = \begin{bmatrix} 1.0020 & 0.3304 & 0.3304 \\ 0.3304 & 1.0000 & 0.1090 \\ 0.3304 & 0.1090 & 1.0000 \end{bmatrix} \text{ and } R_{ML} = \begin{bmatrix} 1.0000 & 0.3333 & 0.3333 \\ 0.3333 & 1.0000 & 0.1112 \\ 0.3333 & 0.1112 & 1.0000 \end{bmatrix} .$$

Clearly, for this example EFA and EFA_BM produce practically identical solutions. Thus, solving a more complicated problem by EFA_BM is not justified.

5.7   EFA as a low-rank-plus-sparse matrix decomposition

In this section we briefly demonstrate that the classical EFA can be considered as a special case of matrix completion problem also known as *robust* PCA (RPCA). We see that their goals overlap to some extent. We employ here a well-known small dataset to illustrate the similarities and differences between EFA and RPCA.

It was discussed in Section 5.2 that the EFA model definition and the related EFA constraints make it possible to derive a model $R_{XX} = \Lambda\Lambda^\top + \Psi^2$ for the EFA correlation matrix/structure (5.7). Then, the classical EFA problem finds Λ and Ψ such that the model correlation structure R_{XX} fits the sample correlation matrix $R(= X^\top X)$ as closely as possible in some sense. This problem is well known from the classical EFA for nearly a century (Harman, 1976; Mulaik, 2010).

Clearly, the EFA correlation structure R is composed of a low-rank matrix $L = \Lambda\Lambda^\top$ and a diagonal matrix Ψ^2, which is, of course, sparse as at least $p^2 - p$ of its entries are zeros. Thus, the classical EFA can be considered as the *oldest* low-rank-plus-sparse matrix decomposition of a correlation matrix. In this sense, the notorious discussion about permitting Ψ^2 to be PD or PSD takes another meaning. The request for PD Ψ^2 implies for an EFA decomposition of R into a low-rank part plus a sparse matrix which should be the *sparsest* matrix of full rank p. By allowing Ψ^2 to be PSD, we actually increase the approximation power of the low-rank part. Implicitly, this makes such EFA solution more PCA-alike.

In the last couple of decades there is considerable interest in developing methods and algorithms for low-rank-plus-sparse matrix decomposition of large data containing noise and/or missing entries, known also as matrix completion. Let $\|S\|_0$ denote the number of non-zero entries in matrix S. A great number of works (Cai et al., 2008; Candès et al., 2009; Chandrasekaran et al., 2011) are concerned with solving the following problem:

$$\min_{L,S} \gamma \, \text{rank}(L) + \|S\|_0, \text{ s.t } L + S = X \,, \tag{5.30}$$

where X is some kind of given data matrix (square or rectangular) and $\|S\|_0$ is the number of the non-zero entries in S. The unknowns involved are a low-rank matrix L and a sparse matrix S, which sum should reconstruct X exactly. The tuning parameter γ controls the importance of L and S. For example, large values of γ encourage L with lower ranks and less sparse S, and vice versa.

The common in this stream of works is that they start with some kind of penalization and "convexification" of the initial problem, and then apply methods from convex analysis, and more precisely—(positive) semi-definite programming (SDP) (Boyd and Vandenberghe, 2009). The potential difficulty in applying this approach is to show that the solution of the convex surrogate is close enough to the original one which we are interested in. We already meat this approach in Section 4, where it is adopted by d'Aspremont et al. (2007) for sparse PCA. The benefits are obvious: convex problems have only global solutions. Thus, the resulting algorithms are stable with sound optimality properties. One of the main challenges is the lack of convex formulation of matrix rank, which requires the introduction of a number of convex surrogates, e.g. nuclear matrix norm (Recht et al., 2007). Luckily, the situation should be much simpler if they are applied to EFA. As L is square symmetric, its rank minimization can be achieved by minimizing its trace, which is convex. As we will see later, this implementation is also not problem-free. The listed above convex formulations require the exact recovery of the data, which is the correlation matrix for EFA. This turns into a weakness if solving EFA problems because reduces the flexibility of the choices for L and S.

The EFA algorithms on matrix manifolds, proposed here, do not have rank defining problems, as the required low rank is embedded in the unknown parameters. However, the related optimization problems are not convex and the algorithms can produce local solutions.

Here, we are interested to see how general methods for low-rank-plus-sparse matrix decomposition can be applied to classical EFA problems. In terms of both of the above approaches, one can write the EFA model correlation matrix as $R = L + S = \Lambda\Lambda^\top + \Psi^2$.

We illustrate the behaviour of these two types of algorithms on a tiny example to acknowledge easily their differences. First, we consider the robust PCA (RPCA) method (Cai et al., 2008; Candès et al., 2009) for solving (5.30). As no specific structure is assumed for L and S, it seems obvious that such a general method will not be very appropriate for solving EFA problems (having unknowns with very specific structure). To make the RPCA solutions comparable to the standard EFA ones, one can adapt RPCA codes utilizing singular value thresholding to produce *diagonal non-negative S*. We have adapted the procedure `RobustPCA` from http://dlaptev.org/#code, based on the *alternating direction method of multipliers* (ADMM). The original codes expect rectangular data (input) matrices. For the EFA purpose they are modified to work with square ones. Additionally, S is replaced by a vector, which is subject to the same thresholding as in the original codes. The modified codes `RPCAsq` are provided at the end of the section.

Then, `RPCAsq` is used to find a low-rank-plus-sparse matrix decomposition of the sample correlation matrix (4.3) of the Five Socio-Economic Variables

(Harman, 1976, p. 14) considered in Section 4.6.2. We ask `RPCAsq` to produce a solution with L of rank two and diagonal S ($\lambda = .0001$), which is reproduced below:

$$
L + S = \begin{bmatrix}
.9802 & .0253 & .9724 & .4310 & .0149 \\
.0253 & .7792 & .1326 & .7011 & .8621 \\
.9724 & .1326 & .9795 & .5228 & .1337 \\
.4310 & .7011 & .5228 & .8009 & .7702 \\
.0149 & .8621 & .1337 & .7702 & .9540
\end{bmatrix}
+ \operatorname{diag}
\begin{bmatrix}
.0198 \\
.2208 \\
.0205 \\
.1991 \\
.0460
\end{bmatrix} .
$$

The achieved fit by this RPCA solution is $\|X^\top X - L - S\| = .0487$. To further compare this solution to the existing EFA solutions, we find the EVD of L, say $L = AD_L^2 A^\top$. Then, the matrix AD_L can be considered an analogue of factor loadings. Table 5.2 depicts these factor loadings AD_L and S, together with the factor loadings QD and Ψ^2, found by LS-EFA from Section 5.5. The fit achieved by this LS-EFA solution (.0445) is pretty much the same as for RPCA.

Table 5.2: Two solutions for Five Socio-Economic Variables.

Variable	LS-EFA		RPCA	
	QD	Ψ^2	AD_L	S
Total population	.6214 .7831	.0037	.6187 .7730	.0198
Median school years	.7018 -.5227	.2343	.7049 -.5314	.2208
Total employment	.7019 .6828	.0409	.7088 .6907	.0205
Misc. profess. services	.8810 -.1448	.2028	.8825 -.1487	.1991
Median value house	.7800 -.6045	.0264	.7719 -.5985	.0460

One can note that the RPCA and the LS-EFA loadings are very close, while some diagonal entries in Ψ^2 and S differ considerably. Particularly, the Total population is not a Heywood-case in the RPCA solution, something never achieved by any classical EFA solution. Apparently, this adapted and simplified version of RPCA can be used to do EFA. Nevertheless, using general RPCA algorithms to do EFA, looks as an overkill. For example, if one drops the restriction for diagonal S, then RPCA achieves perfect fit to $X^\top X$:

$$X^\top X = L + S =$$
$$\begin{bmatrix} .6821 & .0521 & .6818 & .4389 & .0224 \\ .0521 & .7465 & .1378 & .6914 & .8631 \\ .6818 & .1378 & .6915 & .5147 & .1219 \\ .4389 & .6914 & .5147 & .8653 & .7777 \\ .0224 & .8631 & .1219 & .7777 & 1.000 \end{bmatrix} + \begin{bmatrix} .3179 & -.0423 & .2906 & 0 & 0 \\ -.0423 & .2535 & .0164 & 0 & 0 \\ .2906 & .0164 & .3085 & 0 & 0 \\ 0 & 0 & 0 & .1347 & 0 \\ 0 & 0 & 0 & 0 & 0 \end{bmatrix}.$$

This, in turn, could be considered as an indication for a possible future direction of the EFA development by dropping the requirement for diagonal Ψ (Eldén and Trendafilov, 2019).

A method for low-rank-plus-diagonal matrix decomposition is proposed in (Saunderson et al., 2012), and seems more appropriate for EFA. They reformulate the EFA problem $X^\top X \approx L + D$, and propose to solve the following SDP:

$$\min_{L,D} \ \text{trace}(L) \ , \ \text{s.t} \begin{cases} L - \text{semi-definite}, \\ D - \text{semi-definite, diagonal} \ , \\ L + D = X^\top X \end{cases} \qquad (5.31)$$

by some standard SDP software, e.g. CVX (Grant and Boyd, 2016). For example, to solve (5.31), one can use of the following CVX lines:

```
cvx_begin
variable D(p,p) diagonal;
variable L(p,p) symmetric;
L == semi-definite(p);
minimize( sum(diag(L)) )
subject to
R == L+D; diag(D) >= 0
cvx_end
```

We try the Five Socio-Economic Variables data, but the outcome is rather disappointing. The resulting L has rank three, which is not very satisfying dimension reduction for data with five variables. We know, that the first two (largest) eigenvalues of the sample correlation matrix R explain more than 93% of the total variance of the data. Thus, a two-rank solution L would look more appropriate from data analysis point of view.

Concerning the solution itself, as suggested by (5.31), L is obtained by changing some diagonal entries of R, while keeping it (L) PSD:

$$R = L + D = \begin{bmatrix} 1.000 & .0098 & .9724 & .4389 & .0224 \\ .0098 & .8315 & .1543 & .6914 & .8631 \\ .9724 & .1543 & 1.000 & .5147 & .1219 \\ .4389 & .6914 & .5147 & .7951 & .7777 \\ .0224 & .8631 & .1219 & .7777 & .9824 \end{bmatrix} + \text{diag} \begin{bmatrix} .0000 \\ .1685 \\ .0000 \\ .2049 \\ .0176 \end{bmatrix}.$$

As expected, the fit is perfect, however there is no way to vary/decrease the rank of L (as in RPCA), which seems inconvenient for data analysis applications. If necessary, the zero entries of D can be avoided by setting a threshold, say, .005. The CVX crashes for thresholds ≥ 0.01.

Of course, a way out of this situation is to consider a truncated EVD of L, and produce a low-rank part of rank two L_2 as follows:

$$L_2 = \begin{bmatrix} .6179 & -.7818 \\ .7174 & .5361 \\ .7080 & -.6967 \\ .8773 & .1401 \\ .7809 & .6003 \end{bmatrix} \begin{bmatrix} .6179 & .7174 & .7080 & .8773 & .7809 \\ -.7818 & .5361 & -.6967 & .1401 & .6003 \end{bmatrix}.$$

We notice that the resulting "factor loadings" are very similar to those obtained by RPCA in Table 5.2. However, the fit achieved by this solution is $\|R - L_2 - D\| = .0678$, which is worse than the RPCA and the LS-EFA fits. To further improve this decomposition, one can re-calculate the diagonal term by $D_2 = \text{Diag}(R - L_2)$. This results in a slightly improved fit $\|R - L_2 - D_2\| = .0576$.

These experiments show that methods for low-rank-plus-sparse matrix decomposition requiring exact recovery of the correlation matrix are not quite appropriate for EFA purposes. Some recent works on RPCA relax the requirement for an exact fit to the data (Aravkin et al., 2014) and can be adapted for EFA purposes.

```
function [L, S] = RPCAsq(X, lambda, mu, tol, max_iter)
% modified to do EFA, i.e. works with square X and produces diagonal S
    % - X is a square data matrix (of the size N x N) to be decomposed
```

```
4        %   X can also contain NaN's for unobserved values
5        % - lambda - regularization parameter, default = 1/sqrt(max(N,M))
6        % - mu - the augmented lagrangian parameter, default = 10*lambda
7        % - tol - reconstruction error tolerance, default = 1e-6
8        % - max_iter - maximum number of iterations, default = 1000
9
10       N = size(X,1);
11  %        unobserved = isnan(X)
12  %        X(unobserved) = 0;
13       normX = norm(X, 'fro');
14
15       % default arguments
16       if nargin < 2
17           lambda = 1/sqrt(N);
18       end
19       if nargin < 3
20           mu = 10*lambda;
21       end
22       if nargin < 4
23           tol = 1e-6;
24       end
25       if nargin < 5
26           max_iter = 1000;
27       end
28
29       % initial solution
30       L = zeros(N);
31       S = zeros(N,1); %diagonal
32       Y = zeros(N);
33
34       for iter = (1:max_iter)
35           % ADMM step: update L and S
36           L = Do(1/mu, X - diag(S) + (1/mu)*Y);
37           S = So(lambda/mu, X - L + (1/mu)*Y);
38           S = diag(S); S = S.*(S>0);
39           % and augmented lagrangian multiplier
40           Z = X - L - diag(S);
41  %          Z(unobserved) = 0; % skip missing values
42           Y = Y + mu*Z;
43
44           err = norm(Z, 'fro') / normX;
45           if (iter == 1) || (mod(iter, 10) == 0) || (err < tol)
46  %              fprintf(1, 'iter: %04d\terr: %f\trank(L): %d\tcard(S): %d
        \n', ...
47  %                          iter, err, rank(L), nnz(S)); %L, S
48           end
49           if (err < tol) break; end
50       end
51  end
52
53  function r = So(tau, X)
54       % shrinkage operator
55       r = sign(X) .* max(abs(X) - tau, 0);
56  end
57
58  function r = Do(tau, X)
```

```
% shrinkage operator for singular values
[U, S, V] = svd(X, 'econ');
%r = U*So(tau, S)*V';
r = U*diag(So(tau, diag(S)))*V';
end
```

5.8 Sparse EFA

5.8.1 Introduction

The classical strategies to interpret principal components were discussed in Section 4.5 together with their weaknesses. It was mentioned there that the same applies for interpreting the factor loadings. The usual practice in FA is to ignore the factor loadings with small magnitudes, or set to zero loadings smaller than certain threshold value (Mulaik, 2010). This makes the factor loadings matrix *sparse* artificially and subjectively.

The common understanding nowadays is that such interpretation problems are solved by producing sparse solutions, e.g. sparse PCA. Moreover, this strategy is particularly suitable for analyzing large data, because great number of the input variables are in fact redundant. Sparse PCA is a difficult but very important tool for the modern applications. There is a great number of papers on sparce PCA (Trendafilov, 2014). EFA is another popular dimension reduction technique, but there exist very few works dealing with the modernization of the classical EFA parameter estimation in order to produce sparse factor loadings, e.g. Choi et al. (2011); Hirose and Yamamoto (2015); Ning and Georgiou (2011).

The EFA model is considerably more complicated than PCA. Particularly, obtaining sparse factor loadings is considerably more complicated by the presence of other parameters affecting the overall fit of the EFA model. For example, penalizing the ℓ_1 matrix norm of the factor loadings can lead to solutions with large unique variances Ψ. Thus, imposing sparseness inducing constraints on the factor loadings in EFA is less straightforward than with the component loadings in PCA.

In this section, we modify the classical EFA by introducing sparse-inducing constraints on the factor loadings. The main goal is to obtain easily interpretable factor loadings which are sparse in an optimal way.

The currently existing works on sparse EFA (Choi et al., 2011; Hirose and Yamamoto, 2015; Ning and Georgiou, 2011) use penalties to obtain sparse factor loadings. Choi et al. (2011) and Hirose and Yamamoto (2015) enhance the classical EM algorithm with penalties, while Ning and Georgiou (2011) solve the classical ML-EFA with additional LASSO-type penalty.

We also adopt LASSO-like penalties to achieve sparse factor loadings. They can be readily incorporated in the EFA re-parameterization proposed in Section 5.3.2 for arbitrary EFA formulation (5.20) and (5.25). Indeed, the matrix of factor loadings Λ is given as a product of an orthonormal matrix Q and a diagonal matrix D. Then, Q is sparsified with LASSO-type penalties, and the sparseness is preserved after multiplication by diagonal D. Thus, Q takes care for the pattern of sparseness of Λ, i.e. the locations of the zero loadings, while D adjusts the magnitudes of Λ for better fit.

The section is organized as follows. The sparse EFA is defined in Section 5.8.2, and the gradient of the sparse-inducing penalty based on the ℓ_1-norm is derived in Section 5.8.3. The remaining of the section is dedicated to examples illustrating the application of sparse EFA and comparison with the approach/software developed by Hirose and Yamamoto (2015).

5.8.2   Sparse factor loadings with penalized EFA

The EFA parametrization introduced in Section 5.3.2 can be readily adopted to establish a sparse version of the classical EFA (Trendafilov et al., 2017). Indeed, in the new EFA formulation the classical factor loadings Λ are parametrized as QD. This implies that if Λ is supposed/sought to be sparse, then Q should have the same pattern of zero entries, i.e. Λ and Q are identically sparse. To see this, consider parametrization of the following hypothetical 5×2 sparse factor loadings matrix Λ:

$$
\Lambda = \begin{bmatrix} \lambda_{11} & 0 \\ 0 & \lambda_{22} \\ \lambda_{31} & 0 \\ \lambda_{41} & 0 \\ 0 & \lambda_{52} \end{bmatrix} = QD = \begin{bmatrix} q_{11} & 0 \\ 0 & q_{22} \\ q_{31} & 0 \\ q_{41} & 0 \\ 0 & q_{52} \end{bmatrix} \begin{bmatrix} d_1 & 0 \\ 0 & d_2 \end{bmatrix} = \begin{bmatrix} q_{11}d_1 & 0 \\ 0 & q_{22}d_2 \\ q_{31}d_1 & 0 \\ q_{41}d_1 & 0 \\ 0 & q_{52}d_2 \end{bmatrix},
$$

which demonstrates that Q is solely responsible for the locations of the zeros in Λ, and D simply adjusts the magnitudes of the non-zero loadings.

Thus, to achieve sparse factor loadings Λ, one simply needs sparse orthonormal Q, which is already a problem resembling the well-known sparse PCA considered in Section 4.7.

Let q_i denote the ith column of Q, i.e. $Q = (q_1, q_2, ..., q_r)$, and $t = (\tau_1, \tau_2, ..., \tau_r)$ be a vector of tuning parameters, one for each column of Q. We consider a penalized version of FA, where the ℓ_1 norm of each of the columns of Q is penalized, i.e. $\|q_i\|_1 \leq \tau_i$ for all $i = 1, 2, ..., r$. If t has equal entries, i.e. $\tau_1 = \tau_2 = ... = \tau_r = \tau$, we simply use τ rather than $t = (\tau, \tau, ..., \tau)$.

Introduce the following discrepancy vector $q_t = (\|q_1\|_1, \|q_2\|_1, ..., \|q_r\|_1) - t$, which can also be expressed as $q_t = 1_p^\top [Q \odot \text{sign}(Q)] - t$, where $\text{sign}(Q)$ is a matrix containing the signs of the elements of Q, and 1_p is a vector with p unit elements. We adapt the scalar penalty function $\max\{x, 0\}$ used by Trendafilov and Jolliffe (2006) to introduce the following vector penalty function

$$P_t(Q) = [q_t \odot (1_r + \text{sign}(q_t)]/2 . \tag{5.32}$$

Then, the penalized version of ML-EFA can be defined as

$$\min_{Q, D, \Psi} \log(\det(R_{XX})) + \text{trace}((R_{XX})^{-1}R) + \tau P_t(Q)^\top P_t(Q) , \tag{5.33}$$

and for the LS- and the GLS-EFA as

$$\min_{Q, D, \Psi} \|(R - R_{XX})W\|^2 + \tau P_t(Q)^\top P_t(Q) . \tag{5.34}$$

Note, that $P_t(Q)^\top P_t(Q)$ penalizes the sum-of-squares of $\|q_i\|_1 - \tau_i$ for all $i = 1, 2, ..., r$, i.e. precise fit of $\|q_i\|_1$ to its corresponding tuning parameter τ_i cannot be achieved.

5.8.3 Implementing sparseness

The gradients of the ML-, LS- and GLS-EFA objective functions with respect to the unknowns $\{Q, D, \Psi\}$ are already derived in Section 5.4.1 and Section 5.5.1. Now we only need the gradient ∇_Q of the penalty term $P_t(Q)^\top P_t(Q)$ with respect to Q, which should be added to the gradient of the corresponding objective function.

We start with the derivative of the penalty term (5.32):

$$d(P_t(Q)^\top P_t(Q)) = 2d(P_t(Q))^\top P_t(Q) = 2d(P_t)^\top P_t, \tag{5.35}$$

which requires the calculation of $d(P_t)$. At this point we need an approximation of $\text{sign}(x)$, and we employ the one already used in (Trendafilov and Jolliffe, 2006), which is $\text{sign}(x) \approx \tanh(\gamma x)$ for some large $\gamma > 0$, or for short $\text{th}(\gamma x)$. See also (Hage and Kleinsteuber, 2014; Luss and Teboulle, 2013). Then

$$
\begin{aligned}
2(dP_t) &= (dq_t) \odot [1_r + \text{th}(\gamma q_t)] + q_t \odot [1_r - \text{th}^2(\gamma q_t)] \odot \gamma(dq_t) , \\
&= (dq_t) \odot \{1_r + \text{th}(\gamma q_t) + \gamma q_t \odot [1_r - \text{th}^2(\gamma q_t)]\} , \quad (5.36)
\end{aligned}
$$

where 1_r is a vector with r unit entries. The next differential to be found is

$$
\begin{aligned}
dq_t &= 1_p^\top \{(dQ) \odot \text{th}(\gamma Q) + Q \odot [E_{p \times r} - \text{th}^2(\gamma Q)] \odot \gamma(dQ)\} \\
&= 1_p^\top \{(dQ) \odot \{\text{th}(\gamma Q) + (\gamma Q) \odot [E_{p \times r} - \text{th}^2(\gamma Q)]\}\} , \quad (5.37)
\end{aligned}
$$

where $E_{p \times r}(= 1_p 1_r^\top)$ is a $p \times r$ matrix with unit entries.

Now we are ready to find the gradient ∇_Q of the penalty term with respect to Q. To simplify the notations, let

$$
w = 1_r + \text{th}(\gamma q_t) + (\gamma q_t) \odot [1_r - \text{th}^2(\gamma q_t)] , \quad (5.38)
$$

and

$$
W = \text{th}(\gamma Q) + (\gamma Q) \odot [E_{p \times r} - \text{th}^2(\gamma Q)] . \quad (5.39)
$$

Going back to (5.35) and (5.36), we find that

$$
\begin{aligned}
2(dP_t)^\top P_t &= \text{trace}[(dq_t) \odot w]^\top P_t = \text{trace}(dq_t)^\top (w \odot P_t) \\
&= \text{trace}\{1_p^\top [(dQ) \odot W]\}^\top (w \odot P_t) \\
&= \text{trace}[(dQ)^\top \odot W^\top] 1_p (w \odot P_t) \\
&= \text{trace}(dQ)^\top \{W \odot [1_p(w \odot P_t)]\} , \quad (5.40)
\end{aligned}
$$

making use of the identity $\text{trace}(A \odot B)C = \text{trace} A(B^\top \odot C)$. Thus, the gradient ∇_Q of the penalty term with respect to Q is (Trendafilov et al., 2017):

$$
\nabla_Q = W \odot [1_p(w \odot P_t)] . \quad (5.41)
$$

5.8.4   Numerical examples

In this section we consider two small examples from the classical EFA. They demonstrate that the sparse EFA solutions agree well with the old EFA solutions, but are easier and clearer for interpretation (Trendafilov et al., 2017).

First, we illustrate the proposed sparse EFA on the well-known dataset from classical EFA, namely, Harman's Five Socio-Economic Variables (Harman, 1976, p. 14). This small dataset is interesting because the two- and the three-factor solutions from LS- and ML-EFA are "Heywood cases" (Harman, 1976; Mulaik, 2010), i.e. Ψ^2 contains zero diagonal entries, or $\Psi^2 \geq 0$. One-factor solution is not considered interesting as it explains only 57.47% of the total variance.

Table 5.3 contains several sparse LS-EFA solutions of (5.34) starting with $\tau = \sqrt{5} = 2.2361$, which is equivalent to the standard (non-sparse) LS-EFA solution. For all of them we have $\Psi^2 \geq 0$. Clearly, POP, EMPLOY and HOUSE tend to be explained by the common factors only, which is already suggested by the non-sparse solution ($\tau = \sqrt{5}$). Increasing the sparseness of the factor loadings results in variables entirely explained by either a common or unique factor. The presence of loadings with magnitudes over 1 demonstrates the well-known weakness of LS-EFA in fitting the unit diagonal of a correlation matrix. It is well known that ML-EFA does not exhibit this problem which is illustrated by the next example.

VARS	$\tau = \sqrt{5}$		$\tau = 1.824$		$\tau = 1.412$		$\tau = 1$	
	QD	Ψ^2	QD	Ψ^2	QD	Ψ^2	QD	Ψ^2
POP	-.62 -.78	.00	.07 1.0	.00	-.00 1.0	.00	.00 -.99	.00
SCHOOL	-.70 .52	.23	.94 -.20	.07	.85 -.00	.27	-.28 -.00	.92
EMPLOY	-.70 -.68	.04	.19 .87	.21	-.00 1.0	.00	-.00 -.99	.00
SERVICES	-.88 .15	.20	.78 .23	.34	.58 .13	.65	-.18 -.00	.97
HOUSE	-.78 .60	.03	1.0 -.22	.00	1.1 -.07	.00	-1.2 .00	.00

Table 5.3: LS–EFA solutions for Five Socio-Economic Variables (Trendafilov et al., 2017).

Next, we illustrate the sparse EFA on another well-known classical dataset, namely, the Holzinger-Harman Twenty-Four Psychological Tests (Harman, 1976, p. 123). It is widely used to illustrate different aspects of classical EFA (Harman, 1976; Mulaik, 2010).

The correlation matrix (Harman, 1976, p. 124) of these data is non-singular and we apply ML-EFA (5.33). The first five columns of Table 5.4 contain the solution (factor loadings QD and unique variances Ψ^2) of (5.33) with $\tau = \sqrt{24} = 4.899$, i.e. the standard ML-FA solution, which is nearly identical to the ML solution obtained in (Harman, 1976, p. 215). Then, we rotate (with normalization) the factor loadings QD from the first four columns by VARIMAX from MATLAB (MATLAB, 2019), and the result is given in the next four columns of Table 5.4.

	$\tau = \sqrt{24} = 4.899$					Varimax rotated				$\tau = 2.2867$					$\tau = 2.1697$				
	QD				Ψ^2	QD and $T_{.41}$				QD				Ψ^2	QD				Ψ^2
1	.60	.39	-.22	.02	.44	**.69**	.16	.19	.16	-.88				.31	-.83				.41
2	.37	.25	-.13	-.03	.78	**.44**	.12	.08	.10	-.25				.86					1.0
3	.41	.39	-.14	-.12	.64	**.57**	.14	-.02	.11	-.53				.70	-.39				.76
4	.49	.25	-.19	-.10	.65	**.53**	.23	.10	.08	-.55				.69	-.55				.67
5	.69	-.28	-.03	-.30	.35	.19	**.74**	.21	.15		.82			.35		.81			.36
6	.69	-.20	.08	-.41	.31	.20	**.77**	.07	.23		.84			.32		.84			.32
7	.68	-.29	-.08	-.41	.28	.20	**.81**	.15	.07		.86			.29		.86			.29
8	.67	-.10	-.12	-.19	.49	.34	**.57**	.24	.13		.64			.54		.63			.54
9	.70	-.21	.08	-.45	.26	.20	**.81**	.04	.23		.87			.27		.86			.28
10	.48	-.49	-.09	.54	.24	.12	.17	**.83**	.17		.17	.91		.29		.07	.89		.33
11	.56	-.14	.09	.33	.55	.12	.18	**.51**	.37			.63		.59			.61		.59
12	.47	-.14	-.26	.51	.44	.21	.02	**.72**	.09			.72		.50			.75		.48
13	.60	.03	-.30	.24	.49	**.44**	.19	**.53**	.08	-.29		.47		.51	-.13		.47		.58
14	.42	.02	.41	.06	.65	.20	.08	**.55**					.46	.75				.37	.79
15	.39	.10	.36	.09	.70	.12	.12	.07	**.52**				.53	.71				.49	.73
16	.51	.35	.25	.09	.55	.41	.07	.06	**.53**				.56	.68				.50	.72
17	.47	-.00	.38	.20	.60	.06	.14	.22	**.57**				.72	.54				.77	.50
18	.52	.15	.15	.31	.59	.29	.03	.34	**.46**				.65	.61				.68	.59
19	.44	.11	.15	.09	.76	.24	.15	.16	.37				.33	.82				.18	.89
20	.61	.12	.04	-.12	.59	.40	.38	.12	.30				.33	.77				.31	.78
21	.59	.06	-.12	.23	.58	.38	.17	**.44**	.22			.50		.68			.37		.75
22	.61	.13	.04	-.11	.60	.40	.37	.12	.30				.28	.80	-.60				.64
23	.69	.14	-.10	-.04	.50	**.50**	.37	.24	.24	-.59	.02			.64	-.70				.55
24	.65	-.21	.02	.18	.50	.16	.37	**.50**	.30			.63		.59			.61		.60
(5.20)		14.28											16.71					17.08	

Table 5.4: ML-FA solutions for Twenty-Four Psychological Tests with their values for the ML objective function (5.20) (Trendafilov et al., 2017).

The classical EFA approach to interpret the rotated loadings QD repeats the steps outlined in Section 4.5.3 in the PCA context: sets up a threshold which cuts off the loadings with lesser magnitudes. It is helpful to sort the loadings magnitudes and look for jumps indicating possible cut-off values. The procedure of choosing the interpretation threshold, found to be .41, is described in detail in (Trendafilov et al., 2017). The hypothetical matrix of thresholded loadings is denoted by T_{41} and is formed by the bold loadings of the second four columns of Table 5.4 and zeros elsewhere. Then, following the strategy from Section 4.5.3, the interpretation is based on the normalized $T_{.41}$ (that has columns with unit maximal magnitude).

This is a lengthy and completely subjective procedure. For large loadings matrices, such an approach would be simply impossible to apply. The sparse EFA provides a reasonable alternative by directly producing sparse matrix of factor loadings. However, the sparse solutions produce worse fit than the classic ones. In general, the increase of the loadings sparseness worsens the fit. Thus, one needs to find a compromise between fit and sparseness, i.e. to optimize the value of the tuning parameter τ in (5.33).

In sparse PCA, tuning parameters as τ are usually found by cross-validation for large applications, or by employing information criteria (Trendafilov, 2014). For small applications, as those considered in the paper, the optimal tuning parameter τ can be easily located by solving the problem for several values of τ and compromising between sparseness and fit. Another option is to solve (5.33) for a range of values of τ and choose the most appropriate of them based on some index of sparseness. The following one is adopted in (Trendafilov et al., 2017):

$$\mathrm{IS}(\tau) = \frac{\text{original fit}}{\text{fit for } \tau} \times \left(\frac{\#_0}{pr}\right)^2 , \qquad (5.42)$$

where $\#_0$ is the number of zeros among all pr loadings of $\Lambda = QD$. IS increases with the sparseness and when the fit of the sparse solution of (5.33) is close to the original one, i.e. with $\tau = 0$.

Trendafilov et al. (2017) solve (5.33) for 100 values of τ from $\tau = \sqrt{24} = 4.899$ to 1. The maximum of IS is for $\tau = 2.1697$ and is $\mathrm{IS}(2.1697) = 0.4409$. The corresponding sparse factor loadings QD and unique variances Ψ^2 are depicted in the last five columns of Table 5.4. This matrix of factor loadings has 25 non-zero loadings (26%). The ML fit is 17.08. It turns out, that it is closer to the hypothetical $T_{.46}$ in LS sense than to $T_{.41}$. The sparse matrix obtained with $\tau = 2.2867$ (the second largest $\mathrm{IS}(2.2867) = 0.4032$) is also depicted in Table 5.4. It may look more like $T_{.41}$, but in fact, it is also closer to $T_{.46}$ in LS sense than to $T_{.41}$. Thus, the adopted index of sparseness (5.42) implies that the choice of the cut-off point .41 is incorrect, and should have been set to .46.

5.9 Comparison to other methods

For short, the proposed procedure for sparse EFA will be referred to as **SEFA**. In this section we compare the performance of **SEFA** with the method proposed in Hirose and Yamamoto (2015) developed with the same purpose. In short, Hirose and Yamamoto (2015) use ML-EFA and a sparsity inducing penalty controlled by a regularization parameter $\rho \geq 0$. They can switch between three types of penalties: LASSO, SCAD (Fan and Li, 2001) and MC+ (Zhang, 2010). They utilize expectation–maximization (EM)-like algorithm enhanced with coordinate descent to compute the entire solution path. Their codes in R are available online as the package `fanc`, which also will be used for short reference to their work.

Now, we will demonstrate first how `fanc` finds sparse factor solutions for the Five Socio-Economic Variables, (Harman, 1976, p. 14). The results will be compared with the performance of `SEFA`.

The solution with $\rho = .001$ (and less) is identical with the non-constrained solution ($\rho = 0$) depicted in Table 5.3. We find several (`fanc`) solutions with increasing values of ρ. They are reproduced in Table 5.5:

VARS	$\rho = .005$			$\rho = .01$		$\rho = .05$			$\rho = .1$	
	$\Lambda = QD$	Ψ^2		Λ	Ψ^2	QD	Ψ^2		QD	Ψ^2
POP	.991	.001	.005	.983		.005	.938		.005	.916	-.002	.005
SCHOOL		.891	.193		.882	.194		.828	.195		.802	.196
EMPLOY	.966	.117	.036	.959	.114	.036	.913	.107	.036	.890	.101	.036
SERVICES	.424	.783	.185	.418	.774	.185	.388	.727	.186	.374	.706	.186
HOUSE	.006	.955	.074	.004	.947	.074		.895	.073		.870	.073
Value of (5.20)	-1.072			-1.071		-1.052			-1.032	

Table 5.5: Four `fanc` solutions with LASSO penalty ($\gamma = \infty$) for Five Socio-Economic Variables (Trendafilov et al., 2017).

By looking at Table 5.5, one can conclude that the solution with $\rho = .05$ is the best: it has three exact zero loadings, and its fit is better than the next one with $\rho = .1$, which has two zero loadings. Hirose and Yamamoto (2015) provide a number of ways to evaluate the quality of their solutions. Some of them are provided in Table 5.6 for completeness.

ρ_0	ρ	Goodness-of-fit			Criteria			
		GFI	AGFI	SRMR	AIC	BIC	CAIC	EBIC
.001	.0010784	.7039321	$-\infty$	.0123146	17.13224	24.40584	39.40584	70.45754
.005	.0052776	.7056696	-3.4149556	.0197663	15.13684	21.92553	35.92553	63.37206
.01	.0116752	.7070369	-1.1972236	.0301804	13.14765	19.45144	32.45144	56.29280
.05	.0571375	.7097125	-.4514375	.0983878	11.37472	17.19360	29.19360	49.42979
.1	.0849835	.7081857	-1.1886072	.1327246	13.61488	19.91867	32.91867	56.76003
.15	.1264004	.7033769	-1.2246731	.1759992	14.07468	20.37846	33.37846	57.21982
.2	.1880019	.6916270	-1.3127974	.2264768	14.91420	21.21799	34.21799	58.05935
.7	.6185911	.5735386	-2.1984603	.3974809	22.95663	29.26041	42.26041	66.10178
.8	.9200627	.2063196	-.4881507	.4632527	44.27396	47.66831	54.66831	56.87865

Table 5.6: Quality measures for several `fanc` solutions with LASSO penalty for Five Socio-Economic Variables (Trendafilov et al., 2017).

Note that ρ_0 is the input value for `fanc`, while ρ is the actual value of the tuning parameter used by `fanc` to produce the loadings. According to all goodness-of-fit measures and the information criteria collected in Table 5.6, the solution with $\rho_0 = .05$ seems to be the best one indeed. What seems surprising is that `fanc` is incapable to produce sparser solutions, containing more zeros than three. Instead, the increase of ρ_0 results in loadings containing only two zeros. Such solutions are depicted in Table 5.5 with $\rho = .1$ and in Table 5.7 with $\rho = .7$. As ρ controls the importance of the penalty term, it is logical to expect and desirable to have sparser loadings with larger ρ, which is not the case. Moreover, further increase of ρ ($\geq .8$) simply results in invalid solutions containing one zero column. The remedy proposed by Hirose and Yamamoto (2014) is to replace the LASSO constraint by the MC+ one, which needs one more parameter γ. The best solution we found is with $\rho = 1.6$, $\gamma = 1.5$ and is depicted in Table 5.7. It has five zero loadings and ML fit 1.159163. In order to get sparser solutions Hirose and Yamamoto (2014) suggest trying correlated factors. We were unable to identify a pair of parameters (ρ, γ) for which the oblique solution provides better sparseness and/or ML fit. For comparison, the SEFA solution with $\tau = .9$ is depicted in the last three columns of Table 5.7. It has six zero loadings and even better (lower) minimum of the objective function (5.20).

VARS	$\rho = .7$			$\rho = 1.6,\ \gamma = 1.5$			SEFA ($\tau = .6$)		
	Λ		Ψ^2	Λ		Ψ^2	QD		Ψ^2
POP	.705	-.015	.005	.690		.005	.971	-.000	.054
SCHOOL		.554	.214		.684	.007	.000	.760	.266
EMPLOY	.677	.053	.037	.663	.077	.030	1.00	.000	.000
SERVICES	.229	.477	.211			1.00	.000	.000	1.00
HOUSE		.645	.076		.449	.299	.000	1.00	.000
Value of (5.20)	-0.254			1.159			0.758		

Table 5.7: More `fanc` solutions with LASSO penalty for Five Socio-Economic Variables. The solution for $\rho = .8$ has an empty (zero) second column (Trendafilov et al., 2017).

With this simple example we demonstrate that `fanc` has two serious drawbacks to be taken into account for practical use. First, `fanc` is incapable to produce full range of sparse solutions. Second, the relationship between the sparseness and the parameter ρ for its control is not linear. This considerably complicates the location of an optimal ρ (which provides a reasonable fit with a sensible sparseness)—the main difficulty in any sparse analysis of large data. The presence of another parameter γ for MC+, puts additional difficulty in the `fanc` application.

Now, let us consider the Twenty-Four Psychological Tests (Harman, 1976,

p. 123). The standard ML solution (fanc with $\rho = 0$) has four zeros. The solution with $\rho = 0.1$ seems the sparsest possible with LASSO penalty and has 26 zeros. The further increase of ρ produces invalid solutions. For $\rho = 0.14$, the loadings already have one zero column. Then, let us replace the LASSO constraint by MS+. Hirose and Yamamoto (2014) find solution with $\rho = .02$, $\gamma = 4$ which resembles the PROMAX solution. This solution is not satisfactory, it is not sparse enough as it has 70 (73%) non-zero loadings. The best solution they find with MS+ constraint is with $\rho = .14$, $\gamma = 1.1$, which still has plenty of non-zero loadings, 59 (61%). Our solution with the same (ρ, γ) in Table 5.8 is nearly identical to the one reported in (Hirose and Yamamoto, 2014, Table 4.). In order to get sparser solutions, Hirose and Yamamoto (2014) suggest employing correlated factors. For this dataset this strategy pays off. The best solution with oblique factors obtained with $\rho = .21$, $\gamma = 1.1$ is reported in (Hirose and Yamamoto, 2014, Table 4.) and has only 28 non-zero loadings. After several runs of fanc, we are unable to repeat this solution. Our solution with oblique factors and same (ρ, γ) is depicted in Table 5.8 and has 36 non-zero loadings.

| | Orth, $\rho = 0$ | | | | Ψ^2 | Orth, $\rho = .1$ | | | | Ψ^2 | Orth, $\rho = .14$, $\gamma = 1.1$ | | | | Ψ^2 | Oblique $\rho = .21$, $\gamma = 1.1$ | | | | Ψ^2 |
	Λ					Λ					Λ					Λ				
1	.06	.38	.64	.06	.44		.12	.54		.45		.45	.57	.09	.46	.73				.47
2	.03	.25	.40	0	.78		.06	.30		.79		.28	.38		.78	.47				.78
3	.04	.29	.51	-.12	.64		.10	.38	-.07	.68		.29	.52		.65	.55				.69
4	-.02	.38	.45	-.02	.65		.16	.35		.67		.40	.43		.66	.58				.66
5	-.07	.80	.01	.07	.35		.61		.09	.36	-.02	.78	-.17		.36				.80	.35
6	.01	.83		-.08	.31	.03	.64			.32		.77	-.32		.31				.81	.34
7	-.15	.83	-.01		.28	-.04	.65			.30	-.17	.78	-.25		.29				.83	.32
8	-.05	.68	.20	.10	.49		.46	.15	.07	.50		.69	.15		.50	.22			.57	.50
9		.86	-.02	-.11	.26	.00	.68	-.01		.26		.79	-.36		.25			-.08	.90	.24
10	.07	.31	-.07	.81	.24	.07			.73	.23		.52	-.32	.63	.24		-.38	1.0		.21
11	.27	.38	.12	.46	.55	.20	.12	.10	.37	.56	.26	.52		.34	.54			.46	-.28	.57
12	.02	.21	.27	.67	.44		.21		.52	.48			.42	.61	.45			.70		.50
13	-.03	.39	.43	.42	.49		.11	.34	.30	.54		.52	.25	.37	.52	.31		.47		.52
14	.47	.35	.00	.05	.65	.42	.16		.01	.64	.45	.37			.66		-.56			.69
15	.46	.29	.08	.05	.70	.37	.10	.06		.71	.45	.32			.69		-.53			.72
16	.46	.31	.38		.55	.34	.06	.31		.57	.42	.35	.37		.55	.35	-.37			.61
17	.50	.34	.04	.20	.60	.11	.03		.13	.61	.50	.40			.59		-.67			.54
18	.39	.27	.31	.29	.59	.27	.01	.26	.19	.61	.35	.39	.21	.27	.59	.24	-.46			.62
19	.29	.32	.20	.11	.76	.16	.11	.17	.04	.79	.24	.37	.18		.77		-.49			.76
20	.16	.54	.30	.00	.59	.07	.31	.25		.60		.56	.30		.60	.41			.33	.58
21	.12	.38	.36	.35	.58	.12	.33	.26		.58		.51	.26	.32	.57	.34		.42		.57
22	.17	.53	.30	.01	.60	.05	.30	.26		.61		.55	.30		.61	.40			.33	.59
23	.09	.56	.41	.11	.50		.30	.36	.06	.49		.61	.36		.50	.51			.30	.49
24	.16	.54	.10	.41	.50	.05	.28	.11	.36	.51		.65	.26		.51	.51			.29	.51
(5.20)					14.28					14.95					14.41					14.66

Table 5.8: Several fanc solutions with LASSO ($\gamma = \infty$) and MS+ penalty for Twenty-Four Psychological Tests (Trendafilov et al., 2017).

The correlations among the oblique factors for our (.21, 1.1)-fanc solution are found to be

$$\Phi = \begin{bmatrix} 1.0 & .51 & -.52 & .50 \\ .51 & 1.0 & -.51 & .47 \\ -.52 & -.51 & 1.0 & -.49 \\ .50 & .47 & -.49 & 1.0 \end{bmatrix}.$$

Clearly, the solution with oblique factors looks better, than the orthogonal one. Indeed, the corresponding values of the index of sparseness (5.42) are .3805 and .1472, respectively. Nevertheless, it looks again that fanc is unable to produce solutions with arbitrary sparseness. The fanc problems to achieve reasonable sparseness with LASSO, further continue when applying the MC+ penalty. For some data, fanc can achieve better sparseness by employing correlated (oblique) factors. This looks as a way to boost the performance of the numerical method without clear benefit for the EFA solution, e.g. better fit. As illustrated in Table 5.4, the same problem can be solved by SEFA very satisfactory. In fact, the new EFA parametrization absorbs the classical EFA model with correlated factors (5.4). This follows immediately by taking the EVD of $\Lambda\Phi\Lambda^\top = QDQ^\top$ in (5.13).

The reader is encouraged to add penalty terms in the codes EFA from Section 5.6.1 and repeat the small numerical experiments from the last two sections. Both approximations of $\|\|_{\ell_1}$ discussed in Section 4.6.3 can be tried for this purpose, either (4.30) or the one involving tanh().

5.10   Conclusion

Section 5.2 introduces the population and the sample versions of the classical EFA fundamental equation. It shows that EFA models the sample correlation matrix as a sum of low-rank and PSD/PD diagonal matrix. Section 5.3 briefly considers the classical approach to the EFA parameter estimation that parametrizes the low-rank matrix as $\Lambda\Lambda^\top$, where Λ should have exactly rank r. We believe that this is a serious weakness of the classical EFA causing unnecessary complications, both algorithmic and conceptual. For example, this includes optimization on the non-compact manifold of rank r matrices in $\mathcal{M}(p, r)$ and introduction of correlated and uncorrelated factors. Alternatively, we propose to parameterize the low-rank matrix through its truncated EVD, say $QDQ^\top$ for some $Q \in \mathcal{O}(p, r)$ and PD $D \in \mathcal{D}(r)$. Working with compact $\mathcal{O}(p, r)$ and cone of positive diagonal matrices is an easier task.

The new parameterization is adopted to obtain maximum likelihood and least squares estimations of the classical EFA, respectively, in Section 5.4 and Section 5.5. Thus, the EFA estimation becomes an optimization problem on the product of matrix manifolds. Manopt codes are provided in Section 5.6 for dealing with most popular EFA cost functions.

Section 5.7 relates the classical EFA to the modern matrix completion problems. We observe that the classical EFA is, in fact, the oldest method of such kind. This is illustrated by comparing classical EFA and robust PCA solutions. The results suggest investigating potential EFA generalization by relaxing the constraint for diagonal/square Ψ.

The new EFA parameterization introduced in Section 5.3 is utilized in Section 5.8 to develop a EFA modification that produces sparse factor loadings. Next, in Section 5.9, this sparse EFA is applied to several well-known datasets. The results are compared to their classical EFA solutions, as well as to other existing methods for sparse EFA. In particular, it is demonstrated that the new EFA parameterization makes the old concept of considering separately correlated and uncorrelated factors obsolete.

5.11   Exercises

1. For what values of x the following matrix is PSD, i.e. it is a well-defined correlation matrix (Yanai and Takane, 2007):

$$\begin{bmatrix} 1 & 0 & x & x \\ 0 & 1 & x & -x \\ x & x & 1 & 0 \\ x & -x & 0 & 1 \end{bmatrix}.$$

   Find by hand a two-factor EFA decomposition of the form (5.7) for this correlation matrix?

2. *Principal (axes) factor analysis* (PFA) is one of the first and frequently used in the past methods for FA (Harman, 1976, 8), (Mardia et al., 1979, 9.3). The EFA model (5.7) suggests to choose some initial Ψ_0^2, e.g. estimated or random *communalities* collected in a diagonal matrix $H_0^2 (= I_p - \Psi_0^2)$. Then, we form a residual matrix $R_1 = R_{XX} - \Psi_0^2$ and

find its truncated EVD, say, $R_1 = Q_r D_r^2 Q_r^\top = \Lambda_1 \Lambda_1^\top$, where r is the number of desired factors. Then, new/updated unique variances Ψ_1^2 are found as $\Psi_1^2 = \text{diag}(R_{XX}) - \text{diag}(\Lambda\Lambda^\top)$, which gives the next residual $R_2 = R_{XX} - \Psi_1^2$, and so on. PFA is notorious with producing Heywood cases (Ψ^2 with zero or negative entries). Elaborate the PFA algorithm to produce proper and/or sparse solutions.

3. (Sundberg and Feldmann, 2016) Assuming $\Psi^2 > 0$, the following approach to EFA divides its defining equation (5.7) by Ψ^{-1}, which gives:

$$\Psi^{-1} R_{XX} \Psi^{-1} = \Psi^{-1} \Lambda \Lambda^\top \Psi^{-1} + I_p = \Psi^{-1}\Lambda(\Psi^{-1}\Lambda)^\top + I_p \,,$$

or simply $R_{YY} = \Lambda_Y \Lambda_Y^\top + I_p$, where $Y = X\Psi^{-1}$. In this construction, $\Lambda_Y^\top \Lambda_Y = \Lambda^\top \Psi^{-2}\Lambda$, already known from (5.11) and let us assume it is diagonal, say, $\Lambda^\top \Psi^{-2}\Lambda = \Phi \in \mathcal{D}(r)$. Check that

(a) $\Lambda_Y(:= \Lambda_Y(\Lambda_Y^\top \Lambda_Y)^{-1/2})$ contains the first r (normalized) eigenvectors of R_{YY} and $\Phi + I_r$ contains the corresponding eigenvalues;

(b) thus, the EFA parameters Λ and Ψ^2 are found by starting with random $\Psi_0^2 > 0$ and EVD of R_{YY}, then updating Ψ^2 and EVD of the updated R_{YY}, and so on until convergence;

(c) the updating formulas are given by $\Lambda = \Psi\Lambda_Y(\Phi - I_r)^{1/2}$ and $\Psi^2 = \text{diag}(R_{XX}) - \text{diag}(\Lambda\Lambda^\top)$.

4. The aim of the classical *MINRES* method for EFA of a given $p \times p$ correlation matrix R_{XX} is to find factor loadings Λ such that

- $\Lambda\Lambda^\top$ fits the off-diagonal entries of R_{XX}, i.e. $(E_p - I_p) \odot R_{XX}$, and

- has diagonal entries not exceeding unity.

Formally, the classical MINRES solves (Harman, 1976, 9.3):

$$\min_{\substack{\Lambda \\ \text{diag}(\Lambda\Lambda^\top) = I_p}} \|\text{off}(R_{XX} - \Lambda\Lambda^\top)\|_F \,.$$

(a) Can you write Manopt codes to solve this problem on the oblique manifold of *row-wise* normalized $p \times r$ matrices?

(b) Apply your codes from (a) to the sample correlation matrix (4.3). One such solution Λ is given in the first three columns of Table 5.9. Compare it with the provided there classical MINRES and PCA solutions (by plotting the five variables on the plane).

VARS	Oblique Λ		(Harman, 1976, Table 9.2)		PCA	
	Λ	Diag($\Lambda\Lambda^\top$)	Λ	Diag($\Lambda\Lambda^\top$)	AD	Diag($AD^2A^\top$)
POP	.8301 -.5574	.9998	.621 -.783	.999	.5810 -.8064	.9878
SCHOOL	.5405 .8410	.9994	.701 .522	.764	.7670 .5448	.8851
EMPLOY	.8902 -.4561	1.001	.702 -.683	.959	.6724 -.7260	.9793
SERVICES	.9002 .4358	1.000	.881 .144	.797	.9324 .1043	.8802
HOUSE	.5518 .8340	1.000	.781 .605	.976	.7912 .5582	.9375
SS	2.889 2.111	5.000	2.756 1.739	4.495	2.873 1.797	4.670

Table 5.9: Three two-factor decompositions of the sample correlation matrix of Five Socio-Economic Variables, (Harman, 1976, p.14).

(c) The solution from (b) shows that there is nothing left for Ψ^2, which is supposed to be $I_p - \text{Diag}(\Lambda\Lambda^\top)$. Can you modify somehow MINRES EFA to achieve Ψ^2 with non-negative diagonal entries, i.e. PSD Ψ^2 ($\Psi^2 \geq O_p$).

5. Check that the total differential of the ML-EFA objective function (5.20), as a function of $R_{XX}(= QD^2Q^T + \Psi^2)$ is given by

$$\text{trace}[R_{XX}^{-1}(R_{XX} - R)R_{XX}^{-1}](dR_{XX}) ,$$

and that the second differential is:

$$2\text{trace}[R_{XX}^{-1}(dR_{XX})R_{XX}^{-1}RR_{XX}^{-1}(dR_{XX})] - \text{trace}[R_{XX}^{-1}(dR_{XX})R_{XX}^{-1}(dR_{XX})] .$$

6. The *minimum-trace* EFA (MTFA) (Riccia and Shapiro, 1982; ten Berge and Kiers, 1981; Watson, 1992) aims at addressing the problems with the MINRES EFA. For a given $p \times p$ sample correlation matrix R_{XX}, MTFA solves the following problem:

$$\min_{\substack{\Theta \geq O_p \\ R_{XX} = \Theta + \Psi^2 \\ \Psi^2 \geq O_p}} \text{trace}(\Theta) ,$$

where $\Psi^2 \geq O_p$ means that Ψ^2 is PSD. Recently MTFA (and other classical EFA metods) is re-cast as a *semi-definite programming* (SDP) problem (Bertsimas et al., 2017). We already know that the adopted EFA treatment on manifolds naturally addresses the requirements for $\Theta \geq O_p$ and $\Psi^2 \geq O_p$. Compare the Manopt solutions with the corresponding EFA solutions from their Matlab/R codes[1] utilizing the CVX package (Grant and Boyd, 2016). For example, for the Five Socio-

---

[1] https://github.com/copenhaver/factoranalysis

Economic Variables (Harman, 1976, p. 14) one finds solution with PSD Θ, rank$(\Theta) = 2$ and $\|R_{XX} - \Theta - \Psi_\theta^2\| = 0.0678$. Its EVD $\Theta = AD_\theta^2 A^\top$ is used to produce factor loadings (AD_θ), which are depicted in the first three columns of Table 5.10. This solution is compared to the LS-EFA solution from Table 5.2 (also reproduced for convenience in the last three columns of Table 5.10). The fit of the LS-EFA solution is $\|R_{XX} - QD^2Q^\top - \Psi^2\| = 0.0445$, which is better than the one achieved by SDP/CVX. From the SS values, one concludes that Θ explains more variance than $QD^2Q^\top$, namely $4.541 > 4.495$. The factor loadings are pretty similar, but the individual variances differer.

VARS	SDO/CVX			LS–EFA, Table 5.2		
	AD_θ		Ψ_θ^2	QD		Ψ^2
POP	.6179	.7818	.0000	.6214	.7831	.0037
SCHOOL	.7174	-.5361	.1685	.7018	-.5227	.2343
EMPLOY	.7080	.6967	.0000	.7019	.6828	.0409
SERVICES	.8773	-.1401	.2049	.8810	-.1448	.2028
HOUSE	.7809	-.6003	.0176	.7800	-.6045	.0264
SS	2.777	1.764	4.541	2.756	1.739	4.495

Table 5.10: Two rank-2 factor decompositions of the sample correlation matrix of Five Socio-Economic Variables, (Harman, 1976, p.14).

Produce such comparative results for the datasets used in the numerical experiments in (Bertsimas et al., 2017), and possibly for larger ones. Is Manopt always faster and producing better fit, but explaining less common variance than SDP/CVX, i.e. trace$(\Theta) >$ trace(D^2)? More importantly, compare the behaviour, scalability, etc. of the two types of solvers.

7. (a) (Lai et al., 2020, 6.3) Making use of $Q = 2AA^\top - I_p$, change the variables in the classical r-factor LS-EFA problem:

$$\max_{A,\Psi} \|X^\top X - AA^\top - \Psi^2\|_F \ .$$

Construct and explore an alternating algorithm for LS-EFA on the Grassmann manifold by solving:

$$\max_{Q \in \mathcal{G}(p,r)} \|S - Q\|_F \text{ for some } S \in \mathcal{S}(p) \ ,$$

and followed by update of Ψ^2, and so on until convergence;

(b) Construct and explore a simultaneous LS-EFA solution on $\mathcal{G}(p,r) \times \mathcal{D}(p)$, utilising Manopt and possibly (3.19).

8. The classical χ^2 contingency measure (4.15) is adapted for EFA rotation criterion by taking $F \equiv (\Lambda Q) \odot (\Lambda Q)$ (Knüsel, 2008). Write Manopt codes to find orthogonal and oblique rotation matrices Q, i.e. $Q \in \mathcal{O}(p, r)$ and $Q \in \mathcal{OB}(p, r)$, that maximize the χ^2 rotation criterion: trace $\left[F^\top (I_p \odot F 1_r 1_p^\top)^{-1} F (I_r \odot 1_r 1_p^\top F)^{-1} \right]$.

9. In the classical EFA, the factor loadings $\Lambda = [\lambda_1, \lambda_2, \ldots, \lambda_r] \in \mathcal{M}(p, r)$ are not normalized column-wise, which complicates their comparison among columns (factors). Modify the classical VARIMAX (4.19), to produce loadings with equal sum-of-squares (SS), i.e. $\|\lambda_1\|_2 = \|\lambda_2\|_2 = \ldots = \|\lambda_r\|_2$. A way to achieve this is to additionally penalize Λ. Let $B = \Lambda Q$ and $C = B \odot B$. Check that

   (a) the minimization of $1_p^\top C C^\top 1_p$ forces the column sums of squared loadings rotated (by VARIMAX) to become equal. Hint: Consider *Lagrange's identity*:

$$r \sum_{i=1}^{r} x_i^2 = \left(\sum_{i=1}^{r} x_i \right)^2 + \sum_{1 \le i < j \le r} (x_i - x_j)^2 , \qquad (5.43)$$

   where $x_1, \ldots, x_r$ are non-negative numbers. Apparently, if $\sum_{i=1}^{r} x_i$ is constant, then $\sum_{i=1}^{r} x_i^2$ is minimized when $x_i = x_j$ for any $1 \le i < j \le r$, and vice verse. The result follows by substituting $x_i = b_i^\top b_i$ in (5.43), where $B = [b_1, \ldots, b_r]$.

   (b) $1_p^\top C C^\top 1_p$ penalizes the total deviation of all column sums of squares of B from their mean value. Hint: prove that

$$\sum_{j=1}^{r} \left(b_{\cdot j}^2 - b_{\cdot \cdot}^2 \right)^2 = 1_p^\top C C^\top 1_p - \frac{1}{r} \left(\text{trace} \Lambda^\top \Lambda \right)^2 ,$$

   where $b_{\cdot j}^2 = \sum_i^p b_{ij}^2$ and $b_{\cdot \cdot}^2 = \frac{1}{r} \sum_i^r b_{\cdot j}^2 = \frac{1}{r} \sum_j^r (\sum_i^p b_{ij}^2)$.

   (c) write Manopt codes to solve the *penalized* VARIMAX:

$$\min_{Q \in \mathcal{O}(r)} \text{trace } C^\top J_p C - \tau 1_p^\top C C^\top 1_p ,$$

   where $\tau > 0$, and compare with standard VARIMAX solutions.

Chapter 6

Procrustes analysis (PA)

© Springer Nature Switzerland AG 2021

N. Trendafilov and M. Gallo, *Multivariate Data Analysis on Matrix Manifolds*,
Springer Series in the Data Sciences, https://doi.org/10.1007/978-3-030-76974-1_6

6.1   Introduction

Broadly speaking, Procrustes analysis (PA) is about transforming some configuration of points in the plane/space to fit into a given specific pattern of points. PA is named after the character Procrustes from the Greek mythology, a robber dwelling somewhere in Attica (and son of Poseidon). Procrustes had an iron bed (or, even two beds) on which he forced his victims to lie. If a victim was shorter than the bed, he stretched him to fit. Alternatively, if the victim was longer than the bed, he cut off his legs to make the body fit the bed's length.

In this sense, the Procrustes problems are specific type of matrix regression problems and have the following general form:

$$\min_{Q} \|AQ - B\| , \tag{6.1}$$

where $A \in R^{m \times n}$ and $B \in R^{m \times n}$ are given matrices which arise from observed or measured data. The purpose of PA is to find an $n \times n$ matrix Q that solves (6.1). It is usual for the norm in (6.1) to be the Frobenius norm, and in this sense (6.1) is an example of a LS fitting problem. However in Section 6.4, some robust alternatives of the Frobenius norm are considered.

The problem of matching data matrices to maximal agreement by orthogonal or oblique rotations arises in many areas, e.g. factor analysis (Mulaik, 2010), multidimensional scaling (Gower, 1985), morphometrics (Rohlf and Slice, 1990), shape analysis (Dryden and Mardia, 1998). Gower and Dijksterhuis (2004) gives a concise and instructive discussion of a number of classical algorithms for PA which roots go to psychometrics. They also provide a brief historical review and relations of PA to other MDA techniques.

The section is organized as follows. First, Section 6.2 considers PA on the Stiefel manifold and on the orthogonal group. Then, Section 6.3 briefly discusses PA on the oblique manifold, which is historically related to the classical EFA interpretation (Mulaik, 2010, Ch12). Section 6.4 revisits the PA problems when the LS fit is replaced by more robust discrepancy measures. The multi-mode PA is a rather new topic and is related to analysis of multidimensional data arrays and tensor data. It is considered in Section 6.5 in relation to PCA of such multi-mode data. Section 6.6 discusses several less standard PA-like problems. The final Section 6.7 shows how PA helps to develop an alternative approach to EFA which makes it comparable to

PCA. This results in a fast SVD-based EFA algorithm applicable to large data including data with more variables than observations.

6.2 Orthonormal PA

6.2.1 Orthogonal Penrose regression (OPR)

We start with the simpler case when the unknown Q of the Penrose regression problem is an orthogonal matrix, i.e. $Q \in \mathcal{O}(p)$. For short, this problem is abbreviated to OPR. For given matrices $A \in \mathcal{M}(m \times p), C \in \mathcal{M}(p \times n)$ and $B \in \mathcal{M}(m \times n)$, the OPR problem is defined as

$$\min_{Q \in \mathcal{O}(p)} \|AQC - B\|_F . \tag{6.2}$$

The Euclidean gradient ∇_Q of the cost function in (6.2) with respect to the Frobenius inner product is given by the matrix

$$\nabla_Q = A^\top (AQC - B)C^\top . \tag{6.3}$$

Its projection onto the tangent space $\mathcal{T}_Q \mathcal{O}(p)$ can be computed explicitly:

$$g(Q) = \frac{Q}{2}(Q^\top \nabla_Q - \nabla_Q^\top Q) . \tag{6.4}$$

Now, we are ready to solve OPR problem (6.2) numerically. The reader can use the code for SPARSIMAX from Section 4.6.3 as a template. One simply needs to replace the cost function (6.2) of the problem and its projected (Stiefel) gradient (6.4). Alternatively, one can use the construction illustrated in Section 5.6.2, where the cost function and its Euclidean gradient (6.3) are needed only.

The projection formula (6.4) is also interesting because provides information about the first-order optimality condition for the solution of the OPR problem (6.2). Obviously Q is a stationary point of (6.2) only if $g(Q) = O_p$. Specifically, a necessary condition for $Q \in \mathcal{O}(p)$ to be a stationary point of (6.2) is that the matrix $C(AQC - B)^\top AQ$ must be symmetric.

In the particular case of $p = n$ and $C = I_p$, the necessary condition further simplifies to $B^\top AQ$ being symmetric. This necessary condition is well known in the psychometric literature (Gower, 1985; ten Berge, 1977).

6.2.2   Projected Hessian

The projected gradient (6.4) provides useful information for the stationary points of the OPR problem (6.2). It is desirable to have similar kind of "second-order" information to further identify them. This can be achieved by the so-called *projected Hessian* proposed in (Chu and Driessel, 1990).

Consider the following construction producing projected Hessian of a general equality constraint optimisation problem:

$$\text{Minimize} \quad f(x) \tag{6.5}$$

$$\text{Subject to} \quad c(x) = 0_q, \tag{6.6}$$

where $x \in \mathbb{R}^p, f : R^p \longrightarrow R$ and $c : \mathbb{R}^p \longrightarrow \mathbb{R}^q$ with $q < p$ are sufficiently smooth functions. Suppose the constraint $c(x) = [c_1(x), \ldots c_q(x)]^\top$ is regular in the sense that the (gradient) vectors $\nabla c_i(x), i = 1, \ldots, q$, are linearly independent. The feasible set (6.6), rewritten as

$$\mathcal{C} := \{x \in \mathbb{R}^p | c(x) = 0_q\}$$

forms an $(p - q)$-dimensional smooth manifold in $\mathbb{R}^p$. We may express the (gradient) vector $\nabla f(x)$ as the sum of its projection $g(x)$ onto the tangent space $\mathcal{T}_x\mathcal{C}$ and a linear combination of vectors from the normal space $\mathcal{N}_x\mathcal{C}$ of $\mathcal{C}$ at the same x. In other words, we can express $\nabla f(x)$ as follows:

$$\nabla f(x) = g(x) + \sum_{i=1}^{q} \xi_i(x) \nabla c_i(x) , \tag{6.7}$$

for some appropriate scalar functions $\xi_i(x)$.

Assuming that $g(x)$ is smooth in x, we can differentiate both sides of (6.7). Then, for every $x, v \in \mathbb{R}^p$, we can write

$$v^\top \nabla g(x) v = v^\top \left[\nabla^2 f(x) - \sum_{i=1}^{q} \xi_i(x) \nabla^2 c_i(x) \right] v - v^\top \left[\sum_{i=1}^{q} \nabla c_i(x)(\nabla \xi_i(x))^\top \right] v . \tag{6.8}$$

In particular, if $x \in \mathcal{C}$ and $v \in \mathcal{T}_x\mathcal{C}$, then (6.8) reduces to

$$v^\top \nabla g(x) v = v^\top \left[\nabla^2 f(x) - \sum_{i=1}^{q} \xi_i(x) \nabla^2 c_i(x) \right] v \tag{6.9}$$

since $v \perp \nabla c_i(x)$. We note from (6.9) that the condition $v^\top \nabla g(x) v \geq 0$ for every $v \in T_x \mathcal{C}$ is precisely the well-known second-order necessary optimality condition for problem (6.5)—(6.6). The above technique works only if the projected gradient $g(x)$ is explicitly known.

Now, we can apply this technique to calculate the projected Hessian of the OPR problem (6.2) and further identify its stationary points.

From (6.3) and (6.4), observe that the action of the Fréchet derivative of g at Q on a general $H \in \mathcal{M}(p)$ is given by

$$
\begin{aligned}
(\nabla_Q g) H \;=\; & \frac{H}{2} \left[Q^\top A^\top (AQC - B)C^\top - C(AQC - B)^\top AQ \right] + \\
& \frac{Q}{2} \left[H^\top A^\top (AQC - B)C^\top + Q^\top A^\top AHCC^\top - \right. \\
& \left. CC^\top H^\top A^\top AQ - C(AQC - B)^\top AH \right].
\end{aligned} \tag{6.10}
$$

At a stationary point $Q \in \mathcal{O}(p)$, the quantity in the first brackets above is zero. We also know from Section 3.4.2, that a tangent vector $H \in T_Q \mathcal{O}(p)$ has the form $H = QK$, for some skew-symmetric $K \in \mathcal{K}(p)$. Then, after substitution in (6.10), one obtains that the Hessian action on a tangent vector $H \in T_Q \mathcal{O}(p)$ can be calculated as follows:

$$
\langle (\nabla_Q g) QK, QK \rangle = \left\langle C(AQC - B)^\top AQ, K^2 \right\rangle + \left\langle Q^\top A^\top AQKCC^\top, K \right\rangle,
$$

where $\langle , \rangle$ is the Euclidean inner product (2.9), and the assertion follows from its adjoint property $\langle XY, Z \rangle = \langle Y, X^\top Z \rangle$.

It follows from the standard optimization theory (Luenberger and Ye, 2008, Ch 11), that a second-order necessary condition for a stationary point $Q \in \mathcal{O}(p)$ to be a minimizer of the OPR problem (6.2) is that

$$
\left\langle C(AQC - B)^\top AQ, K^2 \right\rangle + \left\langle Q^\top A^\top AQKCC^\top, K \right\rangle \geq 0 \tag{6.11}
$$

for all (non-zero) skew-symmetric $K \in \mathcal{K}(p)$. If the inequality in (6.11) is *strict*, than it becomes also sufficient.

The second-order necessary condition (6.11) can be transformed into a more meaningful form. Let $K = U\Sigma W^\top$ be the SVD of $K \in \mathcal{K}(p)$, where $U, W \in \mathcal{O}(p)$, and $\Sigma = \mathrm{diag}\{\sigma_1, \ldots, \sigma_q\}$ contains the singular values of K with the property $\sigma_{2i-1} = \sigma_{2i}$ for $i = 1, \ldots, \lfloor \frac{p}{2} \rfloor$, and $\sigma_p = 0$ if p is odd. Then, it follows that $K^2 = -U\Sigma^2 U^\top$, which in fact is the spectral decomposition of K^2. We know that $C(AQC - B)^\top AQ$ is necessarily symmetric at any stationary point Q. Let $C(AQC - B)^\top AQ = V\Lambda V^\top$ be its EVD. Similarly for $A^\top A = T\Phi T^\top$ and $CC^\top = S\Psi S^\top$. Note, that the diagonal entries in Φ and Ψ are non-negative, and those in Λ are real.

Denote $P := V^\top U$ and $R := T^\top QKS$ and let $v(Z)$ be the vector containing the main diagonal of the square matrix Z. Direct calculations show that $v(V\Lambda V^\top) = (V \odot V)v(\Lambda)$. In these notations, we can rewrite (6.11) as

$$-\left\langle V\Lambda V^\top, U\Sigma^2 U^\top \right\rangle + \langle \Phi R\Psi, R \rangle =$$
$$-v^\top(\Lambda)(P \odot P)v(\Sigma^2) + v^\top(\Psi)(R \odot R)v(\Phi) \geq (>)0 , \qquad (6.12)$$

where $\odot$ denotes the Hadamard element-wise matrix product .

Now, consider the sufficiency first, i.e. let (6.12) be strictly positive. The second term $v^\top(\Psi)(R \odot R)v(\Phi) \geq 0$ and the vector $(P \odot P)v(\Sigma^2)$ has only positive coordinates. These facts imply that $\Lambda \leq 0$ with at least one strictly negative, as we assume $C(AQC - B)^\top AQ \neq 0$. In other words, the sufficient condition for Q to be a solution for the OPR problem (6.2) is that the (non-zero) matrix $C(AQC - B)^\top AQ$ should be negative semi-definite, i.e. $C(B - AQC)^\top AQ$ be PSD. This sufficient condition for the OPR problem (6.2) is new.

It is impossible to prove that this is also a necessary condition for the OPR problem (6.2). However, this can be achieved for the special case with $C = I_p$. Then, the OPR projected Hessian (6.12) is reduced to

$$\langle (\nabla_Q g)QK, QK \rangle = -\left\langle B^\top AQ, K^2 \right\rangle$$
$$= -v^\top(\Lambda)(P \odot P)v(\Sigma^2) , \qquad (6.13)$$

where here $B^\top AQ = V\Lambda V^\top$. This clearly shows that the condition is also necessary. In other words, a second-order necessary and sufficient condition for a stationary point $Q \in \mathcal{O}(p)$ to be a solution of the OPR problem with $C = I_p$ is that the (non-zero) matrix $B^\top AQ$ be positive semi-definite. This necessary and sufficient condition is obtained first in (ten Berge, 1977).

The expressions (6.12) and (6.13) make it clear why the negative semi-definiteness of $C(AQC-B)^\top AQ$ cannot be elevated to a necessary condition. Another way to see this is to note that

$$(AQ - B)^\top AQ \leq O_p \Longleftrightarrow Q^\top A^\top AQ \leq B^\top AQ .$$

Since $Q^\top A^\top AQ$ is PSD by construction, $(AQ - B)^\top AQ \leq O_p$ certainly implies $O_p \leq B^\top AQ$, but not the converse.

6.2.3 Orthonormal Penrose regression (OnPR)

Here we consider the general case of the Penrose regression problem where the unknown Q is orthonormal, i.e. $Q \in \mathcal{O}(p, q)$. Let $A \in \mathcal{M}(m \times p), C \in \mathcal{M}(q \times n)$ and $B \in \mathcal{M}(m \times n)$ be given matrices. Then, the orthonormal PR (OnPR) is defined as

$$\min_{Q \in \mathcal{O}(p,q)} \|AQC - B\|_F . \tag{6.14}$$

The Euclidean gradient ∇_Q of the OnPR cost function (6.14) with respect to the Frobenius inner product is given by the matrix

$$\nabla_Q = A^\top (AQC - B)C^\top , \tag{6.15}$$

which is identical to the gradient (6.3) for the orthogonal case. However, its projection onto the tangent space $\mathcal{T}_Q \mathcal{O}(p, q)$ is different and is computed explicitly as

$$g(Q) = \frac{Q}{2}(Q^\top \nabla_Q - \nabla_Q^\top Q) + (I_p - QQ^\top)\nabla_Q . \tag{6.16}$$

The OnPR problem (6.14) can be solved numerically straight away following the instructions for the OPR problem in Section 6.2.1.

The projection formula (6.16) provides the first-order optimality condition of the solution of (6.14). Indeed, since the two terms in (6.16) are mutually perpendicular, then (6.16) is necessarily zero if both terms in it are zero. Then, for $Q \in \mathcal{O}(p, q)$ to be a stationary point of the OnPR problem (6.14), the following two conditions must hold simultaneously:

$$C(AQC - B)^\top AQ \text{ be symmetric} , \tag{6.17}$$
$$(I_p - QQ^\top)A^\top (AQC - B)C^\top = O_{p \times q} . \tag{6.18}$$

The second condition (6.18) reflects the difference between the OPR and the OnPR problems being defined on $\mathcal{O}(p)$ and $\mathcal{O}(p, q)$, respectively.

Now, we can demonstrate that the classical Lagrangian approach leads to same necessary conditions. Indeed, consider the OnPR Lagrangian:

$$l(Q, S) = \frac{1}{2} \langle AQC - B, AQC - B \rangle + \left\langle I_q - Q^\top Q, S \right\rangle ,$$

where $Q \in \mathbb{R}^{p \times q}$ and $S \in \mathcal{S}(q)$. Then one can easily find that

$$\nabla_Q l(Q, \Lambda) = A^\top (AQC - B)C^\top - QS = O_{p \times q} \qquad (6.19)$$

is a necessary condition for Q to solve the OnPR.

Multiplying (6.19) by $Q^\top$ we obtain

$$Q^\top A^\top (AQC - B)C^\top = S \in \mathcal{S}(q) ,$$

and substituting it in (6.19) gives

$$(I_p - QQ^\top)A^\top (AQC - B)C^\top = O_{p \times q} ,$$

which are identical to the necessary conditions (6.17)–(6.18) obtained from the projected gradient.

To further identify the OnPR stationary points, we can construct its Hessian by differentiation of the Lagrangian gradient (6.19) to obtain:

$$\nabla_{QQ}^2 l(Q, \Lambda) H = A^\top AHCC^\top - HS ,$$

for some $H \in \mathcal{T}_Q \mathcal{O}(p, q)$. Then, following the standard constrained optimization theory (Luenberger and Ye, 2008, Ch 11) we can state that

$$\langle H, \nabla_{QQ}^2 l(Q, \Lambda) H \rangle \geq 0 \qquad (6.20)$$

is a second-order necessary condition for $Q \in \mathcal{O}(p, q)$ to be a minimizer of OnPR. If (6.20) holds with the *strict* inequality, then it is also a sufficient condition that $Q \in \mathcal{O}(p, q)$ is a local minimizer for the OnPR.

The second-order optimality condition (6.20) can obtain a more specific form if we look back in Section 3.4.3 and note that any $H \in \mathcal{T}_Q \mathcal{O}(p, q)$ can be expressed as $H = QK + (I_p - QQ^\top)W$, for some skew-symmetric $K \in \mathcal{K}(q)$ and arbitrary $W \in \mathcal{M}(p \times q)$. After substitution in (6.20), the second-order necessary condition for a stationary point $Q \in \mathcal{O}(p, q)$ to be a minimizer of OnPR is given by the inequalities:

$$\left\langle Q^\top A^\top AQKCC^\top, K \right\rangle + \left\langle C(AQC - B)^\top AQ, K^2 \right\rangle \geq 0 \qquad (6.21)$$

for any non-zero $K \in \mathcal{K}(q)$ and

$$\left\langle A(I - QQ^\top)WC, A(I - QQ^\top)WC \right\rangle \geq \tag{6.22}$$
$$\left\langle (I - QQ^\top)WQ^\top A^\top (AQC - B)C^\top, (I - QQ^\top)W \right\rangle ,$$

for arbitrary $W \in \mathcal{M}(p)$. The same second-order optimality conditions as (6.21)–(6.22) can be derived making use of the projected Hessian approach from Section 6.2.2 (Chu and Trendalov, 2001, Th. 4).

6.2.4   Ordinary orthonormal PA

The ordinary orthonormal PA (OPA) is frequently used in practice and deserves special attention. The OPA problem is, of course, a particular case of OnPR with $n = q$ and $C = I_q$, and it is defined as

$$\min_{Q \in \mathcal{O}(p,q)} \|AQ - B\|_F , \tag{6.23}$$

for given $A \in \mathcal{M}(m \times p)$ and $B \in \mathcal{M}(m \times q)$.

In this case, according to (6.16), the projected gradient becomes

$$g(Q) = \frac{Q}{2}(B^\top AQ - Q^\top A^\top B) + (I_p - QQ^\top)A^\top (AQ - B),$$

and the first-order optimality conditions for OPA simplify to

$$B^\top AQ \text{ be symmetric} ,$$
$$(I_p - QQ^\top)A^\top (AQ - B) = O_{p \times q} .$$

Now, we can obtain second-order optimality conditions for the OPA problem by simply adjusting (6.21)–(6.22) from the general OnPR problem, such that $n = q$ and $C = I_q$. This results in the following two conditions to hold simultaneously:

$$\left\langle B^\top AQ, K^2 \right\rangle \geq 0 \tag{6.24}$$

for any non-zero $K \in \mathcal{K}(q)$ and

$$\left\langle A(I_p - QQ^\top)W, A(I_p - QQ^\top)W \right\rangle \geq \tag{6.25}$$
$$\left\langle (I_p - QQ^\top)WQ^\top A^\top (AQ - B), (I_p - QQ^\top)W \right\rangle ,$$

for arbitrary $W \in \mathcal{M}(p \times q)$.

The first condition (6.24) boils down to $B^\top AQ$ being PSD, which is equivalent to the well known already necessary and sufficient condition (6.13), for the OPR problem (6.2) with orthogonal unknown Q. The new element here is that (6.24) should hold in conjunction with (6.25).

The second condition (6.25) is new and reflects the fact that here, in OPA, we work with $Q \in \mathcal{O}(p,q)$. However, numerical experiments show that the left-hand side of (6.25) is considerably dominating. This implies that it is probably not quite important for practical reasons, suggesting that the first condition (6.24) is the decisive one.

6.3   Oblique PA

6.3.1   Basic formulations and solutions

The oblique PA (ObPA) relates to another important class of PA problems originating in psychometrics (Cox and Cox, 2001; Gower, 1985; Mulaik, 1972). As its name suggests, ObPA involves *oblique* rotations and is defined as

$$\min_{Q \in \mathcal{OB}(p,q)} \|AQ - B\|_F . \tag{6.26}$$

The rows of $A \in \mathcal{M}(n \times p)$ stand for n observed/given points in $\mathbb{R}^p$. The columns in $Q \in \mathcal{OB}(p,q)$ stand for the direction cosines of q oblique axes in $\mathbb{R}^p$ relative to p orthogonal axes. It is assumed that $n \geq p \geq q$. The goal is to select those oblique axes so that the rows of the oblique transformation AQ are as close as possible to n given points in $\mathbb{R}^q$ which are represented by the rows of $B \in \mathcal{M}(n \times q)$. In factor analysis the problem (6.26) is referred to as oblique Procrustes rotation to a specified factor-structure matrix (containing the covariances/correlations between observed variables and factors). Similarly, the problem

$$\min_{Q \in \mathcal{OB}(p,q)} \|AQ^{-\top} - B\|_F \tag{6.27}$$

is referred to as oblique Procrustes rotation to the specified factor-pattern matrix (containing the weights assigned to the common factors in factor analysis model). Further details can be found in (Mulaik, 1972).

The special case of the ObPA with $p = q$ does not enjoy a closed-form solution as its orthogonal counterpart. To our knowledge, problem (6.26) appears first in explicit form in (Mosier, 1939) where its first approximate solution is given also. The ObPA finds its application in many areas of MDA including EFA for common-factor extraction (MINRES) (Mulaik, 1972), maximal degree structure fitting, and various multidimensional scaling techniques (Cox and Cox, 2001; Gower, 1985; ten Berge, 1991). It also occurs in the context of LS minimization with a quadratic inequality constraint (Golub and Van Loan, 2013, 6.2).

A common feature of the "classical" solutions of (6.26) is that they take advantage of the fact that the constraint $\mathrm{diag}(Q^\top Q) = I_q$ is equivalent to q copies of the unit sphere S^{p-1} in $\mathbb{R}^p$, i.e. the problem can be split up into q separate problems for each column of Q. The main two approaches to ObPA differ in using EVD of $A^\top A$ (Browne, 1967; Cramer, 1974) and SVD of A (Golub and Van Loan, 2013; ten Berge and Nevels, 1977; ten Berge, 1991). Both approaches lead to identical secular equations, which in principle, can be solved through the application of any standard root-finding technique, such as Netwon's method. The convergence can be very slow and also the iteration can diverge, due to the fact that one solves the equation close to a pole. An improved numerical solution of the secular equation by bidiagonalization of A is proposed in (Eldén, 1977).

The second problem (6.27) is less popular. It has been solved column-wise by (Browne, 1972; Browne and Kristof, 1969) making use of planar rotations and in (Gruvaeus, 1970) by using penalty function approach.

Employing methods for matrix optimization on $\mathcal{OB}(p, q)$ turns the solution of (6.26) and (6.27) into a routine task which does not require solving secular equations. Particularly, problem (6.26) can also be solved by solving q separate vector optimization problems on S^{p-1} for each column of Q.

With no further efforts, one can tackle even more complicated oblique versions of the OnPR problem (6.14):

$$\min_{Q \in \mathcal{OB}(p,q)} \|AQC - B\|_F ,$$

and

$$\min_{Q \in \mathcal{OB}(p,q)} \|AQ^{-\top}C - B\|_F ,$$

which cannot be solved by the classical approaches listed above. More complicated problems involving oblique rotation are considered in the next Section 6.5.

Finally, the readers can exercise and write Manopt codes for solving (6.26) and (6.27). They are advised to use as a template the codes provided in Section 11.1.7. Note, that ObPA is a simpler problem as it requires minimization on a single oblique manifold.

6.4   Robust PA

6.4.1   Some history remarks

In this section, we consider the PA where the usual LS objective function is replaced by a more robust discrepancy measure, based on the ℓ_1 norm or smooth approximations of it.

To our knowledge the first consideration of a Procrustes problem employing a robust fit is given in (Siegel and Benson, 1982; Siegel and Pinkerton, 1982) for two- and three-dimensional orthogonal rotations, i.e. for $Q \in \mathcal{O}(p)$ with $p = 2$ or 3. The method is known as the *repeated medians* method (Rousseeuw and Leroy, 1987). See also (Rohlf and Slice, 1990) for a brief review and some generalizations. There is no explicit goodness-of-fit measure to be optimized in this approach which makes it difficult to judge the quality of the results. Explicit robust loss functions are employed in (Verboon, 1994), namely, the ℓ_1, Huber and Hampel loss functions (Rousseeuw and Leroy, 1987), and the algorithms provided solve the *orthogonal* Procrustes problem for arbitrary p.

There are two underlying assumptions here. The first of these is that the errors or noise in the data follows a Gaussian distribution so that the use of the LS criterion (sum-of-squares) is appropriate. However, it is frequently the case that the data contain outliers, that is observations which appear to be inconsistent with the rest of the data. Indeed these occur in almost every true series of observations. There are different ways to take account of outliers (Barnett and Lewis, 1978). Some rejection rules have been derived in order to eliminate them, but their performance is often very poor. It is usually necessary to accommodate outliers in some way, and in particular the use of estimators which are robust is very important, that is estimators which are insensitive to deviations from the standard assumptions. The LS criterion (ℓ_2 norm) is not robust, and the presence of outliers introduces a systematic bias into the least squares estimate that is greater than is introduced by a more robust procedure.

There are many robust estimators available (Huber, 1981), for example, the ℓ_1 norm, where the sum of absolute values of the errors is minimized, or the Huber M-estimator. The use of the ℓ_1 norm can be traced back at least as far as Laplace in 1786. See Chapter 1 in (Rousseeuw and Leroy, 1987) for a comprehensive introduction and further details; historical information is given in many articles and books: see for example (Farebrother, 1987, 1999; Stigler, 1986).

The second underlying assumption in (6.1) is that errors are only present in the matrix B, and the matrix A is error-free. For a conventional linear data fitting problem, the analogous situation is that errors are only present in the dependent variable values. For this case, if large errors are also present in the explanatory or independent variable values, the (usual) ℓ_1 norm criterion is no longer a reasonable robust alternative to least squares (Rousseeuw and Leroy, 1987). The *least median of squares* (LMS) and *least trimmed squares* (LTS) are known to work better in such situations (Rousseeuw and Leroy, 1987). Numerically, they can be treated as special cases of the *weighted least squares* analysis (Rousseeuw and Leroy, 1987).

6.4.2 Robust OnPR

Here we briefly reconsider the OnPR problem (6.14) where the Frobenius matrix (ℓ_2) norm is replaced by the ℓ_1 matrix norm defined as $\|A\|_{\ell_1} = \sum |a_{ij}|$. Let $A \in R^{m \times p}, B \in R^{m \times n}$ and $C \in R^{q \times n}$ be given and consider the following OPR problem:

$$\min_{Q \in \mathcal{O}(p,q)} \|AQC - B\|_{\ell_1} . \tag{6.28}$$

The objective function of problem (6.28) can be rewritten as

$$\|AQC - B\|_{\ell_1} = \mathrm{trace}[(AQC - B)^{\top} \mathrm{sign}(AQC - B)] . \tag{6.29}$$

In order to solve (6.28) by optimization methods on $\mathcal{O}(p,q)$ we need to find a suitable smooth approximation of the term $\mathrm{sign}(AXC - B)$ in (6.29). This situation resembles the problem for smooth approximation of the LASSO constraint in sparse PCA/EFA considered in Chapter 4 and Chapter 5. For example, Trendafilov and Watson (2004) adopted

$$\text{sign}(AQC - B) \approx \tanh[\gamma(AQC - B)] \tag{6.30}$$

for some sufficiently large γ, which is already mentioned in (4.31).

For *orthogonal* rotation Q $(p = q)$, one can show (see Exercise 6, Section 6.9), that the gradient of the objective function (6.29) differs from $A^\top \text{sign}(AXC - B)C^\top$ in a small neighbourhood of zero (matrix), whose diameter depends on the magnitude of γ. Moreover, the following first-order necessary condition for stationary point identification readily holds :

$$C\text{sign}(AQC - B)^\top AQ \ \ \text{must be symmetric,}$$

outside a small neighbourhood of zero residuals $AXC - B$ depending on the value of γ (Trendafilov and Watson, 2004). Note that the corresponding necessary condition for the LS-OPR from Section 6.2.3 looks the same but without the sign function applied to the residual matrix $AXC - B$.

6.4.3   Robust oblique PA

The general form of the oblique PA was considered in Section 6.3.1. The LS fitting problems (6.26) and (6.27) problem considered there will be reformulated as follows. Let $A \in \mathbb{R}^{m \times p}, B \in \mathbb{R}^{m \times n}$ and $C \in \mathbb{R}^{q \times n}$ be given and consider the problems:

$$\min_{Q \in \mathcal{OB}(p,q)} \|AQC - B\|_{\ell_1} \, , \tag{6.31}$$

and, respectively,

$$\min_{Q \in \mathcal{OB}(p,q)} \|AQ^{-\top}C - B\|_{\ell_1} \, . \tag{6.32}$$

One notes that the only difference with Section 6.3.1 is that the ℓ_1 matrix norm replaces the standard Frobenius (ℓ_2) norm.

As in the previous section, one can rewrite the ℓ_1 matrix norm in the objective functions of (6.31) and (6.32) in the following form:

$$\|AQC - B\|_{\ell_1} = \text{trace}[(AQC - B)^\top \text{sign}(AQC - B)] \, ,$$

and approximate $\text{sign}(AXC-B)$ by $\tanh(\gamma(AXC-B))$ for some sufficiently large γ, e.g. $\gamma = 1000$.

Using this approximation, the gradients of (6.31) and (6.32) can be calculated by standard matrix differentiation (see Exercise 6.9). First-order optimality conditions for identification of the stationary points of (6.31) and (6.32) are readily obtained by replacing their gradients ∇ into the following general formula:

$$\nabla = Q\text{diag}(Q^\top\nabla) \ .$$

To illustrate the effect of using robust PA we consider here a small example from EFA considered in (Harman, 1976, Table 15.8). This is a typical situation in which PA is used in EFA. We are given some initial factor loadings (A) depicted in the first two columns of Table 6.1. One can hypothesize or wish that the loadings have certain pattern, and postulates their believe by stating specific target loadings matrix (B). In this particular example, the target is given in the second pair of columns of Table 6.1. One wants to rotate A with the orthogonal or oblique matrix Q, such that the rotated factor loadings AQ are as close as possible to B. In other words, one needs to solve the following Procrustes problems:

$$\min_{Q\in\mathcal{O}(p)} \|AQ - B\| \quad \text{and} \quad \min_{Q\in\mathcal{OB}(p)} \|AQ - B\| \ ,$$

where $\|\|$ stands here for the ℓ_1 and ℓ_2 matrix norms.

Table 6.1: Orthogonal and oblique Procrustes solutions (with l_1 and l_2 fit) for Eight Physical Variables (Harman, 1976, Table 15.8)

Vars	Initial loadings 1	2	Target loadings 1	2	Orthogonal rotated loadings l_2 fit 1	2	l_1 fit 1	2	Oblique rotated loadings l_2 fit 1	2	l_1 fit 1	2
x_1	.856	-.324	9	0	.876	.264	.898	.179	.911	.450	.789	.083
x_2	.848	-.412	9	0	.924	.190	.937	.100	.943	.387	.853	.000
x_3	.808	-.409	9	0	.890	.168	.902	.082	.906	.359	.825	-.015
x_4	.831	-.342	9	0	.867	.235	.886	.151	.897	.419	.787	.056
x_5	.750	.571	0	9	.247	.910	.333	.882	.424	.942	.025	.841
x_6	.631	.492	0	9	.201	.775	.274	.752	.352	.800	.012	.718
x_7	.569	.510	0	9	.141	.751	.212	.734	.288	.764	-.041	.707
x_8	.607	.351	0	9	.268	.648	.329	.620	.392	.691	.107	.581
Rotation	-	-	-	-	.793	.609	.848	.530	.899	.768	.627	.437
matrix Q	-	-	-	-	-.609	.793	-.530	.848	-.438	.641	-.779	.899

In the third and the fourth couples of columns in Table 6.1 are given the rotated by orthogonal matrix Q factor loadings. The rotation matrices are given beneath the corresponding loadings. It is difficult to say which of these loadings are superior. Indeed, the small loadings of the first factor of the l_2 fit are better (smaller) than those of the l_1 fit, but for the second factor things are reversed. The same is valid for the large loadings. The oblique loadings are depicted in the last two couples of columns in Table 6.1. It is clear that the l_2 fit produces better large loadings for both factors, but the "zero" loadings are quite disappointing. The l_1 fit produces very good small loadings for both factors. As a whole, one can conclude that the robust (l_1) oblique Procrustes solution is the best for this example. This is because the interpretation is mainly interested in the vanishing ("zero") loadings.

6.4.4   PA with M-estimator

In this section, we are concerned with the treatment of (6.1) while making explicit two fundamental assumptions: (i) only the response data which make up the matrix B are corrupted, and (ii) the errors in B are such that the use of a more robust estimator than least squares is appropriate. In particular, we will consider problem (6.1) when the norm is the ℓ_1 norm.

As already explained, standard differentiable optimization theory does not apply and although, as we will see, this does not inhibit a solution process, nevertheless it may be preferable to smooth the problem. One possibility is to replace it by the Huber M-estimator (Rousseeuw and Leroy, 1987), which has continuous derivatives. Following the accepted notations, the Huber M-estimator for minimizing the $m \times 1$ residual vector $d = Aq - b$ is defined by

$$\psi(d) = \sum_{i=1}^{m} \rho(d_i), \tag{6.33}$$

where

$$\rho(t) = \begin{cases} t^2/2, & |t| \leq \tau \\ \tau(|t| - \tau/2), & |t| > \tau, \end{cases} \tag{6.34}$$

and τ is a scale factor or tuning constant. Then the oblique Procrustes subproblem:

$$\min_{q^\top q = 1} \|Aq - b_i\|_{\ell_1} \tag{6.35}$$

is replaced by

$$\min_{q^\top q=1} \psi(Aq - b_i) . \tag{6.36}$$

The function (6.33) is convex and once continuously differentiable but has discontinuous second derivatives at points where $|d_i| = \tau$ (the tight residuals). Clearly if τ is chosen large enough, then the loss function is just the least squares function, and if τ tends to zero, then limit points of the set of solutions minimize the ℓ_1 norm of d (Clark, 1985), so in a sense the Huber M-estimator represents a compromise between these two. Frequently, the scale factor τ has to be obtained by satisfying an auxilliary equation as part of the estimation process. However, most algorithms have been developed for the case when τ is assumed given. Then, as observed by Huber, the search for the minimum is the search for the correct partition of the components of d, and the correct signs of those with $|d_i| > \tau$. There has been considerable interest in the relationship between the Huber M-estimator and the ℓ_1 problem, see for example Li and Swetits (1998); Madsen et al. (1994). Indeed for standard linear data fitting problems, it has been argued that the right way to solve ℓ_1 problems is by a sequence of Huber problems (Madsen and Nielsen, 1993). In any event, the value of this criterion has been well established in many situations.

Next we consider the oblique Procrustes subproblem when the objective function in (6.35) is replaced by the Huber M-estimator function defined in (6.33)–(6.34). In other words the objective function in (6.35) is approximated as follows:

$$\|Aq - b_i\|_{\ell_1} \approx \psi(Aq - b_i), \tag{6.37}$$

for $i = 1, 2, \ldots, n$.

The gradient of the right-hand side of (6.37) can be calculated as the following $p \times 1$ vector:

$$\nabla_i = A^\top \psi'(Aq - b_i), \tag{6.38}$$

for $i = 1, 2, \ldots, n$, where $\psi'(Aq - b_i)$ denotes an $m \times 1$ vector with components $(\rho'(a_1 x - b_{1,i}), \rho'(a_2 q - b_{2,i}), \ldots, \rho'(a_m q - b_{m,i}))$, where a_j^T is the $1 \times p$ vector of the jth row of A and

$$\rho'(t) = \begin{cases} t, & |t| \leq \tau \\ \tau \operatorname{sign}(t), & |t| > \tau \end{cases} . \tag{6.39}$$

6.5   Multi-mode PA

6.5.1   PCA and one-mode PCA

We already know from Section 4 (and Section 5) that orthogonal and oblique rotations can be used to transform the component loadings and scores without changing the fit. Indeed, from (4.17) we know that

$$X = FDA^\top = FDQQ^{-1}A^\top . \tag{6.40}$$

A common approach in the classical PCA (and EFA) interpretation is to prepare some target matrix of "desirable" loadings A_t and rotate the initial ones A to this target using Procrustes rotations, i.e. find an orthogonal/oblique matrix Q such that $\min_Q \|AQ - A_t\|$.

Now, suppose that we are interested to achieve simultaneously some desirable component loadings A and scores $C(= FD)$ defined by their target matrices A_t and C_t, respectively. Then, a transformation Q should be sought such that A fits A_t and C fits C_t *simultaneously* as well as possible. Thus, the standard Procrustes problem becomes more complicated:

$$\min_Q w_A\|AQ^{-\top} - A_t\|_F^2 + w_C\|CQ - C_t\|_F^2 , \tag{6.41}$$

where $Q \in \mathcal{O}(p)$ or $\mathcal{OB}(p)$, and w_A and w_C are fixed weights which balance the contribution of the component and the score matrices. This modified Procrustes problem is called the *one-mode* Procrustes problem. Of course, if one is interested in a loadings or a score matrix only with certain structure then the inappropriate term can be dropped off consideration.

6.5.2   Multi-mode PCA and related PA problems

The classical PCA ($X = FDA^\top = CA^\top$) can be viewed as one-mode PCA and a particular case of the following *two-mode* PCA (Magnus and Neudecker, 1988) which is defined as: for a given $n \times p$ data matrix:

$$X = A_2CA_1^\top + E, \tag{6.42}$$

where A_1, A_2 and C are $p \times q, n \times r$ and $r \times q$ matrices, respectively. As in PCA, A_1 and A_2 are called (component) loadings, and C is known as the *core*. The two-mode PCA aims reduction of the number of variables p, as well as reduction of the number n of individuals, a quite likely situation in large data applications.

As in the standard PCA, the solution of the two-mode PCA is not unique. As in (6.40), the loadings matrices A_1 and A_2 in (6.42) can be transformed by any non-singular matrices Q_1 and Q_2 which can be compensated by the inverse transformation applied to the core, as follows:

$$X = A_2 C A_1^\top = (A_2 Q_2^{-\top})(Q_2^\top C Q_1)(A_1 Q_1^{-\top})^\top . \qquad (6.43)$$

Similarly to the one-mode PCA, one can be interested in obtaining a solution defined by the targets $A_{1,t}$, $A_{2,t}$ and C_t, respectively. Then, the following *two-mode* Procrustes problem should be considered:

$$\min_{Q_1, Q_2} w_1 \|A_1 Q_1^{-\top} - A_{1,t}\|^2 + w_2 \|A_2 Q_2^{-\top} - A_{2,t}\|^2 + w \|Q_2^\top C Q_1 - C_t\|^2, \quad (6.44)$$

where $(Q_1, Q_2) \in \mathcal{O}(p_1) \times \mathcal{O}(p_2)$ or $\mathcal{OB}(p_1) \times \mathcal{OB}(p_2)$ and w, w_1 and w_2 are fixed weights.

Let vec(X) denote a column vector of length np whose elements are the elements of the $n \times p$ matrix X arranged column after column, known as the major column form of a matrix. This is also called the *vec-operator* (Magnus and Neudecker, 1988, Ch4). The *Kronecker product* of two matrices $A \in \mathcal{M}(m, n)$ and $B \in \mathcal{M}(p, q)$ is a new $mp \times nq$ matrix defined as:

$$A \otimes B = \begin{bmatrix} a_{11} B & \cdots & a_{1n} B \\ \vdots & \ddots & \vdots \\ a_{m1} B & \cdots & a_{mn} B \end{bmatrix} . \qquad (6.45)$$

Then, (6.43) can be rewritten as follows:

$$x = (A_1 \otimes A_2) c = (A_1 Q_1^{-\top} \otimes A_2 Q_2^{-\top})(Q_1^\top \otimes Q_2^\top) \text{vec}(C) , \qquad (6.46)$$

where $x = \text{vec}(X)$ and $c = \text{vec}(C)$ denote the vec-operator, transforming its matrix argument into a vector by stacking its columns one under another. To obtain (6.46), the following useful identities are utilized:

$$\text{vec}(X_1 X_2 X_3) = (X_3^\top \otimes X_1) \text{vec} X_2 , \qquad (6.47)$$

and

$$(X_1 \otimes X_2)^\top = X_1^\top \otimes X_2^\top . \qquad (6.48)$$

Further details, properties and applications of the Kronecker product can be found in (Golub and Van Loan, 2013, 12.4) and (Lancaster and Tismenetsky, 1985, Ch 12).

Then the two-mode Procrustes problem (6.44) can be rewritten as follows:

$$\min_{Q_1,Q_2} w_1\|A_1 Q_1^{-\top} - A_{1,t}\|_F^2 + w_2\|A_2 Q_2^{-\top} - A_{2,t}\|_F^2 + w\|c^\top (Q_1 \otimes Q_2) - c_t^\top\|_F^2 ,$$

subject to $(Q_1, Q_2) \in \mathcal{O}(p_1) \times \mathcal{O}(p_2)$ or $\mathcal{OB}(p_1) \times \mathcal{OB}(p_2)$.

We can extrapolate (6.46) to m-mode PCA, for analyzing m-dimensional data arrays/tensors, see (Magnus and Neudecker, 1988, p. 363):

$$x = (A_1 \otimes A_2 \otimes \ldots \otimes A_m)c = (\otimes_{i=1}^{m} A_i)c,$$

where A_i is an $n_i \times p_i$ matrix for $i = 1, 2, \ldots, m$, and the vectors x and c are the column forms of the tensors X and C of order m. The vec-operator for tensors works the same as for matrices: it expands the tensor by starting from the leftmost mode/index and keeping the rest fixed.

Then the corresponding *m-mode Procrustes problem* becomes:

$$\min_{Q_i} \sum_{i=1}^{m} w_i\|A_i Q_i^{-\top} - A_{i,t}\|_2^2 + w\|c^\top (\otimes_{i=1}^{m} Q_i) - c_t^\top\|_F^2, \tag{6.49}$$

where $(Q_1, Q_2 \ldots Q_m) \in \prod_{i=1}^{m} \mathcal{O}(p_i)$ or $\prod_{i=1}^{m} \mathcal{OB}(p_i)$.

Note that the *orthogonal* m-mode Procrustes problem (6.49) reduces to

$$\min_{Q_i} \sum_{i=1}^{m} w_i\|A_i Q_i - A_{i,t}\|_F^2 + w\|c^\top (\otimes_{i=1}^{m} Q_i) - c_t^\top\|_F^2, \tag{6.50}$$

where $(Q_1, Q_2 \ldots Q_m) \in \prod_{i=1}^{m} \mathcal{O}(p_i)$.

The particular case of problem (6.50) with $w = 0$ and $w_i = 1$ for all $i = 1, 2, \ldots, m$ is well known as *generalized Procrustes problem* (Gower, 1985).

6.5.3   Global minima on $\mathcal{O}(p)$

Not every critical point is a minimum. Except for the maxima and saddles points there may be multiple local minima. Unfortunately, second-order

(Hessian) information is not generally available to allow us an theoretical treatment of the character of these minima.

For the one-mode *orthogonal* Procrustes problem ($m = 1$), we can demonstrate the uniqueness of the local minimum. Problem (6.41) becomes

$$\min_{Q \in \mathcal{O}(p)} w_1 \|AQ - A_t\|_F^2 + w\|CQ - C_t\|_F^2 . \tag{6.51}$$

Now, let $M = w_1 A^\top A_t + wC^\top C_t$. For this particular case direct calculations show that a first-order necessary condition for $Q \in \mathcal{O}(p)$ to be a stationary point of the one-mode orthogonal Procrustes problem (6.51) is that the matrix $Q^\top M = S$ be symmetric. Such Q are readily found by a polar decomposition $M = QS$, where S is a symmetric (not necessarily PD) matrix. The different stationary points are related by $Q_2 = Q_1 Z$ where Z is a matrix of the form $UDU^\top$, where D is a diagonal matrix with entries $\{1, -1\}$ and $U^\top SU$ is diagonal, i.e. the corresponding S's differing only in the signs of their eigenvalues. Assuming S has distinct non-zero eigenvalues, there are two stationary points of interest, corresponding to the PD $S(= S_{pd})$ and to $S(= S_a)$ having the smallest magnitude eigenvalue negative and the rest positive.

For this problem, the constrained Hessian has a simple form. Following the technique described in Section 6.2.2 (Chu and Trendafilov, 1998a), we find that the quadratic form $H_Q(K, K)$ of the projected Hessian at a stationary point $Q \in \mathcal{O}(p)$ for a tangent vector QK is given by $\langle Q^\top M, K^2 \rangle = \langle S, K^2 \rangle$, where K is an arbitrary skew-symmetric and $\langle , \rangle$ denotes the Euclidean/Frobenius inner product (2.9).

Based on the projected Hessian and the fact that the diagonal entries of K^2 are non-positive for all skew-symmetric K, one might conclude that the stationary point with $S = S_{pd}$ is the only stationary point with a negative definite H. However, one can prove the property of all skew-symmetric matrices that no diagonal entry K^2 can be less than the sum of the rest (a consequence of the eigenvalues of K being pairs, $i\omega, -i\omega$, or 0). Thus, the stationary point with $S = S_a$ is another solution with negative definite H (and with some technical work can be shown to be the only other). These two solutions are the only local minima, and correspond to the global minima on each of the two components of $\mathcal{O}(p)$.

For the multi-mode Procrustes problem, we have no such analysis, so it is not guaranteed that the algorithm will find the global minimum. This is pretty common situation in practice, when numerical algorithms are in use.

6.6   Some other PA problems

6.6.1   Average of rotated matrices: generalized PA

Suppose we are given a set of $m(\geq 2)$ matrices $A_i \in \mathcal{M}(n, p_i)$. Then the generalized Procrustes problem is defined as

$$\min_{Q_i} \sum_{1 \leq i < j \leq m} \|A_i Q_i - A_j Q_j\|_F , \tag{6.52}$$

where $Q_i \in \mathcal{O}(p_i, r)$ or $Q_i \in \mathcal{OB}(p_i, r)$ for some $r \leq \min\{p_1, \ldots, p_m\}$.

Problem (6.52) can be readily solved as minimisation on matrix manifolds. Historically, (6.52) is attacked by making use of the following identity:

$$\sum_{1 \leq i < j \leq m} \|A_i Q_i - A_j Q_j\|_F = m \sum_{i=1}^{m} \|A_i Q_i - \bar{A}\|_F , \tag{6.53}$$

where

$$\bar{A} = \frac{1}{m} \sum_{i=1}^{m} A_i Q_i \tag{6.54}$$

is the average of all rotated A_i's and known as the average configuration (Gower and Dijksterhuis, 2004, 9.1).

Then, (6.52) can be solved by minimizing the right-hand side of (6.53) in alternating manner: initialize Q_i, update $A_i Q_i$, update $\bar{A}$, solve m Procrustes problems for Q_i, repeat the previous three steps until convergence. Particularly, if we are interested in $Q_i \in \mathcal{O}(p_i, r)$, the algorithm requires the SVDs of m matrices with sizes $p_i \times r$, $i = 1, \ldots, m$ at every iteration step.

Moreover, direct calculations show that (Gower and Dijksterhuis, 2004, 9.1):

$$\sum_{i=1}^{m} \|A_i Q_i\|_F = \sum_{i=1}^{m} \|A_i Q_i - \bar{A}\|_F + m\|\bar{A}\|_F , \tag{6.55}$$

which suggests that the total sum-of-squares is split into residual sum-of-squares and sum-of-squares attributable to the average configuration.

If $Q_i \in \mathcal{O}(r)$ for all i, i.e. $p_1 = \ldots = p_m = r$, then, the left-hand side of (6.55) is constant and (6.52) is equivalent to maximizing $\|\bar{A}\|_F$.

6.6.2 Mean rotation

Here we briefly consider PA problems on the subset of all orthogonal $p \times p$ matrices with *unit* determinant. As known from Section 3.1.2, this is the special orthogonal group $\mathcal{SO}(p)$. More specifically, we are interested in $\mathcal{SO}(3)$, containing the rotations in $\mathbb{R}^3$ and particularly important in computer graphics, robotics and kinematics.

The average or the arithmetic mean of arbitrary $Q_1, Q_2 \in \mathcal{O}(3)$ is unlikely to be an orthogonal matrix, i.e. $\frac{Q_1 + Q_2}{2} \notin \mathcal{O}(3)$. In order to define a reasonable "average" rotation it is worth recollecting that it is supposed to represent the most typical rotations in the sample. Intuitively, this means that the average rotation should minimise the sum of some kind of distances to all of the given rotations, i.e.

$$\min_{Q \in \mathcal{SO}(3)} \sum_i d(Q, Q_i) , \tag{6.56}$$

which is a generalization of the variational definition of average of positive reals $x_1, \ldots, x_m$ (Moakher, 2002):

$$\min_{x > 0} \sum_i |x - x_i|^2 , $$

with solution $x = \frac{1}{m} \sum_i^m x_i$ and denoted by $\bar{x}$.

That is why it is helpful to think of the average rotation as a specific minimisation problem on $\mathcal{SO}(3)$ (Hartley et al., 2011; Moakher, 2002). For this reason, we consider the *chordal distance* between $Q_1, Q_2 \in \mathcal{SO}(3)$ defined/induced by the Frobenius norm in $\mathcal{M}(3)$:

$$d(Q_1, Q_2) = \|Q_1 - Q_2\|_F , \tag{6.57}$$

which after plugging in (6.56) defines the average rotation Q of $Q_1, \ldots, Q_n$ as

$$\min_{Q \in \mathcal{SO}(3)} \sum_{i=1}^m \|Q - Q_i\|_F^2 = \min_{Q \in \mathcal{SO}(3)} \|Q - \bar{Q}\|_F^2 , \tag{6.58}$$

where $\bar{Q} = \frac{1}{m} \sum_{i=1}^m Q_i$.

The right-hand side of (6.58) clearly indicates that the (unknown) average rotation Q is the projection of $\bar{Q}$ onto $\mathcal{SO}(3)$ (Moakher, 2002, Pr 3.3).

Problem (6.58) is a standard orthogonal Procrustes problem. Its solution is given by the SVD of $\bar{Q} = UDV^\top$, and the unknown $Q = UV^\top$, if $\det(\bar{Q}) > 0$. Otherwise, the solution is given by $Q = U\mathrm{diag}(1, 1, -1)V^\top$ (Moakher, 2002, Pr 3.5).

The average rotation defined by (6.58) is referred to as the Euclidean average and employs the chordal distance (6.57). By changing the distance in (6.56), one can obtain different mean rotation. For example, one can consider the geometric mean of positive reals $x_1, \ldots, x_m$ given by the following variational definition (Moakher, 2002, 3.2):

$$\min_{x>0} \sum_i |\log x - \log x_i|^2 = \min_{x>0} \sum_i |\log x_i^{-1} x|^2 , \tag{6.59}$$

with solution $x = \sqrt[m]{x_1 \ldots x_m}$ and denoted by $\tilde{x}$.

To extend the geometric mean definition (6.59) to matrices in $\mathcal{SO}(3)$ we need to define logarithm from a non-singular matrix argument, i.e. on $\mathcal{GL}(3)$. In general, the question of existence of logarithm of a real $A \in \mathcal{GL}(p)$ involves conditions on its Jordan blocks (Culver, 1966).

Particularly, the matrices $A \in \mathcal{M}(p)$ close to I_p are non-singular, and thus, one can define the so-called *principal logarithm* as follows:

$$\log A = -\sum_{i=1}^{\infty} \frac{(I_p - A)^i}{i} , \tag{6.60}$$

for $A \in \mathcal{M}(p)$, such that $\|I_p - A\|_F < 1$ (Curtis, 1984, Ch 4) (Hall, 2015, Ch 2.3). The infinite series (6.60) is convergent *absolutely*, because the series of norms is convergent, and thus, it is convergent (Dieudonné, 1969a, 5.3.2). Unfortunately, the convergence of (6.60) is rather slow and its convergence region is rather small. Alternative definitions of logarithm with better computational properties are listed in (Higham, 1989, Ch 11).

Luckily the situation simplifies for $p = 3$, when we consider $\mathcal{SO}(3)$. Let us define the following isomorphism between $\mathbb{R}^3$ and $\mathcal{K}(3)$, known as the *hat map*:

$$x = \begin{bmatrix} x_1 \\ x_2 \\ x_3 \end{bmatrix} \longleftrightarrow \hat{x} = \begin{bmatrix} 0 & -x_3 & x_2 \\ x_3 & 0 & -x_1 \\ -x_2 & x_1 & 0 \end{bmatrix} .$$

Then, the exponent of $\hat{x} \in \mathcal{K}(3)$ is given by the explicit Rodrigues' formula (Marsden and Ratiu, 1999, p. 294): for $\|x\| \neq 0$

$$\exp(\hat{x}) = I_3 + \sin\|x\| \left(\frac{\hat{x}}{\|x\|}\right) + (1 - \cos\|x\|)\left(\frac{\hat{x}}{\|x\|}\right)^2 \in \mathcal{SO}(3) , \quad (6.61)$$

and $\exp(\hat{x}) = I_3$, if $\|x\| = 0$. To prove (6.61), it is crucial to note first that

$$\hat{x}^3 = -\|x\|^2\hat{x}, \ \hat{x}^4 = -\|x\|^2\hat{x}^2, \ \hat{x}^5 = \|x\|^4\hat{x}, \ \hat{x}^6 = \|x\|^4\hat{x}^2, \dots ,$$

which helps to split the exponential series in two parts containing even and odd powers as follows:

$$
\begin{aligned}
\exp(\hat{x}) &= I_3 + \hat{x} + \frac{\hat{x}^2}{2!} + \frac{\hat{x}^3}{3!} + \frac{\hat{x}^4}{4!} + \dots \\
&= I_3 + \hat{x} + \frac{\hat{x}^2}{2!} - \frac{\|x\|^2\hat{x}}{3!} - \frac{\|x\|^2\hat{x}^2}{4!} + \frac{\|x\|^4\hat{x}}{5!} + \frac{\|x\|^4\hat{x}^2}{6!} - \frac{\|x\|^6\hat{x}}{7!} - \frac{\|x\|^6\hat{x}^2}{8!} \dots \\
&= I_3 + \hat{x}\left[I_3 - \frac{\|x\|^2\hat{x}}{3!} + \frac{\|x\|^4\hat{x}}{5!} - \frac{\|x\|^6\hat{x}}{7!} + \dots\right] \\
&\quad + \hat{x}^2\left[\frac{1}{2!} - \frac{\|x\|^2\hat{x}}{4!} + \frac{\|x\|^4\hat{x}}{6!} - \frac{\|x\|^6\hat{x}}{8!} \dots\right] \\
&= I_3 + \frac{\sin\|x\|}{\|x\|}\hat{x} + \frac{1 - \cos\|x\|}{\|x\|^2}\hat{x}^2 .
\end{aligned}
$$

The exponential map (6.61) produces a rotation by $\|x\|$ radians around the axis given by x. Note that $\frac{\hat{x}}{\|x\|} \in \mathcal{K}(3)$, $\left(\frac{\hat{x}}{\|x\|}\right)^2 \in \mathcal{S}(3)$ and $\frac{\text{trace}(\hat{x}^2)}{\|x\|^2} = -2$.

Now, let $Q = \exp(\hat{x}) \in \mathcal{SO}(3)$. We are interested in a kind of inverse function (logarithm) that returns $\hat{x} \in \mathcal{K}(3)$ from a given $Q \in \mathcal{SO}(3)$. By applying the trace operation on both sides of (6.61) we find that

$$\text{trace}(\exp(\hat{x})) = \text{trace}(Q) = 3 - 2(1 - \cos\|x\|) = 1 + 2\cos\|x\| ,$$

i.e. for known Q we have

$$\|x\| = \arccos\left(\frac{\text{trace}Q - 1}{2}\right) = \cos^{-1}\left(\frac{\text{trace}Q - 1}{2}\right) . \quad (6.62)$$

The logarithm of Q can be found by expressing $\hat{x}$ from (6.61) as follows. First, observe that if $\|x\| \neq 0$, then

$$Q - Q^\top \left(= \exp(\hat{x}) - \exp(\hat{x})^\top\right) = 2\sin\|x\| \left(\frac{\hat{x}}{\|x\|}\right) ,$$

which gives

$$\hat{x} \, (= \log Q) = \frac{\|x\|}{\sin \|x\|} \left(\frac{Q - Q^{\top}}{2} \right) \in \mathcal{K}(3) \, , \tag{6.63}$$

and $\|x\|$ is available from (6.62). Provided $\|x\| < \pi$, the expression (6.63) defines the desired $\log Q$ of $Q \in \mathcal{SO}(3)$ through its skew-symmetric part. If $\|x\| = 0$, then $\log Q = 0$. Alternative formulas for (6.61) and (6.63) and further aspects can be found in (Cardoso and Leite, 2010) and (Iserles et al., 2000, B). Riemannian exponential and logarithm functions on the Stiefel manifold are considered in (Zimmermann, 2017).

Now we can consider the *Riemannian distance* between $Q_1, Q_2 \in \mathcal{SO}(3)$ (Moakher, 2005):

$$d(Q_1, Q_2) = \frac{1}{\sqrt{2}} \left\| \log(Q_1^{-1} Q_2) \right\|_F = \frac{1}{\sqrt{2}} \left\| \log(Q_1^{\top} Q_2) \right\|_F \, , \tag{6.64}$$

which helps to generalize the geometric mean of positive reals (6.59) to a kind of geometric mean of the rotations $Q_1, \ldots, Q_m \in \mathcal{SO}(3)$ after plugging it in (6.56) as

$$\min_{Q \in \mathcal{SO}(3)} \sum_{i=1}^{m} \left\| \log(Q_i^{\top} Q) \right\|_F^2 \, . \tag{6.65}$$

The optimality condition for the minimiser of (6.65) is given by:

$$\sum_{i=1}^{m} \log(Q_i^{\top} Q) = 0 \, ,$$

which resembles (with $Q^{\top} = Q^{-1}$) the optimality condition for the geometric mean of positive reals (6.59) (Moakher, 2002, p. 8):

$$\sum_{i=1}^{m} \log(x_i^{-1} x) = 0 \, .$$

Another way to generalize the geometric mean of rotations is to make use of the *hyperbolic distance*:

$$d(Q_1, Q_2) = \| \log Q_1 - \log Q_2 \|_F \, ,$$

and define the fitting problem with respect to it as follows:

$$\min_{Q \in \mathcal{SO}(3)} \sum_{i=1}^{m} \| \log Q_i - \log Q \|_F^2 \, . \tag{6.66}$$

The solution of (6.66) is readily obtained from its first-order optimality condition, which gives $\log Q = \frac{1}{m}\sum_i^m \log Q_i \in \mathcal{K}(3)$.

It is worth mentioning briefly an interpolation problem in $\mathcal{SO}(n), n > 2$, which attracts considerable attention in robotics and computer graphics. Let $Q_1, \ldots, Q_m \in \mathcal{SO}(n)$ and $t_1 < \ldots < t_m$ be real numbers. Find a differentiable (matrix-valued) function $Q : [t_1, t_m] \to \mathcal{SO}(n)$, such that $Q(t_i) = Q_i$ for $i = 1, \ldots, m$. A classical way to solve this problem is to find some $K \in \mathcal{K}(n)$, such that $Q(t) = \exp(tK)$, i.e. make use of the exponential map and transform the original problem on $\mathcal{SO}(n)$ into a new one on the vector (linear) space $\mathcal{K}(n)$. However, the line segment in $\mathcal{K}(n)$, say $l(t)$, calculated through the principal logarithm results in $\exp(l(t))$, deviating significantly from $\exp(tK)$ within the chosen t-interval (Shingel, 2009).

An alternative, PA-like, approach to attack this problem, together with a fast algorithm to solve it in higher dimensions $(n > 3)$ is considered in (Escande, 2016). The idea is, for given $Q \in \mathcal{SO}(n)$ and $K \in \mathcal{K}(n)$, to solve:

$$\min_{\substack{X \in \mathcal{K}(n) \\ \exp X = Q}} \|X - K\|_F^2 \; ,$$

which gives the (if unique) element $X \in \mathcal{K}(n)$ closest to K and is a logarithm of $Q \in \mathcal{SO}(n)$. For implementation details, see (Escande, 2016).

The mean rotation problems considered so far are LS problems. We know from Section 6.4 that more robust estimations can be obtained by replacing the Frobenius norm by the ℓ_1 matrix norm. Then, for example, the LS average (6.58) can be replaced by

$$\min_{Q \in \mathcal{SO}(3)} \sum_{i=1}^m \|Q - Q_i\|_{\ell_1} \; , \tag{6.67}$$

which is another (non-smooth) matrix optimisation problem on $\mathcal{SO}(3)$. In fact, (6.67) generalizes the variational definition of median of $x_1, \ldots, x_m \in \mathbb{R}$:

$$\min_x \sum_i |x - x_i| \; .$$

In this sense, the solution Q of (6.67) can be viewed as the median rotation of $Q_1, \ldots, Q_n$. An algorithm for solving (6.67) is proposed in (Hartley et al., 2011), as well as an algorithm for computing the median rotation from (6.65), where the Frobenius norm is replaced by the ℓ_1 matrix norm.

At the end of the section, it is worth reminding that every $Q \in \mathcal{SO}(3)$ is used to be parametrised via the *Euler angels* ϕ, θ, ψ $(0 \le \phi, \psi < 2\pi, 0 \le \theta < \pi)$. Let $\{e_1, e_2, e_3\}$ be the coordinate axes in $\mathbb{R}^3$. The rotations of θ and ϕ radians about e_1 and e_3 can be written in matrix forms as

$$Q_1(\theta) = \begin{bmatrix} 1 & 0 & 0 \\ 0 & \cos\theta & -\sin\theta \\ 0 & \sin\theta & \cos\theta \end{bmatrix} \text{ and } Q_3(\phi) = \begin{bmatrix} \cos\phi & -\sin\phi & 0 \\ \sin\phi & \cos\phi & 0 \\ 0 & 0 & 1 \end{bmatrix}.$$

Then, every $Q \in \mathcal{SO}(3)$ can be expressed as (a product of) three consecutive rotations, by means of the Euler angles as follows:

$$Q = Q_3(\phi)Q_1(\theta)Q_3(\psi) . \tag{6.68}$$

Note, that there is a potential problem with the parametrisation (6.68). If $\theta = 0$, then $Q_1(\theta) = I_3$, and the product (6.68) reduces to

$$Q = Q_3(\phi)Q_3(\psi) = \begin{bmatrix} c_1 & -s_1 & 0 \\ s_1 & c_1 & 0 \\ 0 & 0 & 1 \end{bmatrix} \begin{bmatrix} c_2 & -s_2 & 0 \\ s_2 & c_2 & 0 \\ 0 & 0 & 1 \end{bmatrix} = \begin{bmatrix} c & -s & 0 \\ s & c & 0 \\ 0 & 0 & 1 \end{bmatrix},$$

where $c = c_1 c_2 - s_1 s_2 = \cos(\phi + \psi)$ and $s = s_1 c_2 + c_1 s_2 = \sin(\phi + \psi)$. In other words, the first and the third rotation can be combined into a single rotation about e_3 by an angle $\phi + \psi$. Graphically, this means that we rotate twice about e_3. The problem here is that there are infinite number of ϕ's and ψ's which sum to the same $\phi + \psi$. For further details on such "singular" cases in the Euler angles representation and relations to the Rodrigues' formula (6.61) see http://motion.pratt.duke.edu/RoboticSystems/3DRotations.html.

In order to use the Euler angles ϕ, θ, ψ as local coordinates at every point of $\mathcal{SO}(3)$, and keep its structure of a smooth manifold, one needs to work with $0 < \phi, \psi < 2\pi, 0 < \theta < \pi$ (Arnold, 1989, 30A). This is similar to the situation with the local coordinates ϕ, θ $(0 < \phi < 2\pi, 0 < \theta < \pi)$ on the unit sphere $\mathcal{S}^2 \subset \mathbb{R}^3$, which has an atlas with two charts (see Section 3.2). Using $0 \le \phi < 2\pi, 0 \le \theta < \pi$ for the whole sphere may look like $\mathcal{S}^2$ can be covered by an atlas with a single chart. This is not correct, and thus, this is not a genuine parametrisation of $\mathcal{S}^2$. Details about the connection between orthogonal matrices and rotations can be found in (Artin, 2011, Ch 5.1).

6.7 PA application to EFA of large data

Some problems related to the classical EFA model (5.1) were already mentioned in Section 5.2.2. EFA tries to express p random variables by fewer r new variables and additional adjustment for each variable, i.e. involving another p unique variables (factors).

EFA can be interpreted as an attempt to improve PCA by doing a similar low-rank approximation and polishing it by a slight adjustment of each fitting variable. However, this implies that EFA expresses p observable variables by new unknown $r + p$ variables. Another baffling feature of the EFA model (5.1) is that it includes both random and non-random unknowns. One would expect some hybrid solution obtained by parametric and nonparametric methods. Instead, the classical EFA takes the approach outlined in Section 5.2.1 and gets rid of the unknown random variables f and u.

Alternatively, one can consider the sample EFA definition from Section 5.2.2 as a starting point. Then, the EFA fundamental equation (5.5) can be interpreted as a specific data matrix factorization (Trendafilov and Unkel, 2011). In fact, the idea is to make EFA a more precise (but also more sophisticated) dimension reduction technique, a kind of PCA generalization.

Thus, the data and the factors are *not* considered random variables, which seems natural when factoring a data *matrix*. Let X be a standardized data matrix of n observations on p variables, i.e. $X^\top 1_{n \times 1} = 0_{p \times 1}$ and $X^\top X$ is the sample correlation matrix. Note that for some large data only centring can be a plausible option. Thus, the rank of X is at most $\min\{n, p\} - 1$. The r-factor EFA fundamental equation (5.5) can be formulated as a specific matrix decomposition/factorization of X in the following form:

$$X \approx F\Lambda^\top + U\Psi = [F\ U][\Lambda\ \Psi]^\top , \qquad (6.69)$$

where the parameters F, Λ, U and Ψ (diagonal) are unknown matrices with sizes $n \times r$, $p \times r$, $n \times p$ and $p \times p$, respectively. The fundamental EFA equation (6.69) means that EFA presents the data X as a linear combination of common and unique factors, F and U. The corresponding weights Λ are called factor loadings, and Ψ – uniquenesses, and Ψ^2 contains the variances of U. Usually, the number of common factors $r(\ll \min\{n, p\})$ is unknown and is specified before the analysis.

We stress that F and U are *not random* variables, but fixed unknown ma-

trices. The classical EFA assumptions that the common and unique factors
are uncorrelated and are mutually uncorrelated (Mulaik, 2010, Ch6.2) are
adapted to the new EFA formulation (6.69) by simply requiring $F^\top F = I_r, U^\top U = I_p$ and $F^\top U = 0_{r \times p}$, respectively. Making use of them, we find
from (6.69) that the sample correlation matrix $X^\top X$ is modelled as

$$X^\top X \approx C = \Lambda\Lambda^\top + \Psi^2 . \tag{6.70}$$

The attractiveness of the new EFA approach is that *all* unknown parameters F, Λ, U and Ψ can be found simultaneously without any distributional
assumptions. Indeed, the classic EFA solves (6.70), and thus, finds Λ and Ψ
only. Additional assumptions and efforts are needed to find F and U (Unkel
and Trendafilov, 2010). In this sense, (6.69) is a EFA model that permits
to fit data, rather than correlations. Another important benefit from this
approach is that F and U are found by orthonormal PA which boils down
to computing SVD. This overcomes the algorithmic weakness of the classical
EFA depending on iterative procedures and makes it also applicable to large
data.

6.7.1   The classical case $n > p$

The new EFA formulation (6.69) suggests that the unknown EFA parameters
can be found by solving the following constraint least squares (LS) problem:

$$\min_{F,\Lambda,U,\Psi} \|X - (F\Lambda^\top + U\Psi)\|^2 = \min_{F,\Lambda,U,\Psi} \|X - [F\ U][\Lambda\ \Psi]^\top\|^2, \tag{6.71}$$

subject to $F^\top F = I_r, U^\top U = I_p, F^\top U = O_{r \times p}$ and Ψ diagonal.

Problem (6.71) is solved by alternating minimization over one unknown and
keeping the rest ones fixed. The algorithm is called GEFALS.

First, consider minimizing (6.71) with respect to the diagonal matrix Ψ.
The optimal Ψ is necessarily given by $\Psi = \text{diag}(X^\top U)$. In a similar way, one
can establish that the optimal Λ (F, U and Ψ fixed) is given by $X^\top F$. The
rotational indeterminacy of Λ can be avoided by taking the lower triangular
part of $X^\top F$ (Trendafilov and Unkel, 2011).

Finally, we minimize (6.71) over $[F, U]$, with Λ and Ψ kept fixed (known).
Let form an $n \times (r + p)$ block-matrix $Z = [F, U]$ and a $p \times (r + p)$ block-

matrix $W = [\Lambda, \Psi]$. Straightforward calculations and the EFA constraints show that

$$Z^\top Z = \begin{bmatrix} F^\top F & F^\top U \\ U^\top F & U^\top U \end{bmatrix} = \begin{bmatrix} I_r & O_{r \times p} \\ O_{p \times r} & I_p \end{bmatrix} = I_{r+p} . \tag{6.72}$$

Then, the original EFA problem (6.71) reduces to a standard orthonormal PA (6.2.4) for known X and $W(= [\Lambda, \Psi])$:

$$\min_{Z \in \mathcal{O}(n, r+p)} \|X - ZW^\top\|^2 . \tag{6.73}$$

GEFALS finds (Λ, Ψ) and Z in an alternating manner. To find Z one needs SVD of XW, which is always rank deficient (see Exercise 5.11). Thus, the orthonormal Procrustes problems (6.73) have no unique orthonormal solution Z. This phenomenon is notorious as the *factor indeterminacy*. The above considerations quantify its origin, and thus, clarifies the existing rather philosophical explanations (Mulaik, 2010, Ch13).

6.7.2 The modern case $p \gg n$

The modern applications frequently require analysing data with more (and usually much more) variables than observations ($p \gg n$). Such data format causes considerable problems in many classic multivariate techniques, including EFA. Indeed, the classic EFA problem ($n > p$) is to fit a hypothetical correlation structure of the form (6.70) to the sample correlation matrix $X^\top X$ (Mulaik, 2010, 6.2.2). However, $X^\top X$ is *singular* when $p \gg n$. Of course, one can still fit C to $X^\top X$, but this will differ from the original EFA problem. To see this, recall that the EFA model is originally defined by (6.69) and the assumptions for the involved unknowns, while (6.70) is their derivative. Specifically, when $p \gg n$, the classic constraint $U^\top U = I_p$ *cannot* be fulfilled any more. Thus, the classic EFA correlation structure (6.70) turns into

$$C = \Lambda \Lambda^\top + \Psi U^\top U \Psi . \tag{6.74}$$

The new correlation structure (6.74) coincides with the classic one (6.70), if the more general constraint $U^\top U \Psi = \Psi$ is introduced in place of $U^\top U = I_p$. In other words, an universal EFA definition, valid for any n and p, should assume the constraint $U^\top U \Psi = \Psi$. An important consequence from this new assumption is that Ψ^2 cannot be p.d., when $p \gg n$ (Trendafilov and Unkel, 2011, Lemma 1).

The rest of the classic EFA constraints $F^\top F = I_m$ and $U^\top F = O_{p \times m}$ remain valid. Thus, for $p \gg n$, EFA requires solution of the following constraint LS problem:

$$\min_{F, \Lambda, U, \Psi} \|X - (F\Lambda^\top + U\Psi)\|^2 , \qquad (6.75)$$

subject to $F^\top F = I_m, U^\top U \Psi = \Psi, F^\top U = O_{m \times p}$ and Ψ diagonal. This problem is called generalized EFA (GEFA).

The updating formulas for Λ and Ψ used in the previous Section 7.4.7 are well know in the classical EFA. They are easily obtained by left-hand side multiplication of the fundamental EFA equation (6.69) with $F^\top$ and $U^\top$, respectively:

$$F^\top X \approx F^\top F\Lambda^\top + F^\top U\Psi = \Lambda^\top \qquad (6.76)$$
$$U^\top X \approx U^\top F\Lambda^\top + U^\top U\Psi = \Psi . \qquad (6.77)$$

As Ψ is assumed diagonal, (6.77) becomes $\Psi \approx \operatorname{diag}(U^\top X)$. Simple calculations show that the first-order necessary conditions for minimum of (6.71) with respect to Λ and Ψ are $\Lambda = X^\top F$ and $\Psi = \operatorname{diag}(U^\top X)$.

For $p \gg n$, we have $U^\top U \neq I_p$, and the left-hand side multiplication of (6.69) with $U^\top$ leads to an expression different from (6.77) for the classic EFA with $n > p$:

$$U^\top X \approx U^\top F\Lambda^\top + U^\top U\Psi = U^\top U\Psi . \qquad (6.78)$$

Also, one can easily find that in this case the first-order necessary condition for minimum of (6.71) with respect to Ψ becomes $U^\top U\Psi = U^\top X$, instead of the classical condition $\Psi = U^\top X$ in (6.77).

The requirement for diagonal Ψ, and the fact that $U^\top U$ contains an identity submatrix and zeros elsewhere imply that the new optimality condition $U^\top U\Psi = U^\top X$ can also be rewritten, respectively, as $U^\top U\Psi = \operatorname{diag}(U^\top X)$ as well as $U^\top U\Psi = U^\top U\operatorname{diag}(U^\top X)$ (Trendafilov and Fontanella, 2019).

Thus, the new EFA constraints, requiring diagonal Ψ and $\Psi = U^\top U\Psi$, bring us to *two* possible updating formulas, $\Psi = \operatorname{diag}(U^\top U\Psi)$ and $\Psi = \operatorname{diag}(U^\top X)$. They are equivalent for analytic purposes. However, for finite precision arithmetic $U^\top U \Psi$ and $U^\top X$ are usually equal only approximately. In general, calculating $U^\top U \Psi$ is (much) faster than calculating $U^\top X$. In this sense, working with $\Psi = U^\top U \Psi$ as an updating formula is preferable when analysing large data.

Considering Λ and Ψ fixed, write the GEFA problem (6.75) for the case $p \gg n$ as

$$\min_{\substack{F^\top F = I_m \\ U^\top F = O_{p \times m}}} \|X - F\Lambda^\top - U\Psi\|^2 , \tag{6.79}$$

keeping in mind the new constraint $U^\top U \Psi = \Psi$. Problem (6.79) can be solved by solving the following two problems alternatively: for fixed U,

$$\min_{F^\top F = I_m} \|(X - U\Psi) - F\Lambda^\top\|^2 , \tag{6.80}$$

and for fixed F,

$$\min_{\substack{U^\top U \Psi = \Psi \\ U^\top F = O_{p \times m}}} \|(X - F\Lambda^\top) - U\Psi\|^2 . \tag{6.81}$$

Problem (6.80) is a standard orthonormal Procrustes problem. The solution is $F = VW^\top$, where $VDW^\top$ is SVD of $(X - U\Psi)\Lambda$. However, problem (6.81) needs more attention. Eventually it can also be solved as a Procrustes problem (Trendafilov and Fontanella, 2019).

6.7.3 EFA and RPCA when $p \gg n$

In Section 5.7 we briefly mentioned methods for RPCA and how they can be utilized for EFA of data with classical format of more observations than variables ($n > p$). As they can be applied to any format data, they can also be used for EFA of data with more variables than observations ($p \gg n$). In particular, the procedure RobustPCA from http://dlaptev.org/#code, adapted in Section 5.7 to produce diagonal sparse part RPCAsq, can be readily used for EFA problems with singular correlation matrix ($p \gg n$).

The general procedure RobustPCA can work directly with the data matrix, however there is no reasonable way to compare the results with GEFAN in terms of fit and quality of the estimated parameters.

6.7.4   Semi-sparse PCA (well-defined EFA)

The classical EFA finds only Λ and Ψ by fitting the model $\Lambda\Lambda^\top + \Psi^2$ to $X^\top X$ with respect to some goodness-of-fit criterion, e.g. least squares or maximum likelihood. As Λ and Ψ have $pr+p$ unknown parameters and $X^\top X$ has $p(p+1)/2$ (or $p(p-1)/2$) informative entries, the EFA problem becomes reasonable only for specific values of r. The problem that many possible F exist for certain X, Λ and Ψ and is known as the *factor indeterminacy*. Its detailed analysis shows that the classical EFA model does not make sense for $n > p + r$ (Eldén and Trendafilov, 2019, Th. 2.2.):

- if $n = p + r$, the EFA problem (6.73) with diagonal Ψ, determines U and the subspace spanned by the columns of F;

- if $n > p + r$, U is determined but F is undetermined. At the optimum the parameter Λ always has the value 0.

To see this, consider the QR decomposition of the $n \times p$ data matrix X:

$$X = Q \begin{bmatrix} R \\ O \end{bmatrix} = QY,$$

where $Q \in \mathcal{O}(n)$ is orthogonal and $R \in \mathcal{M}(p)$ is upper triangular and non-singular (due to the full column rank of X). Note, that the dimensions of the zero matrices are omitted in this section to simplify the notations. Thus, we have then, the original EFA problem transforms into a new EFA problem for a $p \times p$ data matrix Z:

$$\min_{F,U,\Lambda,\Psi} \|Y - [FU][\Lambda\ \Psi]^\top\|^2 ,$$

where $F := Q^\top F, U := Q^\top U$ and $[F\ U]^\top[F\ U] = I_{r+p}$. Enlarge $[F\ U]$, such that $[F\ U\ W] \in \mathcal{O}(n)$ is an orthogonal matrix. Then

$$\begin{aligned} \Delta &= \|Y - [F\ U][\Lambda\ \Psi]^\top\|^2 = \left\| \begin{bmatrix} F^\top \\ U^\top \\ W^\top \end{bmatrix} Y - \begin{bmatrix} I_r & O \\ O & I_p \\ O & O \end{bmatrix} [\Lambda\ \Psi]^\top \right\|^2 \\ &= \|F^\top Y - \Lambda^\top\|^2 + \|U^\top Y - \Psi\|^2 + \|W^\top Y\|^2 . \end{aligned} \quad (6.82)$$

It can be proven that if U is the maximizer of the second term of (6.82), then $U_1 \in \mathcal{O}(p)$ and $U_2 = O_{(n-p)\times p}$ (Eldén and Trendafilov, 2019, Lemma

2.1.). Then, we can show that at the minimizer of the second term, both the first and the third term in Δ can be made equal to zero. Thus, we get the overall minimum of Δ by minimizing the second term alone.

Now, let first $r + p = n$. Then Δ becomes

$$\Delta = \|F^\top Y - \Lambda^\top\|^2 + \|U^\top Y - \Psi\|^2,$$

and the first term can always be made equal to zero due to the fact that Λ is unconstrained. However, we will see below that its value is equal to zero at the minimum. The matrix $[F\ U]$ is square and orthogonal and at the minimum for the second term it has the structure:

$$[F\ U] = \begin{bmatrix} O & U_1 \\ F_2 & O \end{bmatrix},$$

thus the subspace spanned by F is determined.

Next, consider the case $n > m + p$, where the objective function is

$$\Delta = \|F^\top Y - \Lambda^\top\|^2 + \|U^\top Y - \Psi\|^2 + \|W^\top Z\|^2.$$

Again, as Λ is unconstrained, the first term can always be made equal to zero. At the minimum for the second term, the orthogonal matrix $[F\ U\ W]$ must have the structure

$$[F\ U\ W] = \begin{bmatrix} O & U_1 & O \\ F_2 & O & W_2 \end{bmatrix}$$

where $F_2^\top W_2 = O$. Therefore, the third term is

$$\|W^\top Y\|^2 = [O\ W_2^\top] \begin{bmatrix} R \\ O \end{bmatrix} = 0.$$

Thus, at the minimum of Δ neither F_2 nor W_2 are determined by the model. In both cases Λ is equal to zero, since $F^\top Y = O$ at the optimum.

For years, the authors are looking for ways to extract well-defined factors by requiring them to possibly satisfy additional utility features. Instead, Eldén and Trendafilov (2019) propose an alternative EFA model:

$$X \approx F\Lambda^\top + U\Psi^\top, \tag{6.83}$$

subject to $[F\ U]^\top [F\ U] = I_{r+k}$ for some k such that $r + k < \min\{n, p\}$. The involved parameters in (6.83) have different sizes from those in (6.69): $F \in \mathcal{O}(n, r), \Lambda \in \mathcal{M}(p, r), U \in \mathcal{O}(n, k), \Psi \in \mathcal{M}(p, k)$. Note, that Ψ is no longer required to be diagonal. Instead, Ψ is required to be *sparse* rectangular:

- with *only one* non-zero element $\psi_i, 1 \leq i \leq p$ in each row,

- but possibly with more than one non-zero elements in its columns.

These assumptions imply that $\Psi^\top \Psi$ is diagonal, but $\Psi\Psi^\top$ cannot be diagonal. As a consequence, the "unique" factors in U are allowed to contribute to more than one observable variable in X. In this sense, U is composed of shared factors, rather than unique factors as in the classical EFA (6.69). In addition, as k is the rank of the factor $U\Psi^\top$, it can be seen as a parameter that determines the degree of dimension reduction or data compression of the model together with r.

The new block-matrix approximation $X \approx [F\ U][\Lambda\ \Psi]^\top$ is a PCA-like factorization of X, with a loadings block-matrix composed of dense and sparse parts: Λ and Ψ, respectively. For this reason, it is called for short the *semi-sparse* PCA (SSPCA).

As its classical EFA analogue (6.70), the first term $\Lambda\Lambda^\top$ of the new decomposition $X^\top X \approx \Lambda\Lambda^\top + \Psi\Psi^\top$ takes care for the low-rank approximation of $X^\top X$. The second term $\Psi\Psi^\top$ provides sparse adjustment of the low-rank part, in order to improve the overall fit to $X^\top X$. The two parts of the new matrix factorization are orthogonal (uncorrelated) as in the classical EFA. They are spanned by vectors taken from orthogonal subspaces. In this sense, they should be clearly distinguished, in the same way "common" and "unique" factors are in the classical EFA. However, it is important to recognize that the unique factors are "unique" because Ψ is assumed diagonal, not because they come from subspace orthogonal to the one of the common factors. The purpose of the two orthogonal subspaces is to deliver a model as a sum of two terms, as in the classical EFA.

The SSPCA algorithm works as follows (Eldén and Trendafilov, 2019, 3.2). It starts with finding F and Λ. Let $s(\geq r)$ be the rank of the data matrix $X \in \mathcal{M}(n,p)$. Then, its SVD can be written as $X = Q\Sigma V^\top$, where $Q \in \mathcal{O}(n,s), \Sigma \in \mathcal{D}(s)$, and $V \in \mathcal{O}(p,s)$. Denote $Y = \Sigma V^\top \in \mathcal{M}(s,p)$. Since the column vectors of X are in the subspace spanned by the columns of Q, we also require that for the original $[\tilde{F}\ \tilde{U}] \in \mathcal{M}(n, r+k)$ be in that subspace. Then, for some $F \in \mathcal{M}(s,r),\ U \in \mathcal{M}(s,k)$ defined by $[F\ U] := Q[F\ U]$, the SSPCA problem (6.83) requires the minimization of $\|Y - [F\ U][\Lambda\ \Psi]^\top\|$, subject to the new F, U, Λ and Ψ. Thus, we choose $k \leq s-r$ (from $s \geq r+k$), for the rank of the product $U\Psi^\top$.

Further on, partition

$$\Sigma = \begin{bmatrix} \Sigma_1 & O \\ O & \Sigma_2 \end{bmatrix} \ , \ V = \begin{bmatrix} V_1 & V_2 \end{bmatrix}, \ \Sigma_1 \in \mathcal{D}(r), \ V_1 \in \mathcal{M}(p, r) \ ,$$

and let

$$Y = \begin{bmatrix} Y_1 \\ Y_2 \end{bmatrix} := \begin{bmatrix} \Sigma_1 V_1^\top \\ \Sigma_2 V_2^\top \end{bmatrix}.$$

We choose F and $\Lambda^\top$ so that $F\Lambda^\top$ is the best rank-r approximation of Y. Due to the definition of Y we have

$$F = \begin{bmatrix} I_r \\ O \end{bmatrix}, \ \Lambda^\top = Y_1 = \Sigma_1 V_1^\top \ ,$$

which implies *uniqueness* of F and Λ in the new EFA model.

Next, the SSPCA algorithm finds U and Ψ. After the elimination of F and Λ, the residual becomes

$$\left\| \begin{bmatrix} O \\ Y_2 \end{bmatrix} - U\Psi^\top \right\|.$$

The requirement $F^\top U = O$ implies $U = \begin{bmatrix} O \\ U_2 \end{bmatrix}$, where $U_2 \in \mathcal{M}(s - r, k)$. As the columns of U are to be orthogonal and normalized, the same applies to U_2. Thus we arrive at the minimization problem,

$$\min_{U_2, \Psi} \| Y_2 - U_2 \Psi^\top \| \ ,$$

subject to $U_2^\top U_2 = I_k$, and Ψ having the prescribed structure.

This problem contains two unknowns and is solved by alternating over U_2 and Ψ. If Ψ is fixed, we have an ordinary orthonormal Procrustes problem (OPA) considered in Section 6.2.4. Its solution is given by the SVD of $Y_2^\top \Psi \in \mathcal{M}(s - r, k)$ (Golub and Van Loan, 2013, Chapter 12).

If U_2 is given/found, we need to update Ψ, i.e. to find the optimal locations of the single non-zero element $\psi_i, 1 \le i \le p$ in every row of Ψ. This minimization problem corresponds to modelling each column of Y_2 as

$$Y_{2,j} \approx \psi_j U_{2,c(j)} \ , \ 1 \le j \le p \ , \tag{6.84}$$

where $c(j)$ is a function that chooses one of the column vectors of U_2:

$$c : \{1, \ldots, p\} \to \{1, \ldots, k\} \ ,$$

and thus,

$$\Psi^\top = [\psi_1 e_{c(1)}, \ldots, \psi_p e_{c(p)}] \ , \ \text{for } e_i \in \mathbb{R}^k \ .$$

We start the computations for determining U_2 and Ψ with k columns in the two matrices. Due to the fact that we allow several vectors from Y_2, say $Y_{2,i}$ and $Y_{2,j}$, to share the same column from U_2, say $U_{2,l}$, i.e. $c(i) = c(j) = l$, it is quite likely that the algorithm finishes with zero columns in Ψ, meaning that some of the $U_{2,l}$'s does not occur in any model (6.84). Such $U_{2,l}$'s are called passive, and those that do occur in the model are called active. The number of active columns may vary during the optimization procedure. When the optimal solution is found, we remove the passive columns in U_2, and delete the corresponding columns of Ψ. This further reduces the effective size k and the rank of $U\Psi^\top$.

The update of the function $c(i)$ and Ψ, given U_2, makes use of the observation that for each column y_j of Y_2 and Ψ, the solution of the LS problem

$$\min_{\psi_j} \|y_j - \psi_j u_l\|, \ 1 \le l \le k \ ,$$

is $\psi_j = y_j^\top u_l$, and the residual is $\|(I_k - u_l u_l^\top) y_j\|$. If the smallest residual is obtained for $l = i$, then $c(j) := i$.

The SSPCA algorithm can be initiated with a random start. However, a good rational start is preferable. For this reason, we note that $Y_2 = \Sigma_2 V_2^\top = I_{s-r} \Sigma_2 V_2^\top$, i.e. the left singular vectors of Y_2 are the canonical unit vectors $e_i \in \mathbb{R}^{s-r}$. Therefore, the identity matrix $I_{s-r,k}$ is a natural choice for U_2.

The SSPCA decomposition is surprisingly resourceful and versatile. With $r = 0$, it can perform sparse PCA and provide additionally the number of useful sparse PC for the particular data. The best existing sparse PCA techniques are generally faster, but lack this feature completely and work with prescribed number of PCs only. SSPCA can also be very successful as a low-rank-plus-sparse matrix decomposition of large correlation matrices compared to a number of existing methods. Most of them target exact fit to the data, which makes them vulnerable to either poor low rank or not sparse enough parts. A recent approach is able to overcome this weakness (Aravkin et al., 2014), but SSPCA can achieve better fit to the data.

6.8 Conclusion

Section 6.2 considers several types of Procrustes problems defined on $\mathcal{O}(p)$ and $\mathcal{O}(p, r)$ and derives their first and second-order optimality conditions, some of which are new in the field of PA. The introduced projected Hessian approach presents an alternative way to study second-order optimality conditions for constraint problems on matrix manifolds.

Section 6.3 briefly lists the corresponding Procrustes problems defined on $\mathcal{OB}(p)$ and $\mathcal{OB}(p, r)$, respectively. With the adopted formalism of optimization on matrix manifolds their solution simply needs a replacement of the constraint manifolds in Manopt from Section 6.2.

Section 6.4 is dedicated to robust versions of the PA problems considered in Section 6.2 and Section 6.3. The LS fit is replaced by either ℓ_1 matrix norm or by Huber M-estimator function.

Section 6.5 considers a class of more complicated PA problems involving simultaneous fit to several data matrices. Such problems naturally arise in multi-way or tensor data analysis discussed further in Section 10.8.

Section 6.6 is left for several more specialized PA problems. Section 6.6.1 considers the so-called *generalized* PA, which boils down to finding the average of several rotated matrices. Section 6.6.2 gives a brief introduction to the problem of mean and median rotation of several rotations in $\mathcal{SO}(3)$.

Section 6.7 is dedicated to the modern treatment of EFA. First, the classical EFA is reformulated as a PA problem in Section 6.7.1. Then, the PA definition of EFA is generalized in Section 6.7.2 for modern data format with more variables than observations. Finally, Section 6.7.4 briefly reminds the factor indeterminacy of the EFA model. The proposed remedy is an alternative extended EFA model compatible to several modern data analysis methods and likely to be more appropriate than the classical EFA model in a number of contemporary applications.

6.9 Exercises

1. The Euclidean gradient of the objective function in (6.26) is given by the matrix $A^{\top}(AQ - B)$. Show that the first-order optimality condition

for $Q \in \mathcal{OB}(p,q)$ to be minimum of (6.26) is

$$A^{\top}(AQ - B) = Q\mathrm{diag}[Q^{\top}A^{\top}(AQ - B)] \ .$$

This optimality condition is even more interesting if written as

$$Q^{\top}A^{\top}(AQ - B) = (Q^{\top}Q)\mathrm{diag}[Q^{\top}A^{\top}(AQ - B)] \ .$$

2. Show that the Euclidean gradient of the objective function in (6.27) is
   $Q^{-\top}(AQ^{-\top} - B)^{\top}AQ^{-\top}$.

3. Every $n \times p$ matrix X is expressed as $X = \sum_{1}^{p} x_j e_j^{\top}$, where x_j and e_j
   denote the jth columns of X and I_p. Use it to prove (6.47).

4. Prove that the differential of the second term in (6.49) is given by

$$w \sum_{1=1}^{m} c^{\top} Q_1^{-\top} \otimes \cdots \otimes Q_i^{-\top} dQ_i^{\top} Q_i^{-\top} \otimes \cdots \otimes Q_m^{-\top} c_t -$$

$$w \sum_{1=1}^{m} c^{\top} Q_1^{-\top} Q_1^{-1} \otimes \cdots \otimes Q_i^{-\top} dQ_i^{\top} Q_i^{-\top} Q_i^{-1} \otimes \cdots \otimes Q_m^{-\top} Q_m^{-1} c \ ,$$

making use of the identity:

$$(A_1 \otimes B_1)(A_2 \otimes B_2) = A_1 A_2 \otimes B_1 B_2 \ , \tag{6.85}$$

provided that the products $A_1 A_2$ and $B_1 B_2$ exist.

5. If A and B are two matrices, then show that $\|A \otimes B\|_F = \|A\|_F \|B\|_F$.

6. Check that the gradient of the approximation (6.30) to the right-hand
   side of (6.29) is given by the following $p \times p$ matrix $\nabla = A^{\top}YC^{\top}$,
   where $Y = \tanh(X) + X \odot [1_p 1_p^{\top} - \tanh^2(X)]$ and $X = \gamma(AQC - B)$.

7. Check that the approximation to the objective function in (6.32) is
   given by the following $p \times p$ matrix: $\nabla = -Q^{-\top}CY^{\top}AQ^{-\top}$, where
   $Y = \tanh(X) + X \odot [E_p - \tanh^2(X)]$ and $X = \gamma(AQ^{-\top}C - B)$.

8. Consider a robust version of (6.52), by solving:

$$\min_{Q_i} \sum_{1 \leq i < j \leq m} \|A_i Q_i - A_j Q_j\|_{\ell_1} \ ,$$

where $Q_i \in \mathcal{O}(p_i, r)$ or $Q_i \in \mathcal{OB}(p_i, r)$ for some $r \leq \min\{p_1, \ldots, p_m\}$.

9. Find the rank of XW defined in Section 6.7.1.

10. Prove that the constraints $F^{\top}F = I_r$ and $U^{\top}F = O_{p\times r}$ of the EFA model (6.71), imply $\mathrm{rank}(U) \leq n - r$.

11. For the GEFA model from Section 6.7.2 ($p \gg n$), prove that the new constraint $U^{\top}U\Psi = \Psi$ implies that Ψ^2 is positive semi-definite, i.e. contains zero diagonal entries. Hint: Assume the opposite, $\Psi^2 > 0$, and find a contradiction.

12. For the GEFA model from Section 6.7.2, prove that if Ψ have m zero diagonal entries, then $\mathrm{rank}(\Psi) = p - m \leq n - r$. Hint: Write $U = [U_1\ O_{n\times(p-n+r)}]$, and respectively, rewrite Ψ as a block-matrix too.

13. The orthonormal Procrustes problem (6.73) have no unique orthonormal solution, because it relies on SVD of a rank deficient matrix.

    (a) Making use of the "economy" SVD (4.9) of an $m \times n$ ($m \geq n$) matrix X of rank r ($\leq n$), prove that any $Q = U_1 V_1^{\top} + U_2 P V_2^{\top}$ is solution of
    $$\max_{Q^{\top}Q=I_n} \mathrm{trace}Q^{\top}X = \mathrm{trace}D \ ,$$
    where $P \in \mathcal{O}(n-r)$ is arbitrary. Hint: You may use the optimality conditions of (6.23) (Trendafilov and Unkel, 2011, Appendix).

    (b) Take advantage of this factor indeterminacy to create an EFA-based procedure for clustering factor scores, thus, combining dimension reduction with *clustering* (Uno et al., 2019).

14. Check that EFA of data matrix $X \in \mathcal{M}(n, p)$ ($n > p + r$) can be performed by solving:
    $$\min_{[F\ U]\in\mathcal{O}(n,p+r)} \|(I_n - FF^{\top})X - U\mathrm{diag}(U^{\top}X)\|_F \ ,$$
    which is the EFA analogue of (4.8) and turns EFA into a projection method of the data X onto the subspace spanned by the columns of $[F\ U]$, the common and unique factors. Solve this modified EFA problem, possibly by producing Manopt codes.

15. (Artin, 2011, 5.1.27) Check that, left multiplication by
    $$Q = \begin{bmatrix} \cos\theta_1 & -\sin\theta_1 & & \\ \sin\theta_1 & \cos\theta_1 & & \\ & & \cos\theta_2 & -\sin\theta_2 \\ & & \sin\theta_2 & \cos\theta_2 \end{bmatrix} \in \mathcal{SO}(4)$$

is the composition of a rotation through the angle θ_1 on the first two coordinates and a rotation through the angle θ_2 on the last two. Such an operation can *not* be realized as a single rotation.

16. (Curtis, 1984, Ch 4) Let $\mathcal{U} = \{A : \|I_n - A\| < 1\}$ and $\mathcal{V}$ be a neighbourhood of 0_n, such that $\exp(\mathcal{V}) \subset \mathcal{U}$. Prove that

    (a) $\exp(\log(A)) = A$, if $A \in \mathcal{U}$;

    (b) $\log(\exp(B)) = B$, if $B \in \mathcal{V}$;

    (c) $\log(AB) = \log(A) + \log(B)$, if $A, B \in \mathcal{U}$ and $\log(A)$ and $\log(B)$ commute. So, if $A \in \mathcal{O}(n) \cap \mathcal{U}$, then $\log(A) \in \mathcal{K}(n)$.

17. (Hall, 2015, Pr 2.9) For all $A \in \mathcal{M}(n)$ with $\|A\| < \frac{1}{2}$, there exists a constant α, such that $\|\log(I_n + A) - A\| \leq \alpha \|A\|^2$.

18. (Bhatia et al., 2019) Let $A, B \in \mathcal{S}_+(n)$ be any two PD matrices.

    (a) Show that

    $$\min_{Q \in \mathcal{O}(n)} \|A^{1/2} - B^{1/2}Q\|_F = d(A, B) ,$$

    where $d^2(A, B) = \text{trace}(A) + \text{trace}(B) - 2\text{trace}[(A^{1/2}BA^{1/2})^{1/2}]$. Hint: consider the SVD of $A^{1/2}B^{1/2} = U\Sigma V^\top = (U\Sigma U^\top)(UV^\top)$.

    (b) Making use of its Procrustes definition, check that $d(A, B)$ defines a metric on $\mathcal{S}_+(n)$, the *Bures-Wasserstein* distance/metric.

Chapter 7

Linear discriminant analysis (LDA)

© Springer Nature Switzerland AG 2021

N. Trendafilov and M. Gallo, *Multivariate Data Analysis on Matrix Manifolds*,

Springer Series in the Data Sciences, https://doi.org/10.1007/978-3-030-76974-1_7

7.1   Introduction

Discriminant analysis (DA) is a descriptive multivariate technique for analyzing grouped data, i.e. the rows of the data matrix are divided into a number of groups that usually represent samples from different populations (Krzanowski, 2003; McLachlan, 2004; Seber, 2004). Recently DA has also been viewed as a promising dimensionality reduction technique (Dhillon et al., 2002; Hastie et al., 2009). Indeed, the presence of group structure in the data additionally facilitates the dimensionality reduction.

The data for DA have a bit different format from the data analysed in the previous sections. As before, n observations are made on p variables which are collected in a data matrix X. The difference is that these n observations are *a-priori* divided into g groups. Thus, if n_i is the number of individuals in the ith group, then we have $n_1 + n_2 + \ldots + n_g = n$. In the classical scenario it is assumed that $n > p$. Usually, the group membership of the individuals is expressed as a $n \times g$ indicator matrix G with $\{0, 1\}$ elements. Each column of G corresponds to one group and has 1 at its ith position if the corresponding ith observation of X belongs to the particular group. Clearly, we have:

$$1_n^\top G = (n_1, \ldots, n_g) \ , \ 1_n^\top G 1_n = n \ , \ G^\top G = N \text{ and } 1_g^\top N 1_g = n \ ,$$

where N is a $g \times g$ diagonal matrix with $(n_1, \ldots, n_g)$ as main diagonal.

The following between-groups and within-groups scatter matrices B and W play very important role in LDA, and are defined as

$$W = \sum_{i=1}^{g} \sum_{j=1}^{n_i} (x_{ij} - \bar{x}_i)(x_{ij} - \bar{x}_i)^\top \ , \tag{7.1}$$

$$B = \sum_{i=1}^{g} n_i (\bar{x}_i - \bar{x})(\bar{x}_i - \bar{x})^\top \ , \tag{7.2}$$

where the within-group means and total mean are

$$\bar{x}_i = \frac{1}{n_i} \sum_{j=1}^{n_i} x_{ij} \text{ and } \bar{x} = \frac{1}{n} \sum_{i=1}^{g} n_i \bar{x}_i \ , \tag{7.3}$$

and x_{ij} denotes a *vector* of the measurements made on the jth observation belonging to the ith group. The expressions (7.1) and (7.2) for W and B

considerably simplify, if the data matrix X is centred, i.e. $\bar{x} = 0_{p \times 1}$. In reality, one usually analyses standardized data, i.e. where the variables are centred and with unit variances, which simplifies the following expressions.

The expressions (7.1) and (7.2) for W and B can be translated in matrix notations. They are usually easier to work with in theoretical derivations, but may be expensive to use for large data. First, we note that the matrix of group means is given by $\bar{X} = N^{-1}G^{\top}X$. It is well known that the total scatter matrix is given by

$$T = X^{\top} J_n X \; , \tag{7.4}$$

where $J_n = I_n - 1_n 1_n^{\top}/n$ is the centring operator. It is convenient to introduce the notation $X_T = J_n X$, because T becomes Gramian matrix $T = X_T^{\top} X_T$. For centred X, (7.4) simplifies to $T = X^{\top}X$.

The between-group scatter matrix B from (7.2) can be readily written in matrix form as follows:

$$
\begin{aligned}
B &= \left(\bar{X} - \frac{E_{g \times n} X}{n} \right)^{\top} N \left(\bar{X} - \frac{E_{g \times n} X}{n} \right) \tag{7.5} \\
&= X^{\top} \left(G N^{-1} - \frac{E_{n \times g}}{n} \right) N \left(N^{-1}G^{\top} - \frac{E_{g \times n}}{n} \right) X \\
&= X^{\top} \left(G N^{-1} G^{\top} - \frac{E_n}{n} \right) X \\
&= X^{\top} G \left(N^{-1} - \frac{E_g}{n} \right) G^{\top} X \; .
\end{aligned}
$$

Clearly, $B = X_B^{\top} X_B$ for $X_B = N^{1/2} \left(\bar{X} - \frac{E_{g \times n} X}{n} \right)$, i.e. B is Gramian. For centred X, (7.5) simplifies to $B = X^{\top} G N^{-1} G^{\top} X = \bar{X}^{\top} N \bar{X}$.

Finally, the within-group scatter matrix W is simply given by

$$
\begin{aligned}
W &= T - B = X^{\top} J_n X - X^{\top} G \left(N^{-1} - \frac{E_g}{n} \right) G^{\top} X \\
&= X^{\top} \left(I_n - \frac{E_n}{n} - G N^{-1} G^{\top} + \frac{G E_g G^{\top}}{n} \right) X \\
&= X^{\top} \left(I_n - G N^{-1} G^{\top} \right) X \; . \tag{7.6}
\end{aligned}
$$

One notes that $I_n - G N^{-1} G^{\top}$ is idempotent, which makes W Gramian by introducing $X_W = (I_n - GN^{-1}G^\top)X = X - G\bar{X}$. Then, we have

$$W = X_W^\top X_W = (X - G\bar{X})^\top (X - G\bar{X}) \, , \qquad (7.7)$$

which is the matrix equivalent of (7.1).

The best known variety of DA is the linear discriminant analysis (LDA), whose central goal is to describe the differences between the groups in terms of discriminant functions (DF), also called canonical variates (CV), defined as linear combinations of the original variables. In this section we mainly deal with a LDA version initially introduced by Fisher (1936) for the case of two groups ($g = 2$). The same (linear) CVs can be obtained if the different populations are assumed to be multivariate Gaussian with a common covariance matrix and the probabilities of misclassification are minimized (Hastie et al., 2009; McLachlan, 2004). Under the same assumptions, CVs appear in one-way MANOVA for best separation of the group means (Rencher, 2002). For example, Fisher's LDA is performed by MATLAB (MATLAB, 2019) using the function for one-way MANOVA. The mathematical equivalent of Fisher's LDA is the *generalized* eigenvalue problem (GEVD) (Golub and Van Loan, 2013, §8.7).

The interpretation of the CVs is based on the coefficients of the original variables in the linear combinations. The problem is similar to the interpretation of PCs discussed in Section4.4: the interpretation can be clear and obvious if there are only few large coefficients and the rest are all close to or exactly zero. Unfortunately, in many applications this is not the case. The interpretation problem is exacerbated by the fact that there are three types of coefficient, raw, standardized and structure, which can be used to describe the canonical variates (Pedhazur, 1982; Rencher, 1992, 2002), where the disadvantages for their interpretation are also discussed. A modification of LDA, aiming at better discrimination and possibly interpretation, is considered in (Duchene and Leclercq, 1988; Krzanowski, 1995). In this approach the vectors of the CV coefficients are constrained orthonormal.

These difficulties are similar to those encountered in Chapter 4 when interpreting PCA. The same interpretation philosophy adopted in sparse PCA can be adapted for use in LDA. This is realized by Trendafilov and Jolliffe (2007) to obtain sparse CVs. The non-zero entries correspond to the variables that dominate the discrimination.

The chapter is divided into two main parts considering separately data with $n > p$ and $p > n$. The former type of data is the well known one from the standard statistics. Here, it will be referred to as *vertical* data. The latter type of data is common for modern applications and will be referred to as *horizontal* data. The classical MDA techniques are developed for vertical data. However, most of them *cannot* be applied, at least, directly to horizontal data.

The classical LDA (of vertical data) and its interpretation are briefly revised in Section 7.2. Section 7.2.2 described a modification of LDA in which the CV is constrained to be orthogonal, and interpretation of its coefficients is compared to that of LDA for a well-known example. In Section 7.3 we consider the classical DA problem subject to additional LASSO constraints. This is the same idea already successfully applied to PCA in Section 4.7. For short, this technique is called DALASS (Trendafilov and Jolliffe, 2006).

LDA of horizontal data is considered in Section 7.4, which is divided into three parts. Section 7.4.5 reviews several approaches to do LDA of horizontal data by replacing the singular within-group scatter matrix W by its main diagonal. Another alternative to avoid the singular W is presented in Section 7.4.7, where sparse LDA is based on minimization of classification error. Section 7.4.8 lists several techniques equivalent to LDA which, however, do not involve inverse W or T, as optimal scaling and common principal components (CPC). Section 7.4.4 briefly reminds the application of multidimensional scaling as discrimination problems. Finally, sparse pattern with each original variable contributing to only one discriminant functions is discussed in Section 7.4.11.

7.2 LDA of vertical data $(n > p)$

In this section we consider the classical LDA well known from the MDA textbooks (Krzanowski, 2003; Mardia et al., 1979). Consider the following linear combinations $Y = XA$ also called discriminant scores (Pedhazur, 1982). This is a linear transformation of the original data X into another vector space. There is interest in finding a $p \times s$ transformation matrix A of the original data X such that the *a-priori* groups are better separated in the transformed data Y than with respect to any of the original variables. Fisher's LDA achieves both goals by finding a transformation A which produces the "best" discrimination of the groups by simultaneous maximization of the between-groups variance and minimization of the within-groups variance of

Y (Fisher, 1936).

The number of "new" dimensions s is typically much smaller than the one of the original variables (p). In this sense, LDA is another dimension reduction technique.

As in PCA, the transformation A can be found sequentially. For a single linear combinations $y = Xa$, Fisher's LDA is expressed formally as the following problem:

$$\max_{a} \frac{a^\top B a}{a^\top W a} . \tag{7.8}$$

It should be noted that because of $T = B + W$ one does not really need to bother calculating W for (7.8). Instead, Fisher's LDA can be defined as

$$\max_{a} \frac{a^\top B a}{a^\top T a} . \tag{7.9}$$

For vertical data $(n > p)$, both LDA definitions work equally well. However, for horizontal data $(p \gg n)$, working with (7.9) may be preferable because T contains information for the singularities of both B and W: the common null space of B and W is spanned by the eigenvectors of T and

$$\mathcal{N}(B) \cap \mathcal{N}(B) \subseteq \mathcal{N}(B + W) , \tag{7.10}$$

where $\mathcal{N}()$ denotes the null space of its argument, e.g.

$$\mathcal{N}(B) = \{x \in \mathbb{R}^p : Bx = 0_p\} .$$

This will be further discussed and utilized in Section 7.4.

The following normalizations give, respectively, the between-groups and within groups covariance matrices:

$$C_B = \frac{B}{g - 1} \text{ and } C_W = \frac{W}{n - g} . \tag{7.11}$$

If the data has a multivariate normal distribution, then Fisher's LDA objective function from (7.8)

$$F(a) = \frac{a^\top B a}{a^\top W a} \tag{7.12}$$

has an F distribution with $g - 1$ and $n - g$ degrees of freedom under the null hypothesis that there is *no* difference among the g group means. Thus, $F(a)$ in (7.12) can be used to test that hypothesis. The larger the $F(a)$ value, the greater the divergence among the group means. The transformation A will successively produce the maximum possible divergence among the group means in its first few dimensions.

An important assumption for a valid LDA is that the population within-group covariance matrices are equal. This can be checked by using the likelihood-ratio test (Krzanowski, 2003, p. 370) to compare each within-group covariance matrix to the common one. If the null hypothesis is rejected in some groups than the results from LDA are considered unreliable. The CPC model was introduced in (Flury, 1988; Krzanowski, 1984) to study discrimination problems with unequal group covariance matrices.

7.2.1 Standard canonical variates (CVs)

Problem (7.8) is equivalent to the following GEVD problem (Krzanowski, 2003):

$$(B - \lambda W)a = 0_p , \qquad (7.13)$$

which can also be written as

$$(W^{-1}B - \lambda I_p)a = 0_p . \qquad (7.14)$$

Thus the maximum of the objective function F in (7.8) is the largest eigenvalue of $W^{-1}B$ and is achieved at the corresponding eigenvector a. The problem looks quite similar to that of PCA but $W^{-1}B$ is not symmetric. Moreover the rank of this matrix is $r \leq \min(p, g - 1)$ and all the remaining eigenvalues are 0s. The number r is called dimension of the canonical variate representation. The number of useful dimensions for discriminating between groups, s, is usually smaller than r, and the transformation A is formed by the eigenvectors corresponding to the s largest eigenvalues ordered in decreasing order. Clearly the $(p \times s)$ transformation A determined by Fisher's LDA maximizes the discrimination among the groups and represents the transformed data in a lower s-dimensional space.

Before moving on, we note that similar to (7.13) and (7.14), the modified LDA problem (7.9) is equivalent to the following GEVD:

$$(B - \mu T)a = 0_p . \qquad (7.15)$$

One can check that the eigenvector a of (7.15) ia also eigenvector of (7.13), and that the eigenvalue λ of (7.13) is related to the eigenvalue μ of (7.15) by:

$$\lambda = \frac{\mu}{1-\mu} \ .$$

The GEVD problem (7.13) can be written in matrix terms as

$$BA - WA\Lambda = O_{p\times r} \ , \tag{7.16}$$

where Λ is the $r \times r$ diagonal matrix of the s largest eigenvalues of $W^{-1}B$ ordered in decreasing order. This is *not* a symmetric EVD (Golub and Van Loan, 2013, §7.7, §8.7) and A is not orthonormal as in the symmetric case of PCA. Instead, $A \in \mathcal{M}(p,r)$ is a full column rank r, such that $A^\top WA = I_s$. The elements of A are called *raw coefficients* (Pedhazur, 1982) and are denoted by A_{raw}. They are, in fact, the coefficients of the variables in the discriminant functions. This normalization makes the within-groups variance of the discriminant scores Y equal to 1.

There is another way to compute the raw coefficients using a symmetric eigenvalue problem as in PCA. To implement this, one needs to rewrite the basic LDA problem (7.8) in the following equivalent form:

$$\max_a a^\top Ba \ \text{subject to} \ a^\top Wa = 1 \ . \tag{7.17}$$

Let $W = U^\top U$ be the Cholesky factorization of W with $U \in \mathcal{U}_+(p)$, i.e. a positive definite upper triangular matrix U. The substitution $a := Ua$ in (7.17) leads to the following symmetric eigenvalue problem:

$$\max_a a^\top U^{-\top} BU^{-1}a \ \text{subject to} \ a^\top a = 1 \ , \tag{7.18}$$

whose (orthogonal) solution A is used in turn to find the raw coefficients $A_{raw} = U^{-1}A$. If W is ill-conditioned then the Cholesky factorization should be replaced by eigenvalue decomposition (Golub and Van Loan, 2013).

One should also keep in mind that Fisher's LDA problem can be solved by using the original data X only, without forming B and W explicitly and solving (7.13) or (7.14). Indeed, based on (7.4) or (7.7) and (7.5), LDA/GEVD solution can be found by the generalized singular value decomposition (GSVD) of X_T or X_W and X_B as follows (Golub and Van Loan, 2013, §8.7.4). There exist $n \times n$ orthogonal matrices Q_1 and Q_2, and a non-singular $p \times p$ matrix P, such that:

$$Q_1^\top X_T P = \mathrm{diag}(\alpha_1, \ldots, \alpha_p) \text{ and } Q_2^\top X_B P = \mathrm{diag}(\beta_1, \ldots, \beta_p) .$$

The values $\{\alpha_1/\beta_1, \ldots, \alpha_p/\beta_p\}$ are called generalized singular values, and the columns of P are called the right generalized singular vectors. Then, the connection between the GSVD of the matrix pair $\left\{ N^{1/2} \left(\bar{X} - \frac{1_{g \times n} X}{n} \right), J_n X \right\}$ and the LDA/GEVD problem (7.15) is evident from

$$P^\top (X_B^\top X_B - \mu X_T^\top X_T) P = \mathrm{diag}(\alpha_1^2 - \mu \beta_1^2, \ldots, \alpha_p^2 - \mu \beta_p^2) . \qquad (7.19)$$

This equation also implies that the right generalized singular vectors P are generalized eigenvectors of $B - \mu T (= X_B^\top X_B - \mu X_T^\top X_T)$ and their eigenvalues are squares of the generalized singular values, i.e. $\mu_i = \alpha_i^2/\beta_i^2$, for $i = 1, \ldots, p$. Particularly, for the LDA/GEVD problem (7.15), we take the largest $g - 1$ values of β_i, because the rest are zeros, and the corresponding $\mu_i = \infty$ for $i = g, \ldots, p$. This finally gives

$$\beta_i^2 B a_i = \alpha_i^2 T a_i \text{ for } i = 1, \ldots, g - 1 . \qquad (7.20)$$

For large data applying this GSVD is rather expensive as it requires $O(n^2 p)$ operations. The method proposed in (Duintjer Tebbens and Schlesinger, 2007) seems a better alternative if one needs to avoid the calculation of large B and W. Further related results can be found in (Shin and Eubank, 2011).

The raw coefficients are considered difficult to interpret when one wants to evaluate the relative importance of the original variables. As in PCA, one can try to identify those raw coefficients that are large in magnitude in a particular discriminant function and conclude that the corresponding variables are important for discrimination between the groups. The problem is that such a conclusion can be misleading in LDA. The large magnitudes may indeed be caused by large between-groups variability, but also can be caused by small within-groups variability (Krzanowski, 2003). This problem with the interpretation of raw coefficients is overcome by an additional standardization of A_{raw} which makes all variables comparable:

$$A_{std} = \mathrm{diag}(W)^{1/2} A_{raw} , \qquad (7.21)$$

and the new coefficients are called the *standardized coefficients* (Pedhazur, 1982).

Finally, the *structure coefficients* are defined as the correlation coefficients between the input variables and the discriminant scores Pedhazur (1982). They are considered by many authors as most appropriate for interpreting the importance of variables for discrimination. Their disadvantage is that they are univariate measures and do not represent the importance of a variable in the presence of other available variables (Rencher, 1992).

The following small example considered in (Trendafilov and Jolliffe, 2007) demonstrates how confusing the LDA interpretation can be with those several coefficients.

Example: Consider the data on 32 Tibetan skulls divided into two groups (1 to 17 and 18 to 32) and discussed and studied in (Everitt and Dunn, 2001, p. 269). On each skull five measurements (in millimetres) were obtained: greatest length of skull (x_1), greatest horizontal breadth of skull (x_2), height of skull (x_3), upper face height (x_4), and face breadth, between outermost points of cheek bones (x_5).

The data are subject to LDA with **SPSS** (SPSS, 2001). There is only one canonical variate ($s = 1$) in this example. The value of Fisher's LDA objective function (7.12) (F value) is 28.012. The output provided in Table 7.1 gives the criterion generally used to assess each variable's contribution to Hotelling's T^2 and their importance for discrimination.

Table 7.1: Tests of Equality of Group Means (Trendafilov and Jolliffe, 2007).

Vars	Wilk's Lambda	F	p
x_1	.651	16.072	.000
x_2	.998	.063	.803
x_3	.947	1.685	.204
x_4	.610	19.210	.000
x_5	.763	9.315	.005

According to Table 7.1 one can conclude that the variable x_4 is the most important for discrimination, followed closely by x_1, and then by x_5 some distance behind. The other two variables do not seem interesting for the problem and can be dropped from further analysis.

No such clear decision can be made if one bases the interpretation on the raw and standardized coefficients—see Table 7.2. According to the raw coefficients magnitudes, x_2 is now the most important variable, which was the least one in Table 7.1. This variable is also important for the standardized coefficients, but x_5 now has the largest coefficient. The structure coefficients imply interpretations similar to those deduced from Hotelling's T^2, but the dominance of x_1 and x_4 is less clear-cut. For two groups, it can be argued that Hotelling's T^2 provides the best way to interpret the single discriminant function. However in case of several groups, Hotelling's T^2 measures the *overall* contribution of each variable to group separation and thus is not helpful for interpretation of any particular discriminant function (Rencher, 2002).

Table 7.2: Canonical Variates for Skull Data (Trendafilov and Jolliffe, 2007).

Vars	Raw Coefficients	Standardized Coefficients	Structure Coefficients
x_1	.090	.367	.759
x_2	-.156	-.578	-.048
x_3	-.005	-.017	.246
x_4	.117	.405	.830
x_5	.117	.627	.578

For comparison, one can apply stepwise logistic regression to this dataset with the group membership considered as the response variable. The best model found includes the greatest length of skull (x_1), greatest horizontal breadth of skull (x_2) and face breadth, between outermost points of cheek bones (x_5). This solution is matching well the interpretation suggested by the raw and the standardized coefficients in Table 7.2, if we take into account that x_1 and x_4 are highly correlated (.755). This example clearly shows that the structure coefficients can be misleading by measuring the overall importance of the variables.

7.2.2   Orthogonal canonical variates (OCVs)

LDA does not provide orthogonal projection of the data, as PCA does, because "the canonical variate space is derived by deforming the axes in the original data space" (Krzanowski, 2003). If orthogonal projections between the original data space are sought for maximal discrimination of the existing groups, then LDA needs to be modified in a PCA fashion. For this reason the basic LDA problem (7.8) in the form (7.13):

$$\max_{a_i} \frac{a_i^\top B a_i}{a_i^\top W a_i} \text{ subject to } a_i^\top W a_i = 1 \text{ and } a_i^\top W a_j = 0, \qquad (7.22)$$

for $i = 1, 2, \ldots, s; i \neq j$, is replaced in (Duchene and Leclercq, 1988; Krzanowski, 1995) by the following blend between PCA and LDA, namely a LDA objective function subject to PCA constraints:

$$\max_{a_i} \frac{a_i^\top B a_i}{a_i^\top W a_i} \text{ subject to } a_i^\top a_i = 1 \text{ and } a_i^T A_{i-1} = 0_{i-1}^\top , \qquad (7.23)$$

where the matrix A_{i-1} is composed of all preceding vectors $a_1, a_2, \ldots, a_{i-1}$, i.e. A_{i-1} is the $p \times (i-1)$ matrix defined as $A_{i-1} = (a_1, a_2, \ldots, a_{i-1})$. The solutions $A = (a_1, a_2, \ldots, a_s)$ are called *orthogonal* canonical variates.

In these notations, the LDA problem (7.22) can be rewritten in a PCA-like form as suggested in (7.18):

$$\max_{a_i} a_i^\top U^{-\top} B U^{-1} a_i \quad \text{subject to } a_i^\top a_i = 1 \text{ and } a_i^T A_{i-1} = 0_{i-1}^\top , \quad (7.24)$$

where $U \in \mathcal{U}_+(p)$ is the positive definite upper triangular matrix from the Cholesky decomposition of W, i.e. $W = U^\top U$. One keeps in mind that the (orthogonal) solution A of (7.24) is used in turn to find the raw coefficients $A_{raw} = U^{-1} A$. This reformulation of (7.22) in a PCA-like format will be used hereafter.

Thus, there are two very similar, PCA-like problems, to be solved. The difference is in the objective functions to be maximized.

The LDA problem (7.24) to find standard canonical variates can be solved by a number of well-known algorithms (Golub and Van Loan, 2013). The algorithm proposed by (Krzanowski, 1995) for finding the orthogonal canonical variates defined in (7.23) is a sequential one inspired by PCA and works in the same manner: find an unit vector a_1 maximizing the objective function, then form the linear subspace orthogonal to a_1 and find an unit vector a_2 from this subspace which maximizes the objective function and so on.

Example: (continued) In Table 7.3 are given the orthogonal canonical variate (raw) coefficients, and the corresponding structure coefficients. The objective function (7.23) at this solution is, as before, 28.012. The raw coefficients again suggest importance of x_2, x_4 (rather than x_1) and x_5, while the structure coefficients are quite similar to those from Table 7.2 and mark x_2 as the least interesting variable.

Table 7.3: Orthogonal Canonical Variates for Skull Data (Trendafilov and Jolliffe, 2007).

Vars	Raw Coefficients	Structure Coefficients
x_1	.290	.851
x_2	-.505	-.066
x_3	-.017	.332
x_4	.574	.900
x_5	.575	.701

7.3 Sparse CVs and sparse OCVs

Following the sparse PCA idea, Trendafilov and Jolliffe (2007) propose to add LASSO constraints to the LDA definition in order to achieve more easily interpretable CVs. In other words, one can impose the constraints:

$$\|a_i\|_1 \leq \tau_i \text{ for } i = 1, 2, \ldots, s \text{ , with } \tau_i \in [1, \sqrt{p}] \text{ ,} \tag{7.25}$$

on the standard LDA problems from the previous Section. Then, the ith CV is found as a solution of either

$$\max_a a^\top U^{-\top} BU^{-1} a \tag{7.26}$$

and/or

$$\max_a \frac{a^\top B a}{a^\top W a} \tag{7.27}$$

both subject to

$$\|a\|_1 \leq \tau \ , \|a\|_2^2 = 1 \text{ and } a^T A_{i-1} = 0_{i-1}^\top \ . \tag{7.28}$$

Trendafilov and Jolliffe (2007) propose to tackle the LASSO inequality constraint (7.28) by introducing an exterior penalty function P into the objective functions to be maximized. The idea is to penalize an unit vector a which does not satisfy the LASSO constraint by reducing the value of the new objective function. Thus, the LDA problems are modified as follows:

$$\max_a \left[a^\top U^{-\top} B U^{-1} a \ - \mu P(\|a\|_1 - t) \right] \tag{7.29}$$

and

$$\max_a \left[\frac{a^\top B a}{a^\top W a} - \mu P(\|a\|_1 - t) \right] \tag{7.30}$$

both subject to

$$\|a\|_2^2 = 1 \text{ and } a^T A_{i-1} = 0_{i-1}^\top \ . \tag{7.31}$$

For short, these penalized LDA problems are called DALASS (**D**iscriminant **A**nalysis via **LASS**O). The exterior penalty function P is zero if the LASSO constraint is fulfilled. It "switches on" the penalty μ (a large positive number) if the LASSO constraint is violated. Moreover, the more severe violations are penalized more heavily. A typical example of an exterior penalty function for inequality constraints is the Zangwill penalty function $P(x) = \max(0, x)$.

The new LDA problems (7.29)–(7.31) and (7.30)–(7.31) require maximization of more complicated objective functions but subject to the well-known constraint from PCA. Because of the PCA-like nature of (7.31) it seems natural to attack the problems sequentially following the PCA tradition. Such an approach was already considered in relation to PCA in Section 4.7.

Then both the standard and orthogonal CVs can be found as s consequent ascent gradient vector flows, each of them defined on the unit sphere in $\mathbb{R}^p$ and orthogonal to all preceding CVs. The loadings for the canonical variates $a_1, a_2, ..., a_s$ can be computed as solutions of s consequent

Example: (continued) DALASS is applied to the skull data with tuning parameter $\tau = 1.2$ and the solution is given in Table 7.4. According to both raw and standardized coefficients one picks up a single variable, greatest length of skull (x_1), as most important for the group discrimination in this dataset. The structure coefficients are not helpful as they suggest several variables to be taken into account and thus complicate the interpretation. The objective function value at this solution is 19.933.

The DALASS orthogonal canonical variates given in Table 7.5 provide even clearer interpretation: the raw coefficients suggest a single variable, the upper face height (x_4), as important for discriminating between the groups in these skull data. The objective function at this solution is 22.050—a smaller drop compared to 28.012. Thus the orthogonal canonical variates can be considered a better solution of the problem. As before the corresponding structure coefficients are clearly more difficult to interpret. DALASS works very well for the skull data and the interpretation of the results can be entirely based on the discriminant function coefficients.

Table 7.4: DALASS Canonical Variates for Skull Data ($\tau = 1.2$) (Trendafilov and Jolliffe, 2007).

Vars	Raw Coefficients	Standardized Coefficients	Structure Coefficients
x_1	.108	.829	.987
x_2	-.005	-.032	.099
x_3	.002	.011	.422
x_4	.053	.228	.848
x_5	.006	.036	.612

Table 7.5: DALASS Orthogonal Canonical Variates for Skull Data ($\tau = 1.2$) (Trendafilov and Jolliffe, 2007).

Vars	Raw Coefficients	Structure Coefficients
x_1	.110	.822
x_2	-.053	.057
x_3	.000	.321
x_4	.992	.993
x_5	.038	.629

Example: Here DALASS is illustrated on the famous Iris data (Fisher, 1936).

This dataset is four-dimensional ($p = 4$): sepal length (x_1), sepal width (x_2), petal length (x_3) and petal width (x_4). It contains 50 observations in each of the three groups of plants: Iris setosa, Iris versicolor and Iris virginica.

In Table 7.6 and Table 7.7 are given the standard and the orthogonal two-dimensional ($s = r = 2$) canonical variates solutions. Fisher's LDA objective function $F(a)$ (7.12) for the first standard CV is 2366.11. The value at the second CV is 20.98 (the third and fourth are 0), so the relative importance of the second CV is less than 1%. The objective values at the orthogonal CVs are 2366.11 and 705.50 (total 3071.61), i.e. the second CV is considerably more important for this analysis.

These two solutions are depicted in Figure 7.1. The three groups are well separated in both of the plots. The latter seems to be superior because the groups are more compact. More objective comparison of the quality of these two discriminations can be achieved by applying them to classify the original observations. The standard CV solution misclassifies three observations: 52 (3), 103 (2) and 104 (2) with error rate of 2%, while the orthogonal CV solution misclassifies four: 40 (2), 78 (3), 104 (2) and 121 (3) with error rate of 2.67%.

Unfortunately, the discriminant functions coefficients for both of the solutions in Table 7.6 and Table 7.7 do not make the interpretation unique and simple.

Table 7.6: Canonical Variates for Fisher's Iris Data (Trendafilov and Jolliffe, 2007).

Vars	Raw Coefficients		Standardized Coefficients		Structure Coefficients	
x_1	-.08	-.00	-.43	-.01	.79	-.22
x_2	-.15	-.22	-.52	-.74	-.53	-.76
x_3	.22	.09	.95	.40	.98	-.05
x_4	.28	-.28	.58	-.58	.97	-.22

DALASS with tuning parameter $\tau = 1.2$ is applied to obtain both standard and orthogonal two-dimensional canonical variates solutions. They are given in Table 7.8 and Table 7.9 and depicted in Figure 7.2.

The standardized coefficients of the DALASS canonical variates suggest that

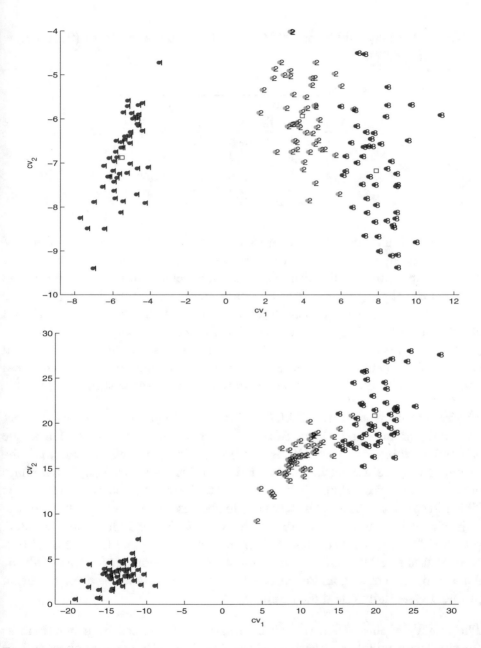

Figure 7.1: *Iris* data plotted against the first two canonical variates. 1 =*Iris setosa*, 2 =*Iris versicolor*, 3 =*Iris virginica*. Squares denote group means. Top: standard canonical variates. Bottom: orthogonal canonical variates (Trendafilov and Jolliffe, 2007).

Table 7.7: Orthogonal Canonical Variates for Fisher's Iris Data (Trendafilov and Jolliffe, 2007).

Vars	Raw Coefficients		Structure Coefficients	
x_1	-.21	-.15	.79	.84
x_2	-.39	.04	-.53	-.48
x_3	.55	.76	.98	.99
x_4	.71	-.62	.97	.91

they are both mainly composed of two of the original variables: the first CV is dominated by sepal length, x_1 and petal length, x_3, and the second CV by sepal length x_1 and sepal width, x_2. One can conclude that the discrimination between the three groups in the *Iris* data can be based on the length of the flowers, and the sepal size. The objective values at the CVs are 1888.10 and 255.60 (total 2143.71), i.e. the second CV is considerably more important (13.5%) for this analysis than for the standard CV from Table 7.6. The total value of the objective function is reasonably high compared to the original 2387.08, so this solution can be considered as quite successful.

The raw coefficients of the DALASS orthogonal canonical variates suggest even simpler interpretations. Both CVs are mainly composed of a single original variable: the first CV is dominated by petal width, x_4, and the second CV by petal length, x_3. According to this result one can discriminate between the three groups in the *Iris* data based on the petal size only. This interpretation is surprisingly simple and also seems reliable. Indeed, the objective values at the DALASS orthogonal CVs are 1430.15 and 1348.91 (total 2779.06, which is only a 10% drop from the original total of 3071.61). Note that the orthogonal CVs are nearly equally important for this analysis. An appropriate orthogonal rotation of the CVs can make the discrimination problem essentially one dimension.

The quality of these DALASS discriminations is assessed by using them to classify the original observations. The DALASS CVs solution misclassifies four observations: 63 (2), 103 (2), 104 (2) and 121 (3) with error rate of 2.67%, while the DALASS orthogonal CVs solution misclassifies five: 9 (3), 50 (2), 78 (3), 104 (2) and 121 (3) with error rate of 3.33%, only marginally worse than for the classical solutions.

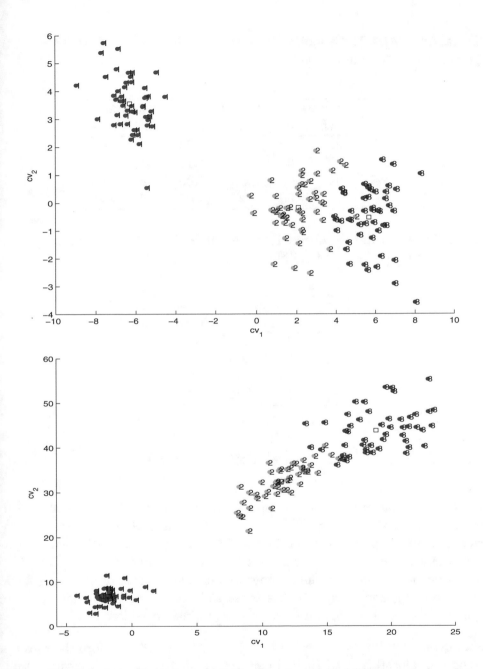

Figure 7.2: *Iris* data plotted against the first two DALASS canonical variates with tuning parameter 1.2. Top: standard CVs. Bottom: orthogonal CVs. (Trendafilov and Jolliffe, 2007).

Table 7.8: DALASS Canonical Variates for Fisher's Iris Data ($\tau = 1.2$) (Trendafilov and Jolliffe, 2007).

Vars	Raw Coefficients		Standardized Coefficients		Structure Coefficients	
x_1	-.21	-.16	-1.09	-.83	.77	-.72
x_2	-.02	.34	-.06	1.15	-.51	.78
x_3	.31	.00	1.35	.00	.98	-.85
x_4	.13	-.00	.27	-.00	.97	-.78

Table 7.9: DALASS Orthogonal Canonical Variates for Fisher's Iris Data ($\tau = 1.2$) (Trendafilov and Jolliffe, 2007).

Vars	Raw Coefficients		Structure Coefficients	
x_1	-.00	-.15	.81	.85
x_2	-.15	-.00	-.44	-.45
x_3	.06	.99	.97	1.00
x_4	.99	-.06	1.00	.96

7.4   LDA of horizontal data ($p > n$)

It turns out that in the modern applications the typical data format is with more variables than observations. Such data are also commonly referred to as the small-sample or horizontal data. The main problem with such data is that the within-groups scatter matrix W is singular and Fisher's LDA (7.8) is not defined. Moreover, the number of variables is usually huge (e.g. tens of thousands), and thus, it makes sense to look for methods that produce sparse CVs, i.e. involving only few of the original variables.

An obvious way to use Fisher's LDA with such data is to ignore their group structure and start with their dimension reduction (PCA). Provided that the within-groups scatter matrix W of the reduced data is not singular, then the classical LDA can be readily applied. One problem with this approach is that W of the reduced data may still be singular. However, the main trouble is that the overall performance depends uncontrollably on the preliminary dimension reduction.

One of the first attempts for systematic study of LDA for horizontal data is made in (Krzanowski et al., 1995). Four approaches are considered there, which to a large extent are still the main routes to deal with the problem:

1. Augmenting W: The aim is to augment W in such a way that "it retains its major characteristics but it becomes non-singular". Related developments are considered in (Bensmail and Celeux, 1996; Friedman, 1989; Guo et al., 2007).

2. Modified canonical analysis: in short, the idea here is to achieve LDA solution (or discrimination in general) by other means (and, thus, avoid dealing with W^{-1}). This is probably the most popular line of research on LDA for $p \gg n$ (Campbell and Reyment, 1981; Clemmensen et al., 2011; Witten and Tibshirani, 2011).

3. Zero-variance discrimination: To restrict the LDA problem on the null subspace of W, i.e. solve (7.8) with additional constraint $Wa = 0_{p \times r}$ (Ye and Xiong, 2006; Zhang et al., 2010).

4. Modelling W: The purpose here is to construct a suitable stochastic model approximating the problem but implying non-singular W. For example, spectroscopic data are modelled by a "moving window" process in (Krzanowski et al., 1995). However, such model/approach is tailored for particular data, and thus, very specific.

7.4.1 LDA through GSVD

It was mentioned that the GSVD computational cost is $O(n^2 p)$, which may be too expensive for large applications with $p \gg n$. However, LDA has some specific features which help to elaborate GSVD for such particular problem. As LDA works with $X_B = J_n G N^{-1} G^\top X$ and $X_T = J_n X$, we can consider the economical SVD $X_T = J_n X = U \Sigma V^\top$, where $U \in \mathcal{O}(n, r), V \in \mathcal{O}(p, r)$ and $\Sigma \in \mathbb{R}^{r \times r}$ is PD diagonal and r is the rank of $J_n X$. Note that $U = J_n U$ is orthonormal and *centred*, i.e. $U \in \mathcal{O}_0(n, r)$. Now, we consider $J_n G N^{-1} G^\top$, which can be at most of rank g. We form the economical SVD $J_n G N^{-1} G^\top U = P \Lambda Q^\top$, where $P \in \mathcal{O}(n, r), Q \in \mathcal{O}(r, r)$ and $\Lambda \in \mathbb{R}^{r \times r}$ is PSD diagonal. Note, that $P = J_n P$, i.e. $P \in \mathcal{O}_0(n, r)$. On can check by direct calculations that:

$$(UQ)^\top (J_n X)(V\Sigma^{-1}Q) = I_r \ ,$$
$$P^\top (J_n GN^{-1}G^\top X)(V\Sigma^{-1}Q) = \Lambda \ .$$

As usually $r > g - 1$, one can simply take $r \equiv g - 1$ in the above SVDs. As before in (7.19), we can write

$$(V\Sigma^{-1}Q)^\top (X_B^\top X_B - \mu X_T^\top X_T)(V\Sigma^{-1}Q) = \mathrm{diag}(\lambda_1^2 - \mu, \ldots, \lambda_{g-1}^2 - \mu) \ ,$$

which could have been derived if full SVDs were taken in the above derivation. Thus, the squares of the diagonal elements of Λ are the generalized singular values and the columns of $V\Sigma^{-1}Q$ are the generalized singular vectors of the corresponding GSVD.

Is there a reasonable way to produce sparse generalized singular vectors remains an open question, i.e. a sparse version of the above GSVD procedure.

7.4.2   LDA and pattern recognition

As a rule, the engineering and computer science literature (Duda et al., 2001; Fukunaga, 1990) treats LDA problems with different criteria, the most popular versions of which are

$$\max_A \mathrm{trace}[(A^\top WA)^{-1}A^\top BA] \ , \tag{7.32}$$

and

$$\max_A \frac{\det(A^\top BA)}{\det(A^\top WA)} \ . \tag{7.33}$$

Such alternative LDA cost functions are influenced by the so-called *clustering indexes* introduced in (Friedman and Rubin, 1967). They are scalar functions of matrix unknowns involving the within- and/or between-groups scatter matrices, W and B, of the data. For example, (7.32) is considered in (Friedman and Rubin, 1967), while (7.33) is a particular case of the index suggested in (Scott and Symons, 1971, p. 396). There is a great number of clustering indexes listed and described in (Charrad et al., 2014).

The LDA problems (7.32) and (7.33) can be readily solved on Stiefel or oblique manifold, $\mathcal{O}(p,r)$ or $\mathcal{OB}(p,r)$, for some $r \leq g - 1$ (making use of

Manopt). The gradients of the objective functions (7.32) and (7.33) do not include W^{-1} or B^{-1}. They involve only $(A^\top W A)^{-1}$, which is a reasonably small matrix, possibly diagonal depending on the adopted constraints. Another positive feature of such LDA reformulations is that sparseness inducing penalties can be readily implemented as in Section 7.4.6. Such type of sparse LDA is considered in (Ng et al., 2011).

A number of other LDA formulations, comparisons of their features, and algorithms for their solutions are found in (Luo et al., 2011; Zhang et al., 2006).

7.4.3 Null space LDA (NLDA)

Roughly, the NLDA problem is

$$\max_{a^\top W a = 0} a^\top B a \,,$$

and requires the computation of the null space of W, which dimensionality is quite high ($\approx p + g - n$) for $p \gg n$.

In general, NLDA proceeds by removing the null space of some PSD matrix, i.e. by projecting it on the subspace which is complement to its null space (Golub and Van Loan, 2013, Corollary 2.4.6.). It turns out, the removal of the null space of T leads to considerable simplification and dimension reduction. It is possible because the null space of T is the intersection of the null spaces of B and W (7.10). This is usually realized by economical SVD of the product factor of $T(= X^\top J_n X)$, which is given by $J_n X = U \Sigma V^\top$, where Σ is the $r \times r$ diagonal matrix containing the non-zero singular values of $J_n X$ in decreasing order. The matrices $U \in \mathcal{O}(n, r)$ and $V \in \mathcal{O}(p, r)$ contain the corresponding left and right singular vectors of $J_n X$. Clearly, r is the rank of $J_n X$ (and $r + 1$ is the rank of the non-centred data matrix X). Now, we can form $T_\pi = V^\top T V = V^\top X^\top J_n X V = \Sigma^2$, which is the projection of T onto the complement of its null space. Then, we can find the corresponding projection of B as $B_\pi = V^\top B V$ and solve (7.15) through a modified much smaller GEVD, which, in fact, transforms into a standard EVD for $b = \Sigma a$, as Σ^2 is diagonal:

$$\max_{b^\top b = 1} b^\top \Sigma^{-1} B_\pi \Sigma^{-1} b \,. \tag{7.34}$$

Note, that now (7.34) should be solved in the null space of the projection $W_\pi (= V^\top W V)$.

The following approach is based on the idea that the null space of W may contain significant discriminant information if the projection of B is not zero in that direction (Ye and Xiong, 2006). For projected B_π and W_π as found above, the following NLDA problem is solved:

$$\max_{A^\top W_\pi A = O} \text{trace}[A^\top B_\pi A] \ . \tag{7.35}$$

To find a matrix which columns span the null space of W_π, we need its EVD, say, $W_\pi = P\Lambda^2 P^\top$, where Λ^2 contains the eigenvalues in *increasing* order. The eigenvectors corresponding to the zero eigenvalues are the ones spanning the null space of W_π. Let collect them in P_0. Then, the solution of (7.35) can be sought in the form $A \equiv P_0 A$, which makes the problem unconstrained:

$$\max_{A} \text{trace}[A^\top P_0^\top B_\pi P_0 A] \ . \tag{7.36}$$

Apparently, the EVD of $P_0^\top B_\pi P_0$ is among the solutions of (7.36), if the orthonormality of A is additionally assumed. Finally, the NLDA CVs are given by $VP_0 A$. Further development and facilitation of this approach is considered in (Chu and Goh, 2010).

Another scenario is considered in (Duintjer Tebbens and Schlesinger, 2007). They argue that NLDA may choose vectors from the common null space of W and B, where both $a^\top W a = 0$ and $a^\top B a = 0$ (which is the null space of T). As a solution, they propose to start with the solution of the standard NLDA problem, i.e.:

$$\max_{\substack{a^\top a = 1, \\ W a = 0_p}} \frac{a^\top B a}{a^\top W a} \ ,$$

and order the CVs according to their between-group variances. If $Ba = 0$ for some a, such CVs are considered not interesting. Thus, NLDA is used to find CVs with non-zero between-group variances. If the number of those CVs is less than $g - 1$, then, the remaining CVs are searched in the complement of the null space of W, i.e. by solving:

$$\max_{\substack{a^\top a = 1, \\ W a \neq 0_p}} \frac{a^\top B a}{a^\top W a} \ .$$

Further theoretical and implementation details of the proposed combined procedure are given in (Duintjer Tebbens and Schlesinger, 2007).

Both GSVD and NLDA methods considered above do not produce sparse CVs, which is an open research area.

7.4.4   LDA with CPC, PLS and MDS

LDA with common principal components (CPC)

Common principal components (CPC) are developed by Flury (1988) and can be used to discriminate several groups of observations with *different* covariance matrices in each group. Zou (2006) already considered briefly such an option. In a simulated study, Flury et al. (1994) demonstrated that even a simpler CPC model with proportional covariance matrices can provide quite competitive discrimination compared to other more complicated methods (Flury, 1988, Ch 5). The CPC version proposed in (Trendafilov, 2010) finds the CPCs one after another. The adopted "power method"-like algorithm can be readily enhanced with additional, say, LASSO constraints to produce sparse CPCs.

LDA with partial least squares (PLS)

Partial least squares (PLS) is another technique which can be used to serve for solving LDA problems. PLS is closely related to canonical correlation analysis (CCA) which is discussed in Section 8.6. It is shown in Section 8.6.1 that LDA can be viewed as a special case of CCA, which closes the loop among those techniques.

Sparse LDA based on metric scaling

Gower (1966) showed that metric scaling of the matrix of Mahalanobis distances between all pairs of groups will recover the canonical variate configuration of group means. However, the Mahalanobis distances use the pooled within-group scatter matrix, and thus, this approach is not applicable for horizontal data. It was mentioned before, that Dhillon et al. (2002) avoided this problem by simply doing PCA of the between-group scatter matrix B to obtain LDA results. Trendafilov and Vines (2009) considered sparse version of this LDA procedure.

The above approach can still be applied if the equality of the population covariance matrices of the groups cannot be assumed. A particularly elegant solution, employing Hellinger distances, can be obtained if the CPC hypothesis is appropriate for the different covariance matrices (Krzanowski, 1990).

Another unexplored option is to consider linear discrimination employing within- and between-group distance matrices (Gower and Krzanowski, 1999), which have sizes $n \times n$. However, the choice of a suitable dissimilarity measure/metric will probably become the major difficulty in implementing this approach.

7.4.5 Sparse LDA with diagonal W

The straightforward idea to replace the non-existing inverse of W by some kind of generalized inverse has many drawbacks and thus is not satisfactory. For this reason, Witten and Tibshirani (2011) adopted the idea proposed by Bickel and Levina (2004) to circumvent this difficulty by replacing W with a diagonal matrix W_d containing its diagonal, i.e. $W_d := I_p \odot W$. Note that Dhillon et al. (2002) were even more extreme and proposed doing LDA of high-dimensional data by simply taking $W = I_p$, i.e. PCA of B. Such LDA version was adopted already to obtain sparse CVs when W is singular (Trendafilov and Vines, 2009).

Probably the simplest strategy can be based on the LDA approach proposed in (Campbell and Reyment, 1981), where LDA is performed in two stages each consisting of EVD of a specific matrix. This approach was already applied in (Krzanowski et al., 1995) with quite reasonable success to LDA problems with singular W. When W_d is adopted, the original two-stage procedure simplifies like this. At the first stage, the original data are transformed as $Y = XW_d^{-1/2}$. Then, at the second stage, the between-groups scatter matrix B_Y of the transformed data Y is formed by (7.2) or (7.5), e.g.:

$$B_Y = \sum_{i=1}^{g} n_i(\bar{y}_i - \bar{y})(\bar{y}_i - \bar{y})^\top ,$$

and followed by some kind of sparse PCA applied to B_Y.

Let the resulting sparse components be collected in a $p \times \min\{p, g - 1\}$ matrix C. Then, the sparse canonical variates are given by $A = W_d^{-1/2}C$.

The sparseness achieved by C, is inherited in A because W_d is diagonal. Note that the calculation of B_Y is not really needed. Following (7.5), the sparse PCA can be performed directly on $(G^\top G)^{-1/2} G^\top Y$.

Krzanowski (1990) proposed a generalization of this two-stage procedure for the case of unequal within-group scatter matrices. He adopted the CPC model for each of the within-group scatter matrices in each group. For horizontal data, this generalized procedure results in a slightly different way of calculating B_Y. Now, $\bar{y}_i = X_i W_{i,d}^{-1/2}$, where X_i is the data submatrix containing the observations of the ith group and $W_{i,d} = I_p \odot W_i$, where W_i is the within-group scatter matrix of the ith group.

7.4.6 Function-constrained sparse LDA

By adopting the simplification $W = W_d$, any method for sparse PCA can be applied for LDA. In particular, the function-constraint reformulation of LDA is straightforward:

$$\min_{\substack{a^\top W_d a = 1 \\ a \perp W_{i-1}}} \|a\|_1 + \tau(a^\top Ba - d)^2 , \tag{7.37}$$

where $W_0 = 0_{p \times 1}$ and $W_{i-1} = W_d[a_1, a_2, ..., a_{i-1}]$, and d is found as a solution of the standard Fisher's LDA problem (7.8) with $W = W_d$. Let $b = W_d^{1/2} a$. Note, that b are in fact, the so-called raw coefficients (Krzanowski, 2003, p.298). As W_d is diagonal, a and b have the same sparseness. Then, the modified Fisher's LDA problem (7.37) to produce sparse raw coefficients is defined as:

$$\min_{\substack{b^\top b = 1 \\ b \perp b_{i-1}}} \|b\|_1 + \tau(b^\top W_d^{-1/2} B W_d^{-1/2} b - d)^2 . \tag{7.38}$$

Thus, problem (7.38) is already known from Section 4.8 as a function-constraint PCA. It can be readily solved by some of the existing software (Boumal et al., 2014; Wen and Yin, 2013). The reader is urged to create Manopt codes for function-constraint LDA by modifying properly the codes from Section 4.8.4.

Instead of solving (7.38), one can consider alternatively

$$\min_{\substack{b^\top b = 1 \\ b \perp b_{i-1}}} \|b\|_1 - \tau b^\top W_d^{-1/2} B W_d^{-1/2} b , \tag{7.39}$$

which numerical solution is considerably faster, but less stable. Moreover, it usually increases the classification error as well.

In experiments with simulated and real data, solving (7.39) is roughly comparable to SLDA (Clemmensen et al., 2011; Merchante et al., 2012) with respect to speed and classification error, having about 5% non-zero entries. Both of the methods (and some other tested) outperform the "Penalized LDA" proposed in (Witten and Tibshirani, 2011).

The SLDA method is briefly outlined in Section 7.4.8. The penalized LDA solves the following problem, with $B = X^\top G (G^\top G)^{-1} G^\top X$ and $W \equiv W_d$ as in (7.39) above:

$$\max_{a^\top W a \leq 1} a^\top B a - P(a) , \tag{7.40}$$

where P is a convex penalty function inducing sparseness of the discriminant vector a, by ordinary or fused LASSO (Tibshirani et al., 2005).

The next a is found also as a solution of (7.40), but with B replaced by

$$B = ((G^\top G)^{-1/2} G^\top X)^\top \Pi ((G^\top G)^{-1/2} G^\top X) ,$$

where $\Pi = I_p - A A^\top$ and the columns of A are the previously found discriminant vectors a, i.e. the next a is orthogonal to $(G^\top G)^{-1/2} G^\top X A$. The penalized LDA problem (7.40) is solved by majorization–minimization (MM) optimization procedure (Hunter and Lange, 2004). Further theoretical and implementation details can be found in (Witten and Tibshirani, 2011).

7.4.7   Sparse LDA based on minimization of the classification error

Fan et al. (2012) argued that ignoring the covariances (the off-diagonal entries in W) as suggested by Bickel and Levina (2004) may not be a good idea. In order to avoid redefining Fisher's LDA for singular W, Fan et al. (2012) proposed working with (minimizing) the classification error instead of Fisher's LDA ratio (7.8). The method is called for short ROAD (from **R**egularized **O**ptimal **A**ffine **D**iscriminant) and is developed for two groups.

Let m_1 and m_2 denote the group means and form $d = \frac{m_1 - m_2}{2}$, and, as before, $T = B + W$. To avoid Fisher's LDA with singular W, (Fan et al., 2012) consider general linear classifier $\delta_a(x) = \mathbb{I}\{a^\top (x - m) > 0\}$, where $m = \frac{m_1 + m_2}{2}$ is the groups mean and $\mathbb{I}$ is indicator function. The misclassification error of $\delta_a(x)$ is $1 - \Phi\left\{\frac{d^\top a}{\sqrt{a^\top T a}}\right\}$. To minimize the classification error of $\delta_a(x)$, one can maximize $\frac{d^\top a}{\sqrt{a^\top T a}}$, or minimize $a^\top T a$ subject to $d^\top a = 1$.

Then, the ROAD problem is to find such a minimizer a, which is moreover *sparse*. Thus, the ROAD minimizer a is sought subject to a LASSO-type constraint introduced as a penalty term:

$$\min_{d^\top a = 1} a^\top T a + \tau \|a\|_1 . \tag{7.41}$$

Further on, Fan et al. (2012) avoid the linear constraint in (7.41) by adding another penalty term to the cost function and solve the following unconstrained problem:

$$\min a^\top T a + \tau_1 \|a\|_1 + \tau_2 (d^\top a - 1)^2 . \tag{7.42}$$

Let us forget for a while for the sparseness of a, and solve (7.42) for the Iris data with $\tau_1 = 0$. Then, the first group *Iris setosa* is perfectly separated by the cloud composed of the rest two groups of *Iris versicolor* and *Iris virginica*. The difference between the means of these two groups is

$$d = (-.1243, .1045, -.1598, -.1537)$$

and the discriminant found is

$$a = (.6623, 1.4585, -5.1101, -.7362) .$$

The constraint $a^\top d = 1$ is fulfilled, but this solution of (7.42) is not convenient for interpretation because one cannot assess the relative sizes of the elements of a. This concern remains even if the LASSO constraint is added.

Thus, it seems reasonable to consider a constrained version of problem (7.42) subject to $a^\top a = 1$. Solution of the following related "dense" problem (with $\tau = 0$)

$$\min_{\substack{d^\top a = 1, \\ a^\top a = 1}} a^\top T a \tag{7.43}$$

is available by Gander et al. (1989). One can consider "sparsifying" their solution to produce unit length ROAD discriminants.

Other works in this direction exploit the fact that the classified error depends on T^{-1} and d only through their product $T^{-1}d$ (Cai and Liu, 2011; Hao et al., 2015). As the ROAD approach, they are also designed for discrimination into two groups. This is helpful for obtaining asymptotic results, however not quite helpful for complicated applications involving several groups.

Finally, a function-constraint reformulation of ROAD may look like this:

$$\min_{a^\top a=1} \|a\|_1 - \tau_1 a^\top T a + \tau_2 (d^\top a - 1)^2 \ ,$$

or, in a more numerically stable form, like this:

$$\min_{a^\top a=1} \|a\|_1 + \tau_1 (\lambda - a^\top T a) + \tau_2 (d^\top a - 1)^2 \ ,$$

where λ is an eigenvalue of T.

Before moving on, it is worth mentioning the way the population case is dealt with when $p > n$ in (Fan et al., 2012). The idea is that nevertheless, in population we still have $n > p$, and all singular covariance matrices eventually behave very well. As a result one ends up with same results as for Fisher's LDA. However, if we always have to analyse data with $p > n$, and deal with the computational consequences, then, isn't more realistic to assume that our population is composed of infinite number of data matrices all of them having $p > n$? There are still very few works making steps towards such "population" thinking, e.g. considering inference under the assumption of $n/p \to \alpha < 1$ in population. In our opinion, extending the ideas considered for supervised problems in (Giraud, 2015) looks more promising.

7.4.8   Sparse LDA through optimal scoring (SLDA)

Clemmensen et al. (2011) make use of the LDA reformulation as optimal scoring problem considered in (Hastie et al., 1994, 1995). The benefit is that the optimal scoring does not require W^{-1}, and thus, is applicable for $p > n$. A very similar approach to LDA is considered in (Ye, 2007).

Let G be the $n \times g$ group indicator matrix known from Section 7.1. Then, the sparse LDA solution is obtained by solving:

$$\min_{a,b} \|Gb - Xa\|^2 + \tau_1 \|a\|_1 + \tau_2 \|a\|_2^2 \ ,$$

subject to $\frac{1}{n}b^{\top}G^{\top}Gb = 1$ and $b^{\top}G^{\top}Gb_{\text{old}} = 0$, where τ_1 and τ_2 are non-negative tuning parameters. The idea is that for $\tau_1 = \tau_2 = 0$, the vectors Xa yield the standard Fisher's CVs (for $n > p$). SLDA solves an elastic net problem (using additional ℓ_2 penalty), which provides some advantages over the LASSO (Zou and Hastie, 2005). Further development of this approach is given in Merchante et al. (2012).

7.4.9 Multiclass sparse discriminant analysis

We already mentioned in the previous Section that the classification error and the CVs depend on T^{-1} and $\mu_1 - \mu_2$ only through their product $a = T^{-1}(\mu_1 - \mu_2)$ (Cai and Liu, 2011; Fan et al., 2012; Hao et al., 2015). This observation is also exploited in (Mai et al., 2016) to derive a procedure for discrimination into more than two groups.

Mai et al. (2016) define and consider the following discriminant directions $a_i = T^{-1}(\mu_1 - \mu_i)$ for $i = 2, \ldots, g$ as unknowns, because T^{-1} is not defined. They claim that the following problem:

$$\min_{a_2,\ldots,a_g} \sum_{i=2}^{g} \frac{1}{2}a_i^{\top}Ta_i - a_i^{\top}(\mu_1 - \mu_i) \tag{7.44}$$

is "convex optimization formulation of the Bayes rule of the multiclass linear discriminant analysis". For large applications, (7.44) is enhanced with additional group LASSO constraints and the problem remains convex. Further details, algorithms and examples can be found in Mai et al. (2016).

Though the examples in Mai et al. (2016) give indication for good results, the explanation how (7.44) becomes "convex optimization formulation of the Bayes rule" seems not quite convincing. For this reason, we provide here an alternative motivation for the convexification of the multiclass LDA.

We start with the observation, that the identity

$$(\mu_1 - \mu_i)^{\top}T^{-1}(\mu_1 - \mu_i) = (\mu_1 - \mu_i)^{\top}T^{-1}TT^{-1}(\mu_1 - \mu_i),$$

makes (7.44) equivalent to

$$-\min_{a_2,\ldots,a_g} \sum_{i=2}^{g} \frac{1}{2}a_i^{\top}(\mu_1 - \mu_i),$$

and, in turn, to

$$\max_{a_2,\ldots,a_g} \sum_{i=2}^{g} a_i^\top (\mu_1 - \mu_i) \,.$$

Now, recall that in case of two groups, $a = T^{-1}(\mu_1 - \mu_2)$ is solution of the LDA problem (7.15), and the corresponding eigenvalue is

$$(\mu_1 - \mu_2)^\top T^{-1}(\mu_1 - \mu_2) = a^\top (\mu_1 - \mu_2) \,. \tag{7.45}$$

However, if $p > n$, then T is singular and $T^{-1}(\mu_1 - \mu_2)$ is not defined. Then, it seems natural to consider a in (7.45) as unknown, and find the best a by maximizing $a^\top (\mu_1 - \mu_2)$, i.e. fulfilling the variational property of the eigenvalue (7.45). To avoid unbounded solutions we additionally impose $a^\top a = 1$.

This idea can be extended for the case of $g(> 2)$ groups by defining the following discriminant directions $a_i = T^{-1}(\mu_1 - \mu_i)$ for $i = 2,\ldots,g$ as in (Mai et al., 2016). Then the best a_i is found by solving the following problem:

$$\max_{\substack{a_2,\ldots,a_g \\ a_i^\top a_i = 1}} \sum_{i=2}^{g} a_i^\top (\mu_1 - \mu_i) \,,$$

which can be rewritten as a problem for maximizing a sum of quadratic forms:

$$\max_{\substack{a_2,\ldots,a_g \\ a_i^\top a_i = 1}} \sum_{i=2}^{g} a_i^\top T a_i \,. \tag{7.46}$$

In the classical case when $n > p$ and T is PD, the largest $g - 1$ eigenvalues a_i of T solve (7.46). The same kind of solution will work for singular T with $p > n$ provided it has at least $g - 1$ positive eigenvalues. For large applications, one can replace $a_i^\top a_i = 1$ by the LASSO constraint $\|a_i\|_{\ell_1} \leq \tau$, which will also make (7.46) a convex problem.

7.4.10   Sparse LDA through GEVD

GEVD is a general framework which accommodates PCA, LDA and canonical correlation analysis (CCA) (Section 8.1). Forgetting about any previous notations, consider a pair of $p \times p$ symmetric matrices $\{U, V\}$, with PD V. The symmetric-definite GEVD is defined as $Ua = \mu Va$. Then, we distinguish the following particular GSVD problems, with specific choices for U and V:

- PCA: $U = X^\top X$ (sample correlation matrix) and $V = I_p$;

- LDA: $U = \bar{X}^\top J_n \bar{X}$ (between-group scatter matrix) and $V = X^\top J_n X$ (total scatter matrix);

- CCA: n observations are measured on p variables, naturally divided into two sets of size p_1 and p_2. Let X_1 denote the $n \times p_1$ matrix that is comprised of the first set of variables, and let X_2 denote the $n \times p_2$ matrix that is comprised of the remaining variables; assume that the columns of X_1 and X_2 have been standardized. Then:

$$U = \begin{bmatrix} O & X_1^\top X_2 \\ X_2^\top X_1 & O \end{bmatrix}, V = \begin{bmatrix} X_1^\top X_1 & O \\ O & X_2^\top X_2 \end{bmatrix} \text{ and } a = \begin{bmatrix} x_1 \\ x_2 \end{bmatrix} .$$

Thus, a method that produces sparse generalized eigenvectors should solve all PCA, LDA and CCA, after specifying the correct corresponding matrix pair $\{U, V\}$. However, in practice, such an universal method does not work equally well in all of the above scenarios.

An example is the universal method proposed in (Sriperumbudur et al., 2011) based on the following sparse GEVD:

$$\max_{a^\top Va \leq 1} a^\top Ua - \lambda P(a) , \qquad (7.47)$$

where $P(a)$ is a penalty term inducing sparseness of a. In contrast to most sparse methods adopting some variant of LASSO, their penalty is different and is supposed to give a tighter approximation than the ℓ_1-based penalties:

$$P(a) = \sum_1^p \log(\varepsilon + |a_i|)$$

where ε is a small non-negative number to avoid problems when some elements a_i of a are zero. Problem (7.47) is then formulated as a difference of convex functions (D.C.) program, giving the name DC-GEV of the method:

$$\min_{\substack{a^\top Va \leq 1 \\ -b \preceq a \preceq b}} \tau\|a\|_2^2 - \left[a^\top (U + \tau I_p)a + \lambda \sum_1^p \log(\varepsilon + |b_i|) \right] ,$$

where the first term is convex and the second term is jointly convex on a and b. The parameter $\tau \geq 0$ is chosen to make $U + \tau I_p$ PSD, e.g. $\tau \geq \max\{0, -\lambda_{\min}(U)\}$. The problem is solved by a local optimization algorithm based on the MM approach (Hunter and Lange, 2004).

With $V = I_p$, DC-GEV switches to sparse PCA, and becomes DC-PCA. The method compares very well even with the best methods for sparse PCA. However, things are more complicated with LDA, DC-FDA, which is the case with $U = B$ and $V = T$. The first weakness of DC-FDA is that it is designed to work with two groups only ($g = 2$). The good news is that in this case a simplified problem is solved:

$$\min_{d^\top a = 1} a^\top T a + \lambda \sum_1^p \log(\varepsilon + |a_i|) \,, \tag{7.48}$$

where $d = \bar{x}_1 - \bar{x}_2$. Further efforts are needed to adapt DC-FDA to work for problems with singular T when $p \gg n$. Little experimental evidence is available for solving such LDA problems with (7.48).

7.4.11   Sparse LDA without sparse-inducing penalty

In this section we consider a new procedure for sparse LDA. The sparseness of the discriminant functions A will be achieved without employing sparse-inducing penalties. Instead, we will look for a solution A with specific pattern of sparseness, with only one non-zero entry in each row of A. The methods are inspired by the works of (Timmerman et al., 2010; Vichi and Saporta, 2009).

The following model represents the original data X by only the group means projected onto the reduced space, formed by the orthonormal discriminant functions A. The model can be formally written as

$$X = G(\bar{X}A)A^\top = G\bar{X}(AA^\top) \,, \tag{7.49}$$

where $\bar{X}$ is the $g \times p$ matrix of group means and G is the $n \times g$ indicator matrix of the groups. We know from Section 7.1 that $\bar{X} = (G^\top G)^{-1}G^\top X = N^{-1}G^\top X$, and the model (7.49) can be rewritten as

$$X = HX(AA^\top) \,, \tag{7.50}$$

where $H = G(G^\top G)^{-1}G^\top$ is a projector. The $p \times r$ orthonormal matrix A contains the orthonormal "raw coefficients" of the problem, and r is the number of required discriminant functions.

We want to find sparse raw coefficients A but without relying on sparseness inducing constrains as in the previous sections. In general, this is unsolvable problem, but it can be easily tackled if we restrain ourselves to a particular pattern of sparseness: e.g. each row of A should posses a *single* non-zero entry. Thus, the total number of non-zero entries in A will be p. To construct A with such a pattern, we introduce a $p \times r$ binary (of 0's and 1's) membership matrix V, indicating which variables have non-zero loadings on each particular discriminant function, i.e. in each column of A. Then, A will be sought in the form of a product $A = \text{Diag}(b)V$, where $\text{Diag}(b)$ is a diagonal matrix formed by the vector b. The ith element of b gives the non-zero value at the ith row of A. In other words, V is responsible for the locations of the non-zero entries in A, while b will give their values. Apparently, the choice of V and b will affect the fit of the model (7.50). Thus, we need to solve the following least squares problem:

$$\min_{V,b} \|X - HX[\text{diag}(b)VV^\top\text{diag}(b)]\| , \qquad (7.51)$$

which is called for short SDP (**S**parse **D**iscriminative **P**rojection).

One can develop possibly a better SDP method if the classification error is minimized instead of fitting the data matrix X or its projection onto the subspace spanned by the discriminant functions. Nevertheless, the main weakness of SDP is that for large p the SDP solutions are not sparse enough, and thus, not attractive for large applications.

7.5 Conclusion

This chapter is divided into two parts depending on the format of the data to be analysed. In the classical situation the number of observations n is greater than the number variables p $(n > p)$. For short, such data matrices are called "vertical". This data format is in contrast with many modern applications, when the number of the involved variable is much bigger than the number of available observations, i.e. $p \gg n$. For short, such data matrices are called "horizontal".

Section 7.1 introduces the basic LDA assumptions and notations common for both data formats. The following Section 7.2 and Section 7.3 consider LDA for vertical data. Section 7.2.1 revisits the standard Fisher's LDA and the ways its solutions (CVs) are interpreted, while in Section 7.2.2 we discuss the less popular but quite useful sometimes option of orthogonal CVs. Section 7.3 proposes sparse versions for those two types of CVs, by penalizing the standard optimization problems.

Section 7.4 is dedicated to a number of approaches to LDA for horizontal data. The first four subsections consider methods dealing solely with the singularity of the within-group scatter matrix W. The remaining six subsections consider methods to obtain sparse CVs for horizontal data.

In particular, Section 7.4.1 treats the horizontal LDA as a generalized SVD (GSVD), i.e. it works with row data rather than with scatter matrices. Section 7.4.2 redefines the LDA cost function to avoid the explicit inverse of W. Alternatively, Section 7.4.3 works with the original LDA cost function, but restricted to the complement of the null space of W/T. Finally, Section 7.4.4 achieves the LDA goals through CPC, PLS or MDS.

Methods for sparse LDA start in Section 7.4.5 with its simplest version when W is (simplified to) diagonal. Section 7.4.6 adopts the function-constrained approach from Section 4.8 to obtain sparse CVs by penalized Fisher's LDA. Section 7.4.7 minimizes the classification error instead of Fisher's LDA ratio (7.8), while Section 7.4.8 rewrites LDA equivalently through optimal scoring. Section 7.4.9 redefines LDA in terms of discriminant directions which requires maximization of a sum of quadratic forms, while Section 7.4.10 considers sparse LDA based on GSVD. Finally, Section 7.4.11 proposes to find a $p \times r$ matrix of CVs with a single non-zero element in every row.

7.6   Exercises

1. Prove that W from (7.1) and B from (7.2) sum up to the following total scatter matrix:

$$T = \sum_{i=1}^{g} \sum_{j=1}^{n_i} (x_{ij} - \bar{x})(x_{ij} - \bar{x})^\top , \qquad (7.52)$$

where x_{ij} denotes the *vector* of measurements made on the jth observation belonging to the ith group, and $\bar{x}$ is the mean of the data.

2. (a) Prove that the solution of (7.13) and (7.14) for two groups is given by:
$$a = W^{-1}(\bar{x}_1 - \bar{x}_2) \ .$$

Hint: Use $n_1\bar{x}_1 + n_2\bar{x}_2 = n\bar{x}$ to express B in terms of $\bar{x}_1 - \bar{x}_2$.

(b) Following the same way of thinking, one can find that the solution of (7.15) is given by
$$a = T^{-1}(\bar{x}_1 - \bar{x}_2) \ .$$

Is there any contradiction with the previous exercise?

3. The LDA constructs rules from the data that help best allocate a new observation to one of them. Suppose we are given data with two groups of observations.

(a) Let the data have a single variable, x, which has normal distribution in each group, $N(\mu_1, \sigma)$ and $N(\mu_2, \sigma)$, respectively. Suppose that a new observation x_0 is allocated to the group where it gets maximal value of the corresponding (normal) density, i.e. allocate x_0 to the first group if $p_{N,1}(x_0) > p_{N,2}(x_0)$, where $p_{N,i}(x) = \frac{1}{\sqrt{2\pi\sigma_i^2}} \exp\left\{-\frac{1}{2}\left(\frac{x - \mu_i}{\sigma}\right)^2\right\}$.

Show that the allocation rule in this scenario is given by
$$|x - \mu_1| < |x - \mu_2| \ ,$$

i.e. the allocation rule is a linear function of x.

(b) Let the data have p variables, i.e. $x \in \mathbb{R}^p$, and the normal densities in each group are $N_p(\mu_1, \Sigma)$ and $N_p(\mu_2, \Sigma)$, respectively. Show that the allocation rule in this case is given by
$$a^\top (x - \mu) > 0 \ ,$$

where $a = \Sigma^{-1}(\mu_1 - \mu_2)$ and $\mu = .5(\mu_1 + \mu_2)$. Thus, the allocation rule is again a linear function of x.

(c) Let the data have p variables, i.e. $x \in \mathbb{R}^p$, and the normal densities in each group are $N_p(\mu_1, \Sigma_1)$ and $N_p(\mu_2, \Sigma_2)$, respectively. Show that the allocation rule is a quadratic function of x:
$$x^\top (\Sigma_1^{-1} - \Sigma_2^{-1})x - 2x^\top (\Sigma_1^{-1}\mu_1 - \Sigma_2^{-1}\mu_2) + c > 0 \ ,$$

where $c = [\ln(\det(\Sigma_1)) + \mu_1^\top \Sigma_1^{-1}\mu_1] - [\ln(\det(\Sigma_2)) + \mu_2^\top \Sigma_2^{-1}\mu_2]$.

4. To avoid W^{-1} in LDA for horizontal data, consider a Procrustes-like version of (7.16) as follows:

$$\min_{\substack{A \in \mathcal{O}(p,r) \text{ or } \mathcal{OB}(p,r) \\ D \in \mathcal{D}(r)}} \|X_B A - X_W AD\|_F . \qquad (7.53)$$

Hint: Problem (7.53) can be split into an orthonormal (or oblique) Procrustes problem, and then, followed by update of D.

5. Consider a sparse version of (7.53) by adding a sparseness inducing penalty to the objective function, e.g. $1_p^\top |A| \leq \tau 1_r^\top$, where $|A|$ is the matrix composed of the absolute values of A and $\tau \geq 0$ is a tuning parameter.

6. Consider a sparsified version of the GSVD of the matrix pair $\{X_B, X_T\}$.

7. (a) Consider LDA for horizontal data based on

$$\max_A \frac{\text{trace}(A^\top BA)}{\text{trace}(A^\top WA)} ,$$

and write Manopt codes for its solution.

(b) Consider a sparse version of this LDA problem by solving

$$\min_A \|A\|_{\ell_1} - \tau \frac{\text{trace}(A^\top BA)}{\text{trace}(A^\top WA)} .$$

8. (Sriperumbudur et al., 2011, p. 30) Let T be PD and b is a given/fixed vector. Suppose that a_1 is the solution to

$$\max_{a^\top Ta} (b^\top a)^2 ,$$

and that a_2 is the solution to

$$\min_{b^\top a = 1} a^\top Ta .$$

Then $a_1 = a_2 \sqrt{b^\top T^{-1} b}$. This shows that the GEVD problem for the pair $\{bb^\top, T\}$ can be solved as a minimization problem. Hint: Use Lagrange multipliers to solve both problems.

9. (Sriperumbudur et al., 2011, p. 33) For $a \in \mathbb{R}^p$ define:

$$\|a\|_\varepsilon = \lim_{\varepsilon \to 0} \sum_1^p |a_i|_\varepsilon = \lim_{\varepsilon \to 0} \sum_1^p \frac{\log\left(1 + \frac{|a_i|}{\varepsilon}\right)}{\log\left(1 + \frac{1}{\varepsilon}\right)} .$$

(a) Show that $\|a\|_0 = \lim_{\varepsilon \to 0} \|a\|_\varepsilon$ and $\|a\|_1 = \lim_{\varepsilon \to \infty} \|a\|_\varepsilon$. Hint: For $\|a\|_1$, denote $\nu = 1/\varepsilon$ and use the L'Hôpital's rule.

(b) For $x \geq 0$ define:

$$x_\varepsilon = \frac{\log(1 + x\varepsilon^{-1})}{\log(1 + \varepsilon^{-1})},$$

and show that for any $x > 0$ and $0 < \varepsilon < \infty$, the value of x_ε is closer to 1 than x is to 1. Hint: Consider separately the cases of $x > 1$ and $1 > x > 0$.

Therefore, for any $\varepsilon > 0$, $\|a\|_\varepsilon$ is a tighter approximation to $\|a\|_0$ than $\|a\|_1$ is.

10. *K*-means clustering (Hartigan, 1975, Ch 4). Let X be a data $n \times p$ matrix with observations divided into $g(\ll n)$ disjoint clusters defined by their indicator $n \times g$ matrix $G = \{g_{ij}\}$, where $g_{ij} = 1$ if ith observation belongs to jth group, and zero otherwise. Let C be the $g \times p$ matrix of cluster centroids. The K-means clustering problem is defined by:

$$\min_{C,G} \|X - CG\|_F^2 .$$

Check that, for known G, the cluster centroids are given by $C = (G^\top G)^{-1} G^\top X$. Then, G is updated to reflect the new C, and so on. For more details see (Gan et al., 2007, Ch 9), (Elkan, 2003).

11. Combine dimension reduction and clustering into a single procedure by adding a clustering index (Friedman and Rubin, 1967) as a penalty term to the original objective function of the PCA formulation (4.8).

(a) Suppose, that g initial groups/clusters are defined/found (say, by K-means clustering) in the data projection matrix F given by their $n \times g$ indicator matrix G. Check that their within-groups scatter matrix is given by:

$$W = I_r - F^\top H F = F^\top (I_n - H)F , \quad H = G(G^\top G)^{-1} G^\top .$$

Hint: Apply (7.6) to the "data matrix" $F \in \mathcal{O}_0(n, r)$.

(b) Combine the PCA formulation (4.8) with the minimisation of $\mathrm{trace}(W)$ (Charrad et al., 2014, 2.19), which leads to:

$$\max_{F \in \mathcal{O}_0(n,r)} \mathrm{trace}\, F^\top (XX^\top + H)F ,$$

where H(and G) is updated at each step based on the new F via K-means clustering.

(c) Explore other clustering updates, as well as other clustering indexes, e.g. maximizing trace($W^{-1}B$) (Banfield and Raftery, 1993).

Chapter 8

Cannonical correlation analysis (CCA)

© Springer Nature Switzerland AG 2021

N. Trendafilov and M. Gallo, *Multivariate Data Analysis on Matrix Manifolds*,

Springer Series in the Data Sciences, https://doi.org/10.1007/978-3-030-76974-1_8

8.1   Introduction

Canonical correlation analysis (CCA) is a natural generalization of PCA when the data contain two sets of variables (Hotelling, 1936). As in PCA, CCA also aims at simplifying the correlation structure between the two sets of variables by employing linear transformations. However, the presence of two sets of variables complicates the problem, as well as the notations.

8.2   Classical CCA Formulation and Solution

CCA is dealing with an $n \times (p_1 + p_2)$ data block-matrix $X = [X_1 \ X_2]$. For simplicity, we assume that X is centred and consider the following scatter matrices $C_{11} = X_1^\top X_1, C_{22} = X_2^\top X_2$ and $C_{12} = X_1^\top X_2$. The classical CCA assumes that $\max\{p_1, p_2\} < n$, which implies that C_{11} and C_{22} are PSD, and that $\operatorname{rank}(C_{12}) \leq \min\{p_1, p_2\}$.

Now, form the following linear combinations $y_1 = X_1 a_1$ and $y_2 = X_2 a_2$. The (Pearson) correlation between them is given by

$$\frac{a_1^\top C_{12} a_2}{\sqrt{a_1^\top C_{11} a_1}\sqrt{a_2^\top C_{22} a_2}} \tag{8.1}$$

and is usually called *canonical correlation coefficient*.

The CCA purpose is to find weights a_1 and a_2, such that the canonical correlation coefficient between y_1 and y_2 is maximized, subject to the condition that y_1 and y_2 have unit variances ($y_1^\top y_1 = 1$ and $y_2^\top y_2 = 1$), i.e.:

$$\max_{\substack{a_1^\top C_{11} a_1 = 1 \\ a_2^\top C_{22} a_2 = 1}} a_1^\top C_{12} a_2 \ . \tag{8.2}$$

The linear combinations y_1 and y_2 are called the first canonical variates. To simplify the notations, let store them in the matrices Y_1 and Y_2, which for now contain a single column. In the same manner, store a_1 and a_2 in the matrices A_1 and A_2.

As in PCA, one can find additional canonical variates $y_1 = X_1 a_1$ and $y_2 = X_2 a_2$ by solving (8.2) under the additional constraints that they are uncorrelated with their predecessors, i.e. $a_1^\top C_{11} A_1 = 0$ and $a_2^\top C_{11} A_2 = 0$. The resulting weight vectors a_1 and a_2 are then inserted as second columns in A_1 and A_2, respectively. Similarly, the new canonical variates y_1 and y_2 are inserted into Y_1 and Y_2 as their second columns. One can repeat this procedure, say, r times. After the rth step, A_1 and A_2 have sizes $p_1 \times r$ and $p_2 \times r$, while Y_1 and Y_2 both have size $n \times r$. CCA assumes that $r < \min\{p_1, p_2\}$.

The CCA problem looks very similar in spirit to PCA, with the exception that the objective function in (8.2) has two unknowns and that C_{12} is rectangular. The classical CCA solution is based on C_{11} and C_{22} being PD, i.e. non-singular and invertible, and is making use of Lagrange multipliers. Consider the Lagrangian function

$$l(a_1, a_2) = a_1^\top C_{12} a_2 - \lambda_1 a_1^\top C_{11} a_1 - \lambda_2 a_2^\top C_{22} a_2 ,$$

which, after differentiation, gives two equations for λ_1 and λ_2:

$$C_{12} a_2 - \lambda_1 C_{11} a_1 = 0_{p_1} \quad \text{and} \quad a_1^\top C_{12} - \lambda_2 a_2^\top C_{22} = 0_{p_2} . \qquad (8.3)$$

Taking into account that $C_{21} = X_2^\top X_1 = (X_1^\top X_2)^\top = C_{12}^\top$, the Lagrange multipliers are found as $\lambda_1 = \lambda_2 = \lambda = a_1^\top C_{12} a_2$. Then, the first equation in (8.3) turns into $C_{21} C_{11}^{-1} C_{12} a_2 - \lambda_1 C_{21} a_1 = 0$, after right-hand side multiplication by $C_{21} C_{11}^{-1}$. The substitution of the second equation from (8.3), finally gives $C_{21} C_{11}^{-1} C_{12} a_2 - \lambda^2 C_{22} a_2 = 0$, or:

$$(C_{22}^{-1} C_{21} C_{11}^{-1} C_{12} - \lambda^2 I_{p_2}) a_2 = 0_{p_2} , \qquad (8.4)$$

which is a (non-symmetric) eigenvalue problem (Golub and Van Loan, 2013, §8.7). Such types of problems are already familiar from (7.13) and (7.14) related to Fisher's LDA considered in Section 7.2.1.

With a similar to (8.3) transformation, one can obtain an eigenvalue problem for a_1:

$$(C_{11}^{-1} C_{12} C_{22}^{-1} C_{21} - \lambda^2 I_{p_1}) a_1 = 0_{p_1} . \qquad (8.5)$$

Let the first r pairs of CVs be collected in A_1 and A_2 with sizes $p_1 \times r$ and $p_2 \times r$, respectively. Then, the equalities $A_1^\top C_{11} A_1 = I_{r_1}$, $A_2^\top C_{22} A_2 = I_{r_2}$, $A_1^\top C_{12} A_2 = 0_{r_1 \times r_2}$ and $A_2^\top C_{21} A_1 = 0_{r_2 \times r_1}$ follow from the eigenvector properties of the CVs.

8.3   Alternative CCA Definitions

The classical CCA assumption for PD C_{11} and C_{22} can be utilized in another way. Let $C_{11} = U_1^\top U_1$ and $C_{22} = U_2^\top U_2$ be the Choleski decompositions with non-zero entries on the main diagonals of the upper triangular matrices U_1 and U_2. Then, one can introduce new unknowns $a_1 \equiv U_1 a_1$ and $a_2 \equiv U_2 a_2$, and the CCA problem (8.2) becomes

$$\max_{\substack{a_1^\top a_1 = 1 \\ a_2^\top a_2 = 1}} a_1^\top (U_1^{-\top} C_{12} U_2^{-1}) a_2 . \tag{8.6}$$

The existence of $U_1^{-\top}$ and U_2^{-1} is guaranteed by the positive definiteness of C_{11} and C_{22} for the classical CCA.

The Lagrangian multipliers approach from the previous section can be readily applied to (8.6). Instead, we prefer to adopt another approach and consider the CCA problem as optimization on a product of two unit spheres. As with LDA in Section 7.2.1, the Choleski transformation helps to simplify the constraints. However, the objective function remains non-symmetric.

Another alternative is to define CCA problem by maximizing the sum of the variation of r canonical correlations collected in Y_1 and Y_2, i.e trace$(Y^\top Y)$. Formally this meant to replace (8.2) by

$$\max_{\substack{A_1^\top C_{11} A_1 = 1 \\ A_2^\top C_{22} A_2 = 1}} \text{trace}(A_1^\top C_{12} A_2) . \tag{8.7}$$

Here it is important to stress that (8.6) and (8.7) are not equivalent: the total variation of the canonical correlations is maximized in (8.7), rather than their individual successive ones as in (8.6). Every solution of (8.6) is solution to (8.7), but the opposite is not true. Similar situation is already observed in (4.1) and (4.2) for PCA in Section 4.2.

Making use of the Choleski decompositions of C_{11} and C_{22} mentioned above we introduce new unknowns $A_1 \equiv U_1 A_1$ and $A_2 \equiv U_2 A_2$. Then, the CCA problem (8.7) becomes

$$\max_{\substack{A_1 \in \mathcal{O}(p_1 \times r) \\ A_2 \in \mathcal{O}(p_2 \times r)}} \text{trace}[A_1^\top (U_1^{-\top} C_{12} U_2^{-1}) A_2] . \tag{8.8}$$

After the solution of (8.8) is found, one should remember to post-multiply A_1 and A_2 by U_1^{-1} and U_2^{-1}, respectively.

Problems (8.7) and (8.8) can be readily upgraded to produce sparse CVs by adding sparseness inducing penalty to their objective functions or by adopting the function-constrained approach used for sparse PCA in Section 4.8 and sparse LDA in Section 7.4.6.

8.4 Singular Scatter Matrices C_{11} and/or C_{22}

The classical CCA heavily depends on the assumption that C_{11} and C_{22} are PD. Unfortunately, in many applications this is not the case. This is especially true nowadays when the number of variables p_1 and p_2 is far bigger than the number of observations n. Keep in mind, that the problem with singular C_{11} and C_{22} can happen even if $p_1 + p_2 < n$. The cause for this is the presence of linearly dependent variables in the data. This phenomenon is well known as *collinearity*.

Similar situation with singular scatter matrices is already faced in Section 7.4 in relation to LDA for horizontal data. We remember that a way to tackle this problem is through regularization. In regularized CCA, a regularization term(s) is added to stabilize the solution, leading to the following modified GEVD:

$$C_{12}(C_{22} + \eta_2 I_{p_2})^{-1} C_{21} a_1 = \lambda^2 (C_{11} + \eta_1 I_{p_1}) a_1,$$

where $\eta_1, \eta_2 > 0$ are regularization parameters.

Another way out is to employ some generalized inverse. However, as with LDA, these two approaches may not be feasible for large data. As with LDA, one can look for alternative CCA definitions that avoid working with C_{11}^{-1} and C_{22}^{-1}.

8.5 Sparse CCA

The CCA solutions (CVs) are notorious with being difficult to interpret. In this sense, producing sparse and easily interpretable solutions is a very important issue for CCA. However, as in PCA, the sparse CVs loose some of their optimal properties: the successive CVs are correlated among the groups of variables. More precisely, let the first r pairs of sparse CVs be collected in A_1 and A_2 with sizes $p_1 \times r$ and $p_2 \times r$, respectively. Then, $A_1^\top C_{11} A_1 = I_{r_1}$ and $A_2^\top C_{22} A_2 = I_{r_2}$ are usually fulfilled as constraints of the methods for sparse CCA. However, $A_1^\top C_{12} A_2 = 0_{r_1 \times r_2}$ and $A_2^\top C_{21} A_1 = 0_{r_2 \times r_1}$ cannot be achieved by the sparse CVs because they are not eigenvectors any more. This resembles the sparse PCA where the sparse PCs cannot be made simultaneously orthonormal and uncorrelated, as the standard PCs (eigenvectors) are. Again, choice should be made for every particular problem to reflect its specific interpretation needs.

8.5.1   Sparse CCA Through Sparse GEVD

The close similarity between CCA and PCA (and LDA) suggests that the available approaches for sparse PCA are likely to be extendable for CCA.

A general approach considering sparse GEVD is proposed in (Sriperumbudur et al., 2011). It can deal with PCA, LDA and CCA as special cases of GEVD and is called for short DC-GEV. The application of the method to sparse LDA was already discussed in Section 7.4.10. Here we briefly outline its adaptation to CCA.

As mentioned in Section 7.4.10, in the CCA scenario, DC-GEV re-names to DC-CCA and solves the following problem:

$$\max_{a^\top V a \leq 1} a^\top U a - \lambda \sum_1^p \log(\varepsilon + |a_i|) , \tag{8.9}$$

where

$$U = \begin{bmatrix} O & X_1^\top X_2 \\ X_2^\top X_1 & O \end{bmatrix}, V = \begin{bmatrix} X_1^\top X_1 & O \\ O & X_2^\top X_2 \end{bmatrix} \text{ and } a = \begin{bmatrix} a_1 \\ a_2 \end{bmatrix} .$$

As the block-matrix U is always singular, the objective functions in (8.9) is replaced by a regularized version $\tau \|a\|_2^2 - a^\top (U + \tau I_p) a$, where $\tau = -\lambda_{\min}(U)$. This choice of the parameter τ makes $U + \tau I_p$ PSD, which makes it possible to solved (8.9) as a D.C. program (Sriperumbudur et al., 2011).

Witten et al. (2009) also attack the problem by adopting a general approach. They propose to replace the standard SVD of the data matrix X by a penalized matrix decomposition (PMD). In this sense, the standard dimension reduction is replaced by *sparse* dimension reduction. The proposal is to approximate the data matrix X with same type of matrix decomposition as SVD but replacing the singular vectors with sparse ones. Such types of methods are available for quite some time (Kolda and O'Leary, 1998), where sparse singular vectors with prescribed cardinality (number of zeros) are obtained.

In other words, PMD considers $X \approx \sum_1^r \lambda_i u_i v_i^\top$, where u and/or v are subject to sparseness inducing penalties, and are found by solving:

$$
\min_{\substack{\lambda \geq 0,\, u,\, v \\ u^\top u = 1\,,\, v^\top v = 1 \\ P_1(u) \leq \tau_1\,,\, P_2(v) \leq \tau_2}} \| X - \lambda u v^\top \|_F \,,
\tag{8.10}
$$

where P_i are convex penalty functions. Witten et al. (2009) take $P_1(u)$ to be the LASSO penalty, i.e. $P_1(u) = 1_p^\top |u|$, while $P_2(v)$ is the fused LASSO, namely, $P_2(v) = \sum_1^p |v_i| + \eta \sum_2^p |v_i - v_{i-1}|$, with $\eta > 0$ (Tibshirani et al., 2005). The solutions u and v of (8.10) also solve

$$
\max_{\substack{u,\, v \\ u^\top u = 1\,,\, v^\top v = 1 \\ P_1(u) \leq \tau_1\,,\, P_2(v) \leq \tau_2}} u^\top X v \,.
\tag{8.11}
$$

To make the PMD problem (8.11) convex, the unit length constraints in (8.11) are replaced by $u^\top u \leq 1$ and $v^\top v \leq 1$. Then (8.11) is solved by a "power method"-like iterations acting in turn on u for fixed v, and then, on v for fixed u, and so on until convergence. The only difference from the classical power method is that u is additionally constrained by LASSO, and v is constrained by fused LASSO.

Now note, that for fixed v, the solution u of

$$
\max_{\substack{u,\, v \\ u^\top u = 1\,,\, v^\top v = 1}} u^\top X v
\tag{8.12}
$$

is given by $\frac{Xv}{\|Xv\|_2}$. Therefore, v that solves (8.12) also solves

$$
\max_{\substack{u,\, v \\ u^\top u = 1\,,\, v^\top v = 1}} v^\top X^\top X v \,,
\tag{8.13}
$$

which is a EVD problem for $X^\top X$. Now, one can ask additionally for sparse v, which transforms (8.13) into a sparse EVD problem.

The attractiveness of the PMD approach is that without any further efforts it applies to CCA (as well as to PCA). Indeed, in our notations, penalized CVs can be obtained by simply inserting penalties in (8.13) which leads to the following PMD problem:

$$\max_{\substack{a_1, a_2 \\ a_1^\top C_{11} a_2 \leq 1 \,, a_2^\top C_{22} a_2 \leq 1 \\ P_1(a_1) \leq \tau_1 \,, P_2(a_2) \leq \tau_2}} a_1^\top C_{12} a_2 \,. \tag{8.14}$$

Both of the considered methods (DC-GEV and PMD) work with horizontal data $(p > n)$. In relation to PMD and other related methods one should note that the power method is the simplest method to adopt from a number of (better) methods for EVD/SVD. It is worth investigating other iterative methods for PMD, e.g. Lanczos bidiagonalisation, which are well described in (Trefethen and Bau, 1997, Ch V, VI) and (Demmel, 1997, Ch 5, 7).

8.5.2   LS Approach to Sparse CCA

Lykou and Whittaker (2010) utilize the LS formulation (8.29) of CCA considered in Section 8.9. The following alternating LS (ALS) algorithm is proposed to find the first CV:

- for given/fixed $y_1 = X_1 a_1$, solve

$$\min_{a_2^\top C_{22} a_2 = 1} \|y_1 - X_2 a_2\|_2 \,,$$

- for given/fixed $y_2 = X_2 a_2$, solve

$$\min_{a_1^\top C_{11} a_1 = 1} \|X_1 a_1 - y_2\|_2 \,.$$

The second CVs, b_1 and b_2, are found subject to the additional constraints $b_1^\top C_{11} a_1 = 0$ and $b_2^\top C_{22} a_2 = 0$, respectively, etc.

Lykou and Whittaker (2010) impose LASSO constraints on a_2 and a_1 at each step of the above ALS algorithm. Moreover, they require that the CVs coefficients do not change their signs while the LASSO shrinking process for the sake of computational advantages. The resulting problem is convex and is solved as a standard quadratic program. The method can be applied to horizontal data as well.

8.6 CCA Relation to LDA and PLS

8.6.1 CCA and LDA

Bartlett (1938) was probably the first to discover the relationship between CCA and LDA. Now, it is well known that when CCA is performed on a dataset with groups and on its indicator matrix representing the group membership, then the CCA solution coincides with Fisher's LDA one. The proof is given as an exercise in Section 8.9.

8.6.2 CCA and PLS

It is very common to say that the fundamental difference between CCA and PLS is that CCA maximizes the correlation while PLS maximizes the covariance. The scenario is the same as in Section 8.2: we are given a centred $n \times (p_1 + p_2)$ data block-matrix $X = [X_1 \ X_2]$, i.e. $1_n^\top X = 0_{p_1+p_2}^\top$. Both CCA and PLS, consider the following two linear combinations $y_1 = X_1 a_1$ and $y_2 = X_2 a_2$. As we know, the CCA goal is to maximize the canonical correlation coefficient between y_1 and y_2 given by (8.1):

$$\frac{a_1^\top C_{12} a_2}{\sqrt{a_1^\top C_{11} a_1} \sqrt{a_2^\top C_{22} a_2}} \ ,$$

which is achieved by solving (8.2):

$$\max_{\substack{a_1^\top C_{11} a_1 = 1 \\ a_2^\top C_{22} a_2 = 1}} a_1^\top C_{12} a_2 \ .$$

In contrast to CCA, the PLS goal is to maximize the *covariance* between y_1 and y_2, given by $a_1^\top C_{12} a_2$, which is achieved by solving:

$$\max_{\substack{a_1^\top a_1 = 1 \\ a_2^\top a_2 = 1}} a_1^\top C_{12} a_2 \ . \tag{8.15}$$

Thus, the relation between CCA and PLS reminds the LDA situation from Section 7.2.2 with classical and orthonormal CVs.

The obvious solution of (8.15) is that a_1 and a_2 are the left and the right singular vectors corresponding to the largest singular value of C_{12}. Alternatively, a_1 is the leading eigenvector of $C_{12}C_{21}$, and a_2 is the leading eigenvector of $C_{21}C_{12}$. Note, that in contrast to CCA, the PLS formulation does not include C_{11} and C_{22}, i.e. PLS is applicable to data with singular C_{11} and/or C_{22}. The approaches for sparse CCA from the previous section can be readily applied to (8.15).

Similar to CCA, there exists a LS equivalent to the PLS problem (8.15). Indeed, one can check by direct calculations that the following LS fitting problem (Barker and Rayens, 2003, Theorem 2):

$$\min_{\substack{a_1^\top a_1 = 1 \\ a_2^\top a_2 = 1}} \|X_1 - X_1 a_1 a_1^\top\|_2^2 + \|X_2 - X_2 a_2 a_2^\top\|_2^2 + \|X_1 a_1 - X_2 a_2\|_2^2 \quad (8.16)$$

is equivalent to (8.15). This alternative PLS formulation is particularly important because (8.16) additionally clarifies the PLS goal: one is interested in two linear combinations y_1 and y_2 which, as in CCA, are as close as possible to each other in LS sense and additionally provide the best rank-one approximations to X_1 and X_2.

The PLS problem (8.15) is frequently considered in its matrix form:

$$\max_{\substack{A_1 \in \mathcal{O}(p_1,r) \\ A_2 \in \mathcal{O}(p_2,r)}} A_1^\top C_{12} A_2 \ . \quad (8.17)$$

In a similar way, one can extend (8.16) as follows:

$$\min_{\substack{A_1 \in \mathcal{O}(p_1,r) \\ A_2 \in \mathcal{O}(p_2,r)}} \|X_1 - X_1 A_1 A_1^\top\|_F^2 + \|X_2 - X_2 A_2 A_2^\top\|_F^2 + \|X_1 A_1 - X_2 A_2\|_F^2 \ .$$

$$(8.18)$$

Instead of using SVD to solve (8.17), Wold (1975) suggested an alternative *non-linear iterative partial least squares* (NIPALS) algorithm. Note, that Eldén (2004) made a brisk warning that, in general, it is not advisable to replace SVD algorithms with NIPALS.

PLS is most frequently employed for regression problems, where the relationship between X_1 and X_2 is not "symmetric", in a sense that, say, X_1 is response and X_2 is predictor. Then, X_1 and X_2 are usually denoted as Y

and X, and the (regression) problem is defined as

$$\min_{A} \|Y - XA\|_F \ , \tag{8.19}$$

which general solution is $A = (X^\top X)^{-1} X^\top Y$, if the inverse of $X^\top X (C_{11})$
exists. Unfortunately, this is frequently not the case, e.g. presence of linearly
dependent variables, known as collinearity, or very few observations. As PLS
is not affected by singular $X^\top X$, it becomes a natural choice for tackling such
problems, along with other approaches as regularization, sparsification, etc.
There exist several variants of PLS regression (Eldén, 2004; Rosipal and
Krämer, 2006).

The notations PLS1 and PLS2 are usually used, if the response Y consists
of a single variable or of a multidimensional block of variables, respectively.
Consider PLS1 regression with $p_1 = 1$. Then, the response $Y (= X_1)$ becomes
a vector y, and (8.17) simplifies to a linear (programming) problem:

$$\max_{A^\top A = I_r} y^\top X A \ ,$$

which can be solved by solving r identical problems for each individual column of A. These sub-problems have the form:

$$\max_{a^\top a = 1} y^\top X a \ ,$$

which solution is

$$a = \frac{X^\top y}{\|X^\top y\|_2} \ .$$

Let the result be stored in a matrix A, column after column. Clearly, the
solution for the next column of A, should be orthogonal to the previous
one(s), i.e. $a^\top A = 0$. This can be achieved if X is replaced/deflated by
$X := X - AA^\top X$, when finding the next column of A. In fact, this is what
NIPALS does in case of PLS1.

8.7 More Than Two Groups of Variables

Probably the most interesting aspect of this class of methods is their relation
to multi-way/tensor data analysis.

8.7.1   CCA Generalizations

It seems that the first attempt to extend CCA to more than two groups of variables is made in (Horst, 1961). A number of CCA generalizations are listed and studied in (Kettenring, 1971), with a warning that the results are difficult to interpret.

The generalized CCA scenario looks as follows. The data containing three groups of variables are given as an $n \times (p_1 + p_2 + p_3)$ data block-matrix $X = [X_1 \ X_2 \ X_3]$. Assuming that X is centred, consider the following scatter block-matrix:

$$X^\top X = \begin{bmatrix} X_1^\top X_1 & X_1^\top X_2 & X_1^\top X_3 \\ X_2^\top X_1 & X_2^\top X_2 & X_2^\top X_3 \\ X_3^\top X_1 & X_3^\top X_2 & X_3^\top X_3 \end{bmatrix} = \begin{bmatrix} C_{11} & C_{12} & C_{13} \\ C_{21} & C_{22} & C_{23} \\ C_{31} & C_{32} & C_{33} \end{bmatrix} .$$

The classical CCA assumes that $\max\{p_1, p_2, p_3\} < n$, which implies that C_{11}, C_{22} and C_{33} are PSD, and that $\mathrm{rank}(C_{12}) \le \min\{p_1, p_2\}$, $\mathrm{rank}(C_{13}) \le \min\{p_1, p_3\}$ and $\mathrm{rank}(C_{23}) \le \min\{p_2, p_3\}$.

The next natural step is to form the following three linear combinations $y_1 = X_1 a_1$, $y_2 = X_2 a_2$ and $y_3 = X_3 a_3$. Probably the simplest way to generalize CCA to deal with three groups of variables is to maximize the sum of their mutual canonical correlation coefficients, subject to the condition that y_1, y_2 and y_3 have unit variances, i.e. $y_i^\top y_i = 1$ for $i = 1, 2, 3$:

$$\max_{\substack{a_1^\top C_{11} a_1 = 1 \\ a_2^\top C_{22} a_2 = 1 \\ a_3^\top C_{33} a_3 = 1}} a_1^\top C_{12} a_2 + a_1^\top C_{13} a_3 + a_2^\top C_{23} a_3 .$$

The CCA purpose is to find weights a_1, a_2 and a_3, such that the sum of the canonical correlation coefficients between y_1, y_2 and y_3 is maximized. This generalized CCA easily extends to m groups of variables, as follows:

$$\max_{a_i^\top C_{ii} a_i = 1} \sum_{1 \le i < j \le m} a_i^\top C_{ij} a_j , \quad i, j = 1, \ldots, m . \tag{8.20}$$

Problem (8.20) is referred to as SUMCOR and can be extended for arbitrary m group of variables. Many other options for generalized CCA are possible, e.g. take the absolute values or the squares of the terms in (8.20). They are known as SABSCOR and SSQCOR, respectively. Of course, for $m = 2$, they all reduce to the classical CCA. Any CCA generalization is supposed to fulfil this consistency condition.

The approach proposed in (Carroll, 1968) is particularly interesting because it can be utilized in several ways, see Exercises 8.9. As above, let $X = [X_1, \ldots, X_m]$ be a data block-matrix with corresponding CVs given by $Y_i = X_i A_i$ for $i = 1, \ldots, m$, and $A_i \in \mathbb{R}^{p_i \times r}$ for some $r \leq \min\{p_1, \ldots, p_m\}$. Consider the following problem:

$$\min_{Y \in \mathcal{O}(n,r)} \sum_{i=1}^{m} \|Y - Y_i\|_F = \min_{Y \in \mathcal{O}(n,r)} \sum_{i=1}^{m} \|Y - X_i A_i\|_F , \tag{8.21}$$

where $Y \in \mathbb{R}^{n \times r}$ is some unknown "target" configuration. In contrast to the standard CCA, there are no constraints on A_i in (8.21). The idea is that at the minimum of (8.21), one has $Y \approx X_i A_i$ for all $i = 1, \ldots, m$. The first very important implication is that $A_i^\top X_i^\top X_i A_i \approx I_r$, which simply follows from $Y \in \mathcal{O}(n,r)$. Second, $\|X_i A_i - X_j A_j\|_F$ are all small, which intuitively suggests that (8.21) gives very much the same results as (8.20).

One can see that at the minimum of (8.21), necessarily

$$Y^\top \left(\sum_{1}^{m} X_i A_i \right) = \left(\sum_{1}^{m} X_i A_i \right)^\top Y , \tag{8.22}$$

i.e. $Y^\top (\sum X_i A_i)$ is a symmetric matrix. If Y is not supposed to be orthonormal, then one can write $Y = \sum X_i A_i$, see also (8.32). However, to accommodate the orthonormal constraint on Y, the general form of $Y^\top (\sum X_i A_i)$ is taken to be $Y^\top S Y$, for some symmetric S, i.e. $SY = \sum X_i A_i$. After substitution of the (unconstrained) solution for $A_i = (X_i^\top X_i)^{-1} X_i^\top Y$, one finds that $S = \sum X_i (X_i^\top X_i)^{-1} X_i^\top$. Thus, Y is composed of the eigenvectors of S, corresponding to the r largest eigenvalues.

Concerning the consistency condition, one can prove directly that for $m = 2$, the solutions of (8.21) and the standard CCA coincide (van de Velden, 2011). For $m > 2$, (8.21) is solved with additional constraints $A_i \in \mathcal{O}(p_i, r)$. Then, the optimality condition (8.22) is replaced by

$$Y^\top X_i A_i = A_i^\top X_i^\top Y , \text{ for } i = 1, \ldots, m . \tag{8.23}$$

Results in similar spirit are obtained in (Gower, 1989), where CCA is related to Procrustes analysis, see also (Gower and Dijksterhuis, 2004, 13.5).

They argue that a CCA generalization ($m > 2$) based on its Procrustes formulation is more adequate one, e.g. preserves CVs orthogonality, see also (8.33). Another group of related works and methods (STATIS, RV coefficient) cast the CCA problem for several groups of variables as simultaneous analysis of several data matrices (Lavit et al., 1994; Sabatier and Escoufier, 1976). Without going into details we only mention that the central idea is to compare the collection of data matrices on the basis of specially configured matrix distances.

Contemporary taxonomy of many (popular) CCA generalizations can be found in (Hanafi and Kiers, 2006). Nowadays, the following MAXBET problem seems one of the most widely used. It is defined as

$$\max_{A} \operatorname{trace}(A^\top X^\top X A) , \tag{8.24}$$

where $X = [X_1 \ldots, X_m] \in \mathbb{R}^{n \times p}$ is a block-data matrix composed of m blocks (of variables), such that $p_1 + \ldots + p_m = p$. It is assumed that $X^\top X$ is PD. The unknown $A \in \mathbb{R}^{p \times r}$ is also a block-matrix composed of m blocks $A_i \in \mathbb{R}^{p_i \times r}$ and $A_i^\top A_i = I_r$, i.e. $A_i \in \mathcal{O}(p_i, r)$.

Hanafi and Kiers (2006) propose a MM-like algorithm for solving (8.24). An extensive study of the problem is proposed in (Liu et al., 2015). It also provides numerical comparison between different iterative solutions of (8.24) and scenarios, including some experiments with Manopt.

8.7.2   CCA Based on CPC

Neuenschwander and Flury (1995) propose an approach based on the CPC idea aiming to overcome the interpretation difficulties of the generalized CCA models. The aims is to produce common CVs (CCV) for each group of variables, but the canonical correlation can vary. The CCV method is restricted to work only for data with equal number p of variables in each group, i.e. $p_1 = \ldots = p_m = p$ and $X = [X_1 \ldots, X_m] \in \mathbb{R}^{n \times mp}$. Another important restriction is that $n \geq mp$. For such data, $X^\top X$ is a block-matrix with square $p \times p$ blocks denoted by C_{ij}. The CCV method assumes that there exists a $p \times p$ non-singular matrix A with unit length columns, i.e. $A \in \mathcal{OB}(p)$, such that $A^\top C_{ij} A = \Lambda_{ij}$ is diagonal for all $i, j = 1, \ldots, m$, say $\Lambda_{ij} = \operatorname{Diag}(\lambda_{ij,1}, \ldots, \lambda_{ij,p})$. This means that the CCVs are the same for all blocks and are collected as columns of A. The diagonal matrices Λ_{ij}

are specific for each block of $X^\top X$. In more compact notations, the CCV definition/model becomes (Neuenschwander and Flury, 1995, p.555):

$$(I_m \otimes A)^\top X^\top X (I_m \otimes A) = \text{diag}_B(\Lambda_{ij}) := \Lambda , \qquad (8.25)$$

where $\Lambda \in \mathcal{M}(mp)$ is a block-diagonal matrix, with diagonal blocks Λ_{ij}.

Note, that the standard CCA applied to each pair $\{X_i, X_j\}$, $i, j = 1, \ldots, m$ of data blocks requires the solution of (8.5), i.e. the EVD of $C_{ii}^{-1} C_{ij} C_{jj}^{-1} C_{ji}$. Then, making use of the CCV assumptions, one finds that

$$C_{ii}^{-1} C_{ij} C_{jj}^{-1} C_{ji} = A\Lambda_{ii}^{-1}\Lambda_{ij}\Lambda_{jj}^{-1}\Lambda_{ji} A^{-1} = A\Xi_{ij}A^{-1} ,$$

where the kth element of the diagonal matrix Ξ_{ij} is $\xi_{ij,k} = \lambda_{ij,k}^2/(\lambda_{ii,k}\lambda_{jj,k})$, for $k = 1, \ldots, p$. This also implies that the columns of A are eigenvectors of $C_{ii}^{-1} C_{ij} C_{jj}^{-1} C_{ji}$ for *any* choice of $\{i, j\}$, $i, j = 1, \ldots, m$.

The CCV estimation problem defined in (Neuenschwander and Flury, 1995) follows closely the normal theory used to derive the maximum likelihood CPC estimators (Flury, 1988, Ch 4). Then, the CCV estimation is formulated as the following constrained optimisation on $\mathcal{OB}(p)$ of a log-likelihood function of the form:

$$\max_{\Lambda, A \in \mathcal{OB}(p)} \log(\det(\Lambda)) - 2\log(\det(Q)) + \text{trace}(\Lambda^{-1}Q^\top X^\top XQ) , \quad (8.26)$$

where $Q = I_m \otimes A$ and Λ as in (8.25). The algorithm proposed to solve (8.26) in (Neuenschwander and Flury, 1995) is generalization of the FG algorithm for the CPC problem considered in the next Chapter 9. Further progress in the CCV approach is made in (Goria and Flury, 1996), but there is still room for much more development. The LS version of the CCV estimation problem is readily defined from (8.25). Of course, both versions of the CCV estimation can be treated as optimisation problems on matrix manifolds.

The next technique is not explicitly related to CPC, but looks for "common" portion of information contained in several data matrices. It has become common to perform multiple analyses on the same set of observations, e.g. patients. This requires analysing simultaneously several datasets each involving different variables measured on the same set of observations. Such type of data is a special case of a three-way data array also called tensor data considered in Section 10.8, but does not really take into account the third mode of the data. In this sense, it can be seen as a method relating several groups of variables.

Lock et al. (2013) propose to analyse such kind of data by extracting the "joint" part of the information carried out by all involved variables and an "individual" part, reflecting the specific information carried out by the variables in each group. The method is called **J**oint and **I**ndividual **V**ariation **E**xplained (JIVE).

For convenience it is assumed here that the data format is changed to variables × observations, because usually $p \gg n$. Thus the data matrices $X_1, X_2, ..., X_m$ have sizes $p_i \times n$ and are collected in a $p \times n$ super-matrix $X = \left[X_1^\top, X_2^\top, \ldots, X_m^\top \right]^\top$, where $p = \sum_1^m p_i$. As in CCA, all X_i are first put on an equal footing by row-centring and making them with unit norm each, i.e. $X_i := X_i / \|X_i\|_F$. Other normalizations are also possible, e.g. the AFM method takes $X_i := X_i / \sigma_i^{\max}$, where $\sigma_i^{\max}$ is the maximal singular value of X_i (Escofier and Pages, 1983).

Then, the JIVE problem is defined as follows. For P, D, Q, U_i, D_i and V_i, ($i = 1, 2, ..., m$) solve the following LS fitting problem:

$$
\min \|X - X_J - X_I\|_F = \min \left\| \begin{bmatrix} X_1 \\ X_2 \\ \vdots \\ X_m \end{bmatrix} - PDQ^\top - \begin{bmatrix} U_1 D_1 V_1^\top \\ U_2 D_2 V_2^\top \\ \vdots \\ U_m D_m V_m^\top \end{bmatrix} \right\|_F ,
$$
(8.27)

where

1. $P \in \mathcal{O}(p, r), Q \in \mathcal{O}(n, r), U_i \in \mathcal{O}(p_i, r)$ and $V_i \in \mathcal{O}(n, r)$,

2. D and D_i are $r \times r$ diagonal;

3. $V_1^\top Q = V_2^\top Q = \ldots = V_m^\top Q = O_r$.

Note that the rank of approximation r can be made different for the joint and the individual terms, X_J ans X_I. The algorithm (8.27) for solving the JIVE problem is based on alternating SVD and is given below.

Lock et al. (2013) compare JIVE to a number of different but related methods as consensus PCA and multiblock PLS (Westerhuis et al., 1998), CCA and PLS. They demonstrate the superior JIVE's performance on artificial and real data analysis problems. Additionally, including sparseness inducing penalties on the JIVE loadings matrices P and $U_i, i = 1, \ldots, m$ in (8.27) is considered and discussed.

Algorithm 8.1 Solving JIVE (8.27).

{Initialization}
Find (truncated) SVD of $X \approx QDP^\top$ and let $X_J = QDP^\top$
{Alternating procedure}
repeat
    Divide $R = X - X_J$ into m blocks R_i of sizes $p_i \times n$
    Find m SVDs of $R_i(I_{n \times n} - PP^\top) = U_i D_i V_i^\top$
    Stack all $U_i D_i V_i$ in X_I
    Update $R = R - X_I$
    Stop if $\|R\|_F < \epsilon$
    Else, find SVD $X - X_I \approx QDP^\top$ and let $X_J = QDP^\top$
until convergence

8.8 Conclusion

Section 8.2 reminds the classical CCA formulation and solution, while Section 8.3 proposes alternative CCA definitions, including translation into matrix notations. Thus, CCA becomes optimization/estimation problem on a product of two Stiefel manifolds. Next, Section 8.4 briefly discusses issues related to the singularity of the involved covariance matrices.

Section 8.5 is dedicated to methods for obtaining sparse canonical variates. Section 8.5.1 describes approaches based on the generalized versions of EVD or SVD, while Section 8.5.2 utilizes the LS regression reformulation of CCA.

Section 8.6.1 briefly reminds the connection between CCA and LDA, while Section 8.6.2 provides alternative definitions of PLS and explains its relation to CCA.

Section 8.7 considers the less developed case of CCA when more than two groups of variables are available. The main difficulty seems to be the great number of possible generalizations which look equally eligible. Several of the most popular and/or accepted ones of them are outlined in Section 8.7.1. Next, in Section 8.7.2 we consider a less standard approach to CCA with several groups of variables based on the CPC from Chapter 9. The aim is to construct *common* CVs for all groups of input variables which requires optimization on an oblique manifold.

8.9   Exercises

1. Let X be a $n \times p$ data matrix which observations are divided into g groups. Suppose G is its indicator matrix representing the group membership of the observations as in Section 7.1. Form a new block-data matrix $[X \ G]$ with size $n \times (p + g)$. Then, the CCA of $[X \ G]$ is related to the following scatter matrix:

$$\begin{bmatrix} X^\top \\ G^\top \end{bmatrix} J_n [X \ G] = \begin{bmatrix} X^\top J_n X & X^\top J_n G \\ G^\top J_n X & G^\top J_n G \end{bmatrix} , \tag{8.28}$$

where the notations from (7.4)–(7.6) are employed. Thus, the CCA problem (8.5) for $[X \ G]$ is equivalent to finding the singular vector of

$$T^{-1/2}(X^\top J_n G)(G^\top J_n G)^{-1/2} ,$$

corresponding to its maximal singular value, or the eigenvector of

$$T^{-1/2}(X^\top J_n G)(G^\top J_n G)^{-1}(G^\top J_n X)T^{-1/2} ,$$

corresponding to its maximal eigenvalue. Note, that $G^\top J_n G$ is singular, and the inverse sign should be understood as a "generalized" one, e.g. $(G^\top J_n G)^{-1/2} = V_r D_r^{-1/2} V_r^\top$, where D_r contains only the largest $r(= g-1)$ non-zero singular values, and V_r collects the corresponding (right) singular vectors. Show that CCA of $[X \ G]$ is equivalent to LDA of X by proving that:

$$B = (G\bar{X})^\top J_n(G\bar{X}) = (X^\top J_n G)(G^\top J_n G)^{-1}(G^\top J_n X) .$$

Hint: Observe that $J_n G = G(GG)^{-1} G^\top J_n G$.

2. Show that the CCA problem (8.2) is equivalent to minimizing the distance (8.29) between the canonical variates y_1 and y_2:

$$\min_{\substack{a_1^\top C_{11} a_1 = 1 \\ a_2^\top C_{22} a_2 = 1}} \|X_1 a_1 - X_2 a_2\|_2 , \tag{8.29}$$

i.e. CCA becomes a LS fitting problem. If C_{11} and C_{22} are PD, then (8.29) becomes a problem on a product of two unit spheres.

3. Show that the CCA problem (8.7) is equivalent to minimizing the LS distance (8.30) between the canonical variates Y_1 and Y_2:

$$\min_{\substack{A_1^\top C_{11} A_1 = I_{r_1} \\ A_2^\top C_{22} A_2 = I_{r_2}}} \|X_1 A_1 - X_2 A_2\|_F \ . \tag{8.30}$$

Thus, CCA becomes a two-sided Procrustes-like problem (Gower and Dijksterhuis 2004, 13.5). If C_{11} and C_{22} are PD, then (8.30) becomes a problem on a product of two Stiefel manifolds, i.e. for $\{A_1, A_2\} \in \mathcal{O}(p_1, r_1) \times \mathcal{O}(p_2, r_2)$. In fact, (8.30) can be extended to more than two, say $m \geq 2$, groups of variables:

$$\min_{\substack{A_i^\top C_{ii} A_i = I_{r_i} \\ A_j^\top C_{jj} A_j = I_{r_j}}} \sum_{1 \leq i < j \leq m} \|X_i A_i - X_j A_j\|_F \ . \tag{8.31}$$

4. Prove the following identity (Gower and Dijksterhuis, 2004, (9.2)):

$$\sum_{1 \leq i < j \leq m} \|X_i A_i - X_j A_j\|_F = m \sum_{i=1}^{m} \|Y - X_i A_i\|_F \ , \tag{8.32}$$

where $Y = \frac{1}{m} \sum_{i=1}^{m} X_i A_i$, and consider possible relations between (8.21) and (8.31). For $m = 2$, make use of (8.32) to show that the standard CCA and (8.21) coincide (Gower and Dijksterhuis, 2004, 13.5).

5. Solve (8.21):

   (a) as an EVD problem of a certain matrix. Hint: For fixed Y, find the unconstrained solution for A_i, and substitute it back in (8.21).

   (b) by making use of an iterative procedure. Make sure that the successive a_i are orthogonal to the previous ones. Hint: initialize (random) Y, and solve linear optimisation for A_i, then update Y, etc.

   (c) Suppose that n observations are made on m categorical variables, and that the ith variable has p_i categories, such that $p_1 + \ldots + p_m = p$. Each categorical variable can be represented by an indicator $n \times p_i$ matrix G_i, assuming that each observation can score on one category only for each variable. For example, the indicator matrix

$$G_i = \begin{bmatrix} 0 & 1 & 0 \\ 1 & 0 & 0 \\ 1 & 0 & 0 \\ 0 & 0 & 1 \end{bmatrix},$$

means that four observations are made on the ith variable which has three categories ($p_i = 3$). The first observation has a score 1 on the second category of the ith variable, and, of course, zeros on the rest categories. The second observation has a score 1 on the first category, and zeros on the rest. The third observation also scores on the first category, etc. Now, if you substitute the indicator matrices G_i of the categorical variables in place of X_i in (8.21), then the resulting problem is equivalent to *multiple correspondence analysis*, or *homogeneity analysis* (Gifi, 1990; ten Berge, 1993). Write Manopt codes that find simultaneous solution $Y, A_1, \ldots, A_m$ of the CCA problem (8.21) subject to $Y \in \mathcal{O}(n, r)$ and $A_i \in \mathcal{O}(p_i, r)$.

6. Prove the following identity (Gower and Dijksterhuis, 2004, (13.28)):

$$\sum_{1 \leq i < j \leq m} \|X_i A_i - X_j A_j\|_F = \text{trace}\{A^\top [m \ \text{diag}_B(X^\top X) - X^\top X]A\} ,$$

(8.33)

where $X \in \mathcal{M}(n, p)$ and $A \in \mathcal{M}(p, r)$ are block matrices composed of m blocks $X_i \in \mathcal{M}(n, p_i)$ and $A_i \in \mathcal{M}(p_i, r)$, respectively, such that $\sum_i p_i = p$. The block-diagonal matrix $\text{diag}_B(X^\top X)$ contains the diagonal blocks of $X^\top X$. Consider algorithms for generalized CCA based on (8.33).

7. Write Manopt codes to solve the CCV problem (8.26). Solve the numerical examples considered in (Neuenschwander and Flury, 1995) and compare your results.

8. Explore the following much faster but less accurate alternative to JIVE from Section 8.7.2:

$$\min \left\| \begin{bmatrix} X_1 \\ X_2 \\ \vdots \\ X_m \end{bmatrix} - PDQ^\top - \begin{bmatrix} U_1 D_1 \\ U_2 D_2 \\ \vdots \\ U_m D_m \end{bmatrix} V^\top \right\|_F ,$$

where

(a) $P \in \mathcal{O}(p, r), Q \in \mathcal{O}(n, r), U_i \in \mathcal{O}(p_i, r)$ and $V \in \mathcal{O}(n, r)$,

(b) D and D_i are $r \times r$ diagonal ,

(c) $Q^\top V = O_r$.

Chapter 9

Common principal components (CPC)

© Springer Nature Switzerland AG 2021
N. Trendafilov and M. Gallo, *Multivariate Data Analysis on Matrix Manifolds*,
Springer Series in the Data Sciences, https://doi.org/10.1007/978-3-030-76974-1_9

9.1   Introduction

The common principal components (CPC) and the proportional principal components (PPC) models are two possible generalizations of the standard PCA for several covariance matrices. The goal of this chapter is to revisit the classical CPC and PPC estimation based on the ML principle, and compare their features to their LS counterparts. The original CPC and PPC models are designed to produce full set of components. Such procedures do not seem appropriate for modern applications involving large number of variables. This problem is addressed by revising the CPC and PPC models and algorithms to produce only few components, and thus, make them suitable for dimension reduction of large data. Finally, we consider few other related problems also involving simultaneous analysis of several (square) matrices, as ICA and INDCSCAL.

The CPC model was introduced and studied by Flury (1988). It is one of many possible generalizations of the standard PCA of several covariance matrices (Jolliffe, 2002). We want to stress that the initial motivation for introducing CPC is to study discrimination problems where the group covariance matrices are not equal as required by LDA, but more generally, share common principal axes (Krzanowski, 1984; Flury, 1988). Later, Flury et al. (1994) demonstrated in a simulated study that even a simpler model than CPC with *proportional* covariance matrices (Flury, 1988, Ch 5) can provide quite competitive discrimination compared to other more complicated methods. Such principal components are called proportional (PPCs). The PPC model is particularly interesting because it admits very simple and fast implementation suitable for large data.

One can appreciate that CPC and PPC can provide a natural alternative to do LDA of small sample (horizontal) data *without* struggling with singular covariance/correlation matrices. Surprisingly, this straightforward option is hardly explored so far (Flury et al., 1994; Zou, 2006). To make this issue more intriguing, we note that CPC can "do" LDA of the Fisher's Iris data and produce three misclassified flowers as the original LDA solution does. Moreover, it can produce solution with only *two* misclassified flowers, which outperforms the classic LDA solution (see Figure 9.1), but is probably due to a randomly accumulated error as it occurs once in, say, 50 runs.

A possible explanation why CPC and PPC are not actively applied to LDA problems could be the fact that their estimation is originally based on

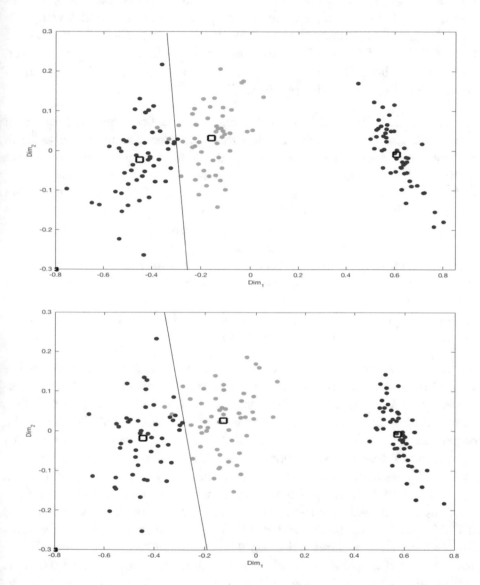

Figure 9.1: CPC for linear discrimination of Fisher's Iris data with three (upper plot) and two misclassified flowers (points).

the assumption of k normal populations with different PD $p \times p$ covariance matrices $\Sigma_i, i = 1, 2, ..., k$. ML procedure for their estimation is proposed in (Flury, 1988, Ch 4). However, such assumptions may not be always appropriate for horizontal data. It makes sense to consider also LS alternatives of the CPC and PPC models and their estimation. For example, the ML procedure for the CPC estimation is realized in the *FG* algorithm based on Jacobi/Givens rotations (Flury, 1988, Ch 9).

The CPC estimation problem can be easily set in LS terms. It becomes a problem for the simultaneous diagonalization of k covariance matrices Σ_i in LS sense. A very similar problem for simultaneous diagonalization of several dis/similarity matrices (usually referred to as INDSCAL) was considered by De Leeuw and Pruzansky (1978) and solved numerically by an algorithm *SUMSCAL* also based on Jacobi rotations. B. Flury pointed out that "...the least-squares procedure...has the apparent advantage of not requiring positive definiteness..." of the covariance matrices (Flury, 1988, p. 206). The main disadvantage is that the INDSCAL aims simultaneous multidimensional scaling and the results "...may be rather unreasonable in PCA" (Flury, 1988, p. 207). There is a great number of other generalizations of PCA, making use of the LS data fitting (Jolliffe, 2002, §13.5). They generalize other (than diagonalization) characteristic PCA features.

The chapter is organized as follows. The CPC estimation problem is formulated in Section 9.2 using both the ML and LS principles. Then, in Section 9.3, the ML and LS problems for fitting the CPC model to the data are reformulated as optimization problems on manifolds. This facilitates the comparison of the ML and LS approaches by inspecting the corresponding optimality conditions. Section 9.4 considers several new procedures for doing CPC. Section 9.4.1 gives a brief reminder of the existing numerical approaches to CPC. Section 9.4.2 shows how the individual eigenvalues/-variances D_i^2 in each group can be found *without* having the ML or LS CPCs. Then, Section 9.4.3 considers CPC reformulations with *known* individual eigenvalues/variances. Section 9.5 considers CPC versions capable to produce only $r(< p)$ CPCs useful for dimension reduction. Section 9.6 deals with the PPCs model. Section 9.6.1 revisits the original PPC model formulation (Flury, 1988, Ch 5) and its LS recasting. Section 9.6.2 is dedicated to dimension reduction with ML and LS PPCs.

We illustrate the considered models and corresponding algorithms by the well-known Fisher's Iris data widely used in (Flury, 1988) and in a number of other research papers.

9.2 CPC estimation problems

The CPC model has been introduced in (Flury, 1988). There are considered k normal populations and it is assumed that their $p \times p$ population covariance matrices $\Sigma_i, i = 1, 2, \ldots, k$ are PD. Then, their ML estimations can be found by minimizing the following log-likelihood function (Flury, 1988, p. 67):

$$\min_{\Sigma_1, \Sigma_2, \ldots, \Sigma_k} \sum_{i=1}^{k} n_i \left[\log(\det(\Sigma_i)) + \text{trace}(\Sigma_i^{-1} S_i) \right] , \tag{9.1}$$

where $S_i, i = 1, 2, \ldots, k$ are given sample covariance matrices.

The hypothesis of the CPC model is that the population covariance matrices can be decomposed simultaneously as

$$\Sigma_i = Q D_i^2 Q^\top , \tag{9.2}$$

where Q is a common $p \times p$ orthogonal matrix for all $\Sigma_i, i = 1, 2, \ldots, k$ and D_i^2 are positive $p \times p$ diagonal matrices. The CPC model assumes that the sample covariance matrices S_i are based on $n_i (\geq p)$ degrees of freedom each. In the LDA context, each group has n_i observations. Let the total number of observations be $n = n_1 + \ldots + n_k$.

The CPC estimation problem is then for given sample $p \times p$ covariance matrices $S_i, i = 1, 2, \ldots, k$ to find their common eigenvectors and corresponding (different) eigenvalues, such that

$$S_i \approx Q D_i^2 Q^\top \tag{9.3}$$

as close as possible in some sense. By substituting (9.2) in (9.1), the ML estimation of the parameters Q and $D_i^2, i = 1, 2, \ldots, k$ is formulated in (Flury, 1988, p. 67) as the following optimization problem (ML-CPC):

$$\min_{\substack{Q \in \mathcal{O}(p) \\ D_i \in \mathcal{D}(p)}} \sum_{i=1}^{k} n_i \left\{ \log \left[\det(Q D_i^2 Q^\top) \right] + \text{trace} \left[(Q D_i^2 Q^\top)^{-1} S_i \right] \right\} . \tag{9.4}$$

Thus, ML-CPC is a constraint optimization problem on the product of the Lie group of all $p \times p$ orthogonal matrices $\mathcal{O}(p)$ and $\mathcal{D}(p)^k = \underbrace{\mathcal{D}(p) \times \ldots \times \mathcal{D}(p)}_{k}$, where $\mathcal{D}(p)$ is the linear subspace of all $p \times p$ diagonal matrices. One notes that such D_i^2 admit zero entries, which, in fact, makes the use of the log-likelihood function (9.4) invalid. The reason for choosing this form is the same as in EFA discussed in Section 5.6.1. For many practical situations, positive D_i^2 can be achieved by working on $\mathcal{D}(p)$, which is simpler than working on the cone of all PD diagonal matrices.

Note that the ML-CPC formulation can be rewritten in a more convenient form, by utilizing that Q is orthogonal matrix and $\det Q = \pm 1$:

$$\min_{\substack{Q \in \mathcal{O}(p) \\ D_i \in \mathcal{D}(p)}} \sum_{i=1}^{k} n_i \left\{ \log \left[\det(D_i^2) \right] + \text{trace} \left[D_i^{-2} \odot (Q^\top S_i Q) \right] \right\} , \qquad (9.5)$$

where $\odot$ denotes the Hadamard (element-wise) matrix product.

The corresponding LS setting of the CPC problem (LS-CPC) is

$$\min_{\substack{Q \in \mathcal{O}(p) \\ D_i \in \mathcal{D}(p)}} \sum_{i=1}^{k} n_i \|S_i - Q D_i^2 Q^\top\|_F^2 , \qquad (9.6)$$

which is very similar to the orthogonal INDSCAL problem (Takane et al., 2010; Trendafilov, 2004). The difference is that INDSCAL is interested in orthonormal matrix $Q \in \mathcal{O}(p, r)$, while LS-CPC works on $\mathcal{O}(p)$. In contrast to the ML definition of CPC (9.4), possible zero entries in D_i^2 do not affect its LS version (9.6).

9.3   ML- and LS-CPC

9.3.1   Gradients and optimality conditions

The Euclidean gradients of the ML-CPC objective function (9.4) with respect to the unknowns Q and D_i are as follows:

$$\nabla_Q = W_i Q D_i^2 , \qquad (9.7)$$

and

$$\nabla_{D_i} = D_i \odot (Q^\top W_i Q) , \qquad (9.8)$$

where $W_i = QD_i^{-2}(D_i^2 - Q^\top S_i Q)D_i^{-2}Q^\top$. After substitution of W_i into (9.7)–(9.8), we arrive at the following gradients:

$$\nabla_Q = QD_i^{-2}(D_i^2 - Q^\top S_i Q) \,, \tag{9.9}$$

and

$$\nabla_{D_i} = D_i^{-1} - D_i^{-3} \odot (Q^\top S_i Q) \,. \tag{9.10}$$

The Euclidean gradients of the LS-CPC objective function (9.6) with respect to the unknowns Q and D_i are as follows:

$$\nabla_Q = -S_i Q D_i^2 \,, \tag{9.11}$$

and

$$\nabla_{D_i} = D_i^3 - D_i \odot (Q^\top S Q) \,. \tag{9.12}$$

The first-order optimality condition for a stationary point of the ML-CPC objective function (9.4) can be found by simply making zero the right-hand sides of (9.9)–(9.10) which leads to the following:

Theorem 9.3.1. *A necessary condition for $(Q, D_1, \ldots, D_k) \in \mathcal{O}(p) \times \mathcal{D}(p)^k$ to be a stationary point of the ML-CPC (9.4) is that the following $k + 1$ conditions must hold simultaneously:*

- $\sum_{i=1}^{k} Q^\top S_i Q D_i^{-2}$ *be symmetric;*
- $(Q^\top S_i Q) \odot D_i^{-3} = D_i^{-1}.$

Similarly the first-order optimality condition for a stationary point of the LS-CPC objective function (9.6) can be described by zeroing the right-hand sides of (9.11)–(9.12):

Theorem 9.3.2. *A necessary condition for $(Q, D_1, \ldots, D_k) \in \mathcal{O}(p) \times \mathcal{D}(p)^k$ to be a stationary point of the LS-CPC is that the following $k+1$ conditions must hold simultaneously:*

- $\sum_{i=1}^{k} Q^\top S_i Q D_i^2$ *be symmetric;*
- $(Q^\top S_i Q) \odot D_i = D_i^3.$

If indeed $D_i, i = 1, ..., k$ do not have zero diagonal elements, then the second optimality conditions in Theorem 9.3.1 and Theorem 9.3.2 coincide and have the form:

$$\text{diag}(Q^\top S_i Q) = D_i^2 . \tag{9.13}$$

Unfortunately, the first optimality conditions in Theorem 9.3.1 and Theorem 9.3.2 are different and thus, in general, one cannot expect equal estimations from ML-CPC and LS-CPC. This is only possible if all S_i are (exactly) simultaneously diagonalizable, which is unrealistic to expect in practical problems. This result is established also in (Beaghen, 1997, Theorem 4.1).

Only in the case when $k = 1$ (i.e., the standard PCA), the first optimality conditions in Theorem 9.3.1 and Theorem 9.3.2 coincide. Then, they present a single optimality condition that $Q^\top S_1 Q$ must be a diagonal matrix, if the entries of D_1^2 are all distinct.

9.3.2   Example: Fisher's Iris data

For illustration, we consider the famous Fisher's Iris data. ML-CPC estimations are obtained by solving (9.4) with Manopt and using the gradients (9.9)–(9.10):

$$Q = \begin{bmatrix} .7367 & -.1640 & -.6471 & -.1084 \\ .2468 & -.8346 & .4655 & .1607 \\ .6047 & .5221 & .5002 & .3338 \\ .1753 & .0628 & .3382 & -.9225 \end{bmatrix},$$

and the diagonals of $D_i^2, i = 1, 2, 3$ as column vectors:

$$d_1^2 = \begin{bmatrix} 48.4592 \\ 5.5394 \\ 7.4688 \\ 1.0140 \end{bmatrix}, \quad d_2^2 = \begin{bmatrix} 69.1495 \\ 7.5369 \\ 6.7125 \\ 5.3645 \end{bmatrix}, \quad d_3^2 = \begin{bmatrix} 14.6461 \\ 12.5066 \\ 2.7526 \\ 1.0169 \end{bmatrix}.$$

The results are nearly identical to those reported in (Flury, 1988, pp. 94–97). As it can be expected, similar ML-CPC estimations are obtained by using the *FG* algorithm (Flury, 1988, Ch 9), which is given in the Appendix for completeness:

$$Q = \begin{bmatrix} -.7363 & -.1645 & .6473 & .1084 \\ -.2469 & -.8343 & -.4660 & -.1607 \\ -.6050 & .5224 & -.4996 & -.3338 \\ -.1754 & .0631 & -.3381 & .9225 \end{bmatrix},$$

and respectively:

$$d_1^2 = \begin{bmatrix} 48.4631 \\ 5.5390 \\ 7.4664 \\ 1.0139 \end{bmatrix}, \quad d_2^2 = \begin{bmatrix} 69.2226 \\ 7.5334 \\ 6.7169 \\ 5.3639 \end{bmatrix}, \quad d_3^2 = \begin{bmatrix} 14.6418 \\ 12.5086 \\ 2.7530 \\ 1.0169 \end{bmatrix}.$$

The minimum of the objective value (9.14) is 35.70.

LS-CPC estimations are obtained by solving (9.6) with Manopt and using the gradients (9.11)–(9.12):

$$Q = \begin{bmatrix} .7269 & .2001 & .6144 & .2313 \\ .2384 & .8200 & -.4515 & -.2582 \\ .6241 & -.5344 & -.4212 & -.3829 \\ .1547 & -.0457 & -.4901 & .8562 \end{bmatrix},$$

and respectively:

$$d_1^2 = \begin{bmatrix} 48.3734 \\ 5.5865 \\ 7.3654 \\ 1.1570 \end{bmatrix}, \quad d_2^2 = \begin{bmatrix} 69.3839 \\ 7.4535 \\ 7.5889 \\ 4.4105 \end{bmatrix}, \quad d_3^2 = \begin{bmatrix} 14.2910 \\ 12.8365 \\ 2.5575 \\ 1.2355 \end{bmatrix}.$$

The estimations of Q obtained by LS-CPC and ML-CPC have same signs, but their magnitudes are similar only remotely.

9.3.3 Appendix: MATLAB code for FG algorithm

MATLAB version of the original *FG* algorithm seems not available on the web. Our MATLAB translation of the original FORTRAN codes published in Flury (1988) is attached here.

```
1    global k epsF epsG n_g
2    A=[];
3
4    %Fisher's Iris Data————————————————————
5    n_g=[49 49 49]
6    A(:,:,3)=[12.4249 9.9216 1.6355 1.0331; 9.9216 14.3690 1.1698
            0.9298;...
7       1.6355 1.1698 3.0159 0.6069; 1.0331 0.9298 0.6069 1.1106];
8    A(:,:,1)=[26.6433 8.5184 18.2898 5.5780; 8.5184 9.8469 8.2653
            4.1204;...
9       18.2898 8.2653 22.0816 7.3102; 5.5780 4.1204 7.3102 3.9106];
10   A(:,:,2)=[40.4343 9.3763 30.3290 4.9094; 9.3763 10.4004 7.1380
            4.7629;...
11      30.3290 7.1380 30.4588 4.8824; 4.9094 4.7629 4.8824 7.5433];
12
13   %————————————————————
14   [p,p1,k]=size(A); epsF=.0001; epsG=.0001;
15   Bold=eye(p); B=orth(rand(p));
16
17   while norm(B-Bold,1)>epsF
18
19   Bold=B;
20   for j=2:p;
21       for m=1:(j-1);
22           T=[]; b=[B(:,m) B(:,j)];
23           for i=1:k
24               T(:,:,i)=b'*A(:,:,i)*b;
25           end
26           [c,s]=G_alg(T); tcs=s/(1+c);
27           B(:,m)=b(:,1)+s*(b(:,2)-tcs*b(:,1));
28           B(:,j)=b(:,2)-s*(b(:,1)+tcs*b(:,2));
29       end
30   end
31   end
32
33   B, disp('Eigenvalues:');
34   d=[];
35   for i=1:k
36       B'*A(:,:,i)*B;
37       d=[d diag(B'*A(:,:,i)*B)];
38   end
39   ld=log(d)*diag(n_g);
40   d
41
42   min=0; for j=1:k;
43       ws=B*diag(d(:,j))*B';
44   %     min=min+n_g(j)*(log(det(ws))+trace(inv(ws)*A(:,:,j)));
45       min=min+n_g(j)*log(det(diag(diag(B'*A(:,:,j)*B))));
46   end; disp('Minimum:'); min
47
48   %════════════════════════
49   function [c,s]=G_alg(T);
50   global k epsF epsG n_g
51   Q=zeros(2); Qnew=eye(2);
52   %while norm(Qnew-Q,'fro')>epsG
53   while norm(Qnew-Q,1)>epsG
```

```
    Q=Qnew;
    S=zeros(2);
for i=1:k
    delta1=Q(:,1)'*T(:,:,i)*Q(:,1); delta2=Q(:,2)'*T(:,:,i)*Q(:,2);
    if delta1*delta2==0; disp('Some delta is 0'); break; end
    w=(delta1-delta2)/(delta1*delta2);
    S=S+n_g(i)*w*T(:,:,i);
end
% find orthogonal Q which diagonalizes S
[c,s]=sym_schur2(S);
end
end

%━━━━━━━━━━━━━━━━━━━━━━━━━━━━━━
function [c,s]=sym_schur2(S);
if S(1,2)~=0
    ratio=(S(1,1)-S(2,2))/(2*S(1,2));
    if ge(ratio,0)
        t=1/(ratio+sqrt(1+ratio*ratio));
    else
        t=-1/(-ratio+sqrt(1+ratio*ratio));
    end
    c=1/sqrt(1+t*t); s=c*t;
else
    c=1; s=0;
end
end
```

9.4 New procedures for CPC estimation

9.4.1 Classic numerical solutions of ML- and LS-CPC

Instead of solving the standard CPC problem (9.5), the classic FG algorithm solves the following problem:

$$\min_{Q \in \mathcal{O}(p)} \sum_{i=1}^{k} n_i \log \left\{ \det \left[\mathrm{diag}(Q^\top S_i Q) \right] \right\} , \qquad (9.14)$$

which equivalence is observed by substituting the ML-CPC optimality condition $D_i^2 = \mathrm{Diag}(Q^\top S_i Q)$ from (9.13) in it (Flury, 1988, pp. 67–71). Once Q is found, one can obtain D_i^2 for all i.

An alternative option is also based on the same ML-CPC optimality condition. One can consider D_i known, and solve

$$\min_{Q \in \mathcal{O}(p)} \sum_{i=1}^{k} n_i \text{trace} \left[D_i^{-2} \odot (Q^\top S_i Q) \right] , \qquad (9.15)$$

followed by updating $D_i^2 = \text{Diag}(Q^\top S_i Q)$. The algorithm starts with a random D_i and alternatively finds Q and D_i at each step until convergence.

9.4.2   Direct calculation of individual eigenvalues/variances

Following (Chu, 1991), we can derive dynamical systems to find *only* the individual eigenvalues for each of the covariance matrices S_i, without paying attention to their CPCs. To see this, assume that $Q(t)$ is a smooth function of t, and $Q(t) \in \mathcal{O}(p)$ for any $t \geq 0$. Then, for $i = 1, 2, ..., k$, define

$$X_i(t) = Q(t)^\top S_i Q(t) ,$$

which after differentiation gives

$$\dot{X}_i = \dot{Q}^\top S_i Q + Q^\top S_i \dot{Q} . \qquad (9.16)$$

Now, to find the individual eigenvalues of the ML-CPC and LS-CPC problems, we have to express $\dot{Q}$ in (9.16) by using the corresponding gradients of Q from (9.9) and (9.11) respectively.

Let us start with the ML-CPC problem, for which we find the following (autonomous) gradient dynamical system:

$$\dot{Q} = Q \sum_{i=1}^{k} n_i \left[D_i^{-2}, Q^\top S_i Q \right] , \qquad (9.17)$$

where $[A, B] = AB - BA$ denotes for short the commutator/bracket of A and B. One can easily check, that the right-hand side of (9.17) is in fact the projection of (9.9) onto $\mathcal{O}(p)$.

The substitution of (9.17) into (9.16) gives

$$
\begin{aligned}
\dot{X}_i &= \dot{Q}^{\top} S_i Q + Q^{\top} S_i \dot{Q} \\
&= \sum_{i=1}^{k} n_i \left[D_i^{-2}, Q^{\top} S_i Q \right]^{\top} Q^{\top} S_i Q + Q^{\top} S_i Q \sum_{i=1}^{k} n_i \left[D_i^{-2}, Q^{\top} S_i Q \right] \\
&= -\sum_{i=1}^{k} n_i \left[D_i^{-2}, Q^{\top} S_i Q \right] Q^{\top} S_i Q + Q^{\top} S_i Q \sum_{i=1}^{k} n_i \left[D_i^{-2}, Q^{\top} S_i Q \right] \\
&= \left[Q^{\top} S_i Q, \sum_{i=1}^{k} n_i \left[D_i^{-2}, Q^{\top} S_i Q \right] \right] \\
&= \left[Q^{\top} S_i Q, \sum_{i=1}^{k} n_i \left[\operatorname{diag}(Q^{\top} S_i Q)^{-1}, Q^{\top} S_i Q \right] \right] \\
&= \left[X_i, \sum_{i=1}^{k} n_i \left[\operatorname{diag}(X_i)^{-1}, X_i \right] \right] .
\end{aligned}
\tag{9.18}
$$

Thus, the simultaneous diagonalization of several symmetric PD matrices $S_i, i = 1, 2, \ldots, k$ with respect to the ML fit can be obtained by solving (9.18) with initial values $X_i(0) = S_i$.

In a similar manner, the gradient dynamical system for the LS-CPC problem is given by

$$
\dot{Q} = -Q \sum_{i=1}^{k} n_i \left[D_i^2, Q^{\top} S_i Q \right] ,
\tag{9.19}
$$

and the right-hand side of (9.19) is the projection of (9.11) onto $\mathcal{O}(p)$. After substitution of (9.19) into (9.16), one finds the corresponding dynamical systems for the LS fit (PD not required):

$$
\dot{X}_i = -\left[X_i, \sum_{i=1}^{k} n_i \left[\operatorname{diag}(X_i), X_i \right] \right], X_i(0) = S_i .
\tag{9.20}
$$

The equations (9.20) are well known in numerical linear algebra (Chu, 1991), and are a generalization of the *double bracket flow* for diagonalization of a single symmetric matrix (Helmke and Moore, 1994). The equation (9.18) is new and reflects the ML fit to the data. The resemblance between the ML and LS dynamical systems (9.18) and (9.20) is striking.

For example, the solution of (9.18) for Fisher's Iris data gives the following ML eigenvalues:

$$d_1^2 = \begin{bmatrix} 48.4601 \\ 5.5393 \\ 7.4688 \\ 1.0143 \end{bmatrix} , \quad d_2^2 = \begin{bmatrix} 69.2233 \\ 7.5365 \\ 6.7126 \\ 5.3643 \end{bmatrix} , \quad d_3^2 = \begin{bmatrix} 14.6442 \\ 12.5066 \\ 2.7526 \\ 1.0170 \end{bmatrix} ,$$

which are very close to the ones obtained by the standard CPCs methods.

Similarly, the solution of (9.20) gives the following LS eigenvalues:

$$d_1^2 = \begin{bmatrix} 48.3722 \\ 5.5867 \\ 7.3665 \\ 1.1571 \end{bmatrix} , \quad d_2^2 = \begin{bmatrix} 69.3819 \\ 7.4538 \\ 7.5902 \\ 4.4109 \end{bmatrix} , \quad d_3^2 = \begin{bmatrix} 14.2907 \\ 12.8367 \\ 2.5575 \\ 1.2354 \end{bmatrix} ,$$

which are also very close to the standard ones considered before.

9.4.3   CPC for known individual variances

The main diagonals of the matrices X_i obtained by (9.18) and (9.20) contain the individual eigenvalues D_i^2 of the covariance matrices $S_i, i = 1, 2, \ldots, k$. One can go one step further and recover the corresponding CPCs. This can be arranged by substituting X_i into the minimization problems (9.4) and (9.6), which leads to solving the following problems: for ML-CPC

$$\min_{Q \in \mathcal{O}(p)} \sum_{i=1}^{k} n_i \text{trace}(D_i^{-2} Q^\top S_i Q) , \tag{9.21}$$

because the first term in (9.4) becomes constant for known D_i^2 and coincides with (9.15), and respectively, for the LS-CPC

$$\min_{Q \in \mathcal{O}(p)} - \sum_{i=1}^{k} n_i \text{trace}(D_i^2 Q^\top S_i Q) . \tag{9.22}$$

It is easy to see, that (9.21) and (9.22) are Procrustes problems. Indeed, solving (9.21) and (9.22) is equivalent to solving

$$\min_{Q \in \mathcal{O}(p)} \sum_{i=1}^{k} n_i \| S_i Q + Q D_i^{-2} \|^2 , \tag{9.23}$$

and respectively

$$\min_{Q \in \mathcal{O}(p)} \sum_{i=1}^{k} n_i \| S_i Q - Q D_i^2 \|^2 \ . \tag{9.24}$$

All of the listed problems can be solved Manopt. Of course, for known diagonal matrices $D_i, i = 1, 2, ...k$, the CPCs Q can be found by integrating the gradient dynamical systems (9.17) and (9.19) on $\mathcal{O}(p)$.

Applied to Fisher's Iris data, this approach gives ML and LS eigenvalues and CPCs, identical to those reported in Section 9.3.2 above. As an exercise, one can solve (9.21) to obtain the corresponding ML-CPC:

$$Q = \begin{bmatrix} .7367 & -.1640 & -.6471 & .1084 \\ .2468 & -.8346 & .4655 & .1607 \\ .6048 & .5221 & .5003 & .3338 \\ .1753 & .0628 & .3381 & -.9225 \end{bmatrix},$$

which correspond well to the standard ones. Similarly, by solving (9.22), one obtains the corresponding LS-CPC as

$$Q = \begin{bmatrix} .7274 & .1999 & .6145 & -.2310 \\ .2385 & .8199 & -.4520 & .2582 \\ .6245 & -.5346 & -.4215 & .3828 \\ .1548 & -.0457 & -.4904 & -.8564 \end{bmatrix},$$

which are also very similar to the LS-CPC obtained before.

9.5 CPC for dimension reduction

The CPC model is designed to find all p components. However, for dimension reduction purposes or simply when the number of variables is high it may not be desirable to find all CPCs, but only r of them. This problem is addressed in (Trendafilov, 2010), where an algorithm is proposed that finds the CPCs sequentially. The great benefit is that their variances are decreasing, at least approximately, in all k groups. Clearly, such CPCs are useful for *simultaneous* dimension reduction.

For illustration, the first two sequential CPCs are

$$Q = \begin{bmatrix} .7467 & -.1044 \\ .4423 & .7998 \\ .4743 & -.5905 \\ .1476 & .0288 \end{bmatrix},$$

and their eigenvalues:

$$d_1^2 = \begin{bmatrix} 46.6777 \\ 7.2244 \end{bmatrix}, \ d_2^2 = \begin{bmatrix} 64.6575 \\ 13.1740 \end{bmatrix}, \ d_3^2 = \begin{bmatrix} 19.0796 \\ 7.8347 \end{bmatrix},$$

which are remotely similar to the first two ML-CPCs considered in Section 9.3.2. The total variance explained by these sequential CPCs is 158.6479, with 130.4148 and 28.2331 for each component. For comparison, the first two ML-CPCs explain less total variance 157.8377, with 132.2548 and 25.5829 for each of them.

This section considers other approaches to achieve CPC procedure with dimension reduction features, i.e. producing only $r(< p)$ CPCs and explaining as much as possible variance in each group.

As before, there are considered k normal populations with same mean vector and it is assumed that their $p \times p$ covariance matrices $\Sigma_i, i = 1, 2, ..., k$ are all different. We can still assume that the CPC hypothesis (9.2) is valid. However, the ML-CPC problem is modified to find only r CPCs such that the covariance matrices Σ_i are presented simultaneously as closely as possible to diagonal matrices D_i^2 with size $r \times r$. In other words, Q is sought on the Stiefel manifold $\mathcal{O}(p, r)$ of all $p \times r$ orthonormal matrices $Q^\top Q = I_r$, i.e. one solves (9.5) with a slightly modified unknown Q.

In the LS case, this modified CPC problem simply reduces to the orthonormal INDSCAL (Takane et al., 2010; Trendafilov, 2004). Applied to Fisher's Iris data, one obtains the following CPCs:

$$Q = \begin{bmatrix} .7292 & -.0876 \\ .2394 & -.8786 \\ .6232 & .4607 \\ .1527 & -.0887 \end{bmatrix},$$

and corresponding individual variances:

$$d_1^2 = \begin{bmatrix} 48.3611 \\ 5.7966 \end{bmatrix}, \ d_2^2 = \begin{bmatrix} 69.3900 \\ 8.5040 \end{bmatrix}, \ d_3^2 = \begin{bmatrix} 14.3304 \\ 12.3864 \end{bmatrix},$$

which look reasonably similar to the first two original LS-CPC. The total variance explained by these CPCs is 158.7684, with 132.0814 and 26.6870 for each component. For comparison, the first two LS-CPCs from Section 9.3.2 explain less total variance 157.9248, with 132.0483 and 25.8765 for each of them.

A possible option to define a dimension reduction version of ML-CPC is to consider (9.5) rather than (9.4). Thus, we avoid the inversion of $Q^\top D_i^2 Q$, but have to keep in mind that $D_i^2 > 0$ should be fulfilled. Of course, for dimension reduction, the minimization of (9.5) with respect to Q should be on the Stiefel manifold $\mathcal{O}(p, r)$ for some pre-specified r.

Another way to attack the ML-CPC problem is by solving (9.21) on $\mathcal{O}(p, r)$, where $D_{0,i}$ are some appropriate initial values. Reasonable choices for $D_{0,i}$ are the r largest eigenvalues of each S_i, or $Q_0^\top S_i Q_0$ for some random $Q_0 \in \mathcal{O}(p, r)$. However, by solving,

$$\min_{Q \in \mathcal{O}(p,r)} \sum_{i=1}^{k} n_i \mathrm{trace}(D_{0,i}^{-2} Q^\top S_i Q) , \qquad (9.25)$$

one finds the CPCs for which the sums of the explained variances are *minimized*. For example, for Fisher's Iris data one obtains by solving (9.25) the following two CPCs:

$$Q = \begin{bmatrix} .6424 & .1716 \\ -.3677 & -.2309 \\ -.5277 & -.3378 \\ -.4167 & .8962 \end{bmatrix} ,$$

and corresponding individual variances:

$$d_1^2 = \begin{bmatrix} 7.4285 \\ 1.0481 \end{bmatrix} , \quad d_2^2 = \begin{bmatrix} 6.6403 \\ 4.7160 \end{bmatrix} , \quad d_3^2 = \begin{bmatrix} 2.7598 \\ 1.1403 \end{bmatrix} ,$$

These CPCs reasonably well correspond to the *last* two ML-CPCs found before. The variances of the first and the second CPCs are, respectively, 16.8285 and 6.9044 with total 23.7329.

This phenomenon is easily explained by remembering that the problem (9.25) is equivalent to (9.23) rewritten in the following form:

$$\min_{Q \in \mathcal{O}(p,r)} \sum_{i=1}^{k} n_i \| Q D_{0,i}^{-2} Q^\top + S_i \|^2 .$$

In order to find the first r CPCs for which the sums of the explained variances are *maximized*, one should solve the opposite problem to (9.25), i.e.:

$$\max_{Q \in \mathcal{O}(p,r)} \sum_{i=1}^{k} n_i \text{trace}(D_{0,i}^{-2} Q^\top S_i Q) \ . \tag{9.26}$$

Indeed, for Fisher's Iris data by solving (9.26), one obtains the following two CPCs:

$$Q = \begin{bmatrix} -.7446 & -.0429 \\ -.4164 & .7992 \\ -.4988 & -.5993 \\ -.1528 & -.0126 \end{bmatrix} \ ,$$

and corresponding individual variances:

$$d_1^2 = \begin{bmatrix} 47.2755 \\ 6.7427 \end{bmatrix} , \quad d_2^2 = \begin{bmatrix} 65.7536 \\ 11.7210 \end{bmatrix} , \quad d_3^2 = \begin{bmatrix} 18.4561 \\ 8.5601 \end{bmatrix} \ .$$

These CPCs reasonably well correspond to the first two ML-CPCs found before. The variances of the first and the second CPCs are, respectively, 131.4852 and 27.0238 with total 158.5090.

The dimension reduction problem (9.26) can be solved my Manopt. The readers are asked to write their own code and apply them to Fisher's Iris data.

Another algorithm can be developed by rewriting the objective function in (9.26) as

$$\sum_{i=1}^{k} n_i \sum_{j=1}^{r} \frac{q_j^\top S_i q_j}{d_{j,i}^2} = \sum_{j=1}^{r} q_j^\top \left(\sum_{i=1}^{k} \frac{n_i S_i}{d_{j,i}^2} \right) q_j \ . \tag{9.27}$$

and observing that the solution of (9.26) can be found by maximizing each of the r terms in the right-hand side of (9.27). Then, the solution of the problem (9.26) can be found by solving sequentially r problems of the form:

$$\max_{\substack{q, q^\top q = 1 \\ q \perp Q_{j-1}}} q^\top \left(\sum_{i=1}^{k} \frac{n_i S_i}{d_{j,i}^2} \right) q \ , \tag{9.28}$$

where $Q_0 := 0$ and $Q_{j-1} = [q_1, q_2, ..., q_{j-1}]$.

Theorem 9.5.1. *The first-order optimality conditions for the problem (9.28) are, for $j = 1, 2, ...r$:*

$$\Pi_j \left(\sum_{i=1}^{k} \frac{n_i S_i}{d_{j,i}^2} \right) q_j = 0_p , \tag{9.29}$$

where Π_j denotes the projector $I_p - Q_j Q_j^\top$.

This first-order optimality condition is nearly identical to (Trendafilov, 2010, Theorem 3.1). This indicates that (9.26) can be solved by a stepwise algorithm very similar to the one outlined in (Trendafilov, 2010, p. 3452).

The adapted stepwise algorithm produces the following two ML-CPCs:

$$Q = \begin{bmatrix} .7453 & .0486 \\ .4212 & -.7968 \\ .4941 & .6022 \\ .1516 & .0124 \end{bmatrix} ,$$

and corresponding individual variances for Fisher's Iris data:

$$d_1^2 = \begin{bmatrix} 47.1683 \\ 6.8356 \end{bmatrix} , \quad d_2^2 = \begin{bmatrix} 65.5591 \\ 11.9278 \end{bmatrix} , \quad d_3^2 = \begin{bmatrix} 18.5765 \\ 8.4426 \end{bmatrix} ,$$

which is very similar to the matrix solution of (9.26). The variances of the first and the second CPCs are, respectively, 131.3038 and 27.2061 with total 158.5099.

In a similar way, one can rewrite the LS version of the CPC problem (9.22) with known individual variances on the Stefel manifold $\mathcal{O}(p, r)$ as

$$\max_{Q \in \mathcal{O}(p,r)} \sum_{i=1}^{k} n_i \text{trace}(D_{0,i}^2 Q^\top S_i Q) ,$$

which can be solved with Manopt.

Alternatively, one can solve it by solving sequentially r problems of the form:

$$\max_{\substack{q, q^\top q = 1 \\ q \perp Q_{j-1}}} q^\top \left(\sum_{i=1}^{k} n_i d_{0j,i}^2 S_i \right) q . \tag{9.30}$$

Theorem 9.5.2. *The first-order optimality conditions for the problem (9.30) are, for $j = 1, 2, ...r$:*

$$\Pi_j \left(\sum_{i=1}^k n_i d_{0j,i}^2 S_i \right) q_j = 0_{p \times 1} . \tag{9.31}$$

The first-order optimality conditions (9.31) indicates that (9.30) can be solved by the same type of stepwise algorithm as in the ML case. The properly adapted stepwise algorithm produces the following two LS-CPCs:

$$Q = \begin{bmatrix} .7328 & -.1602 \\ .2676 & .8728 \\ .6063 & -.2839 \\ .1542 & .3632 \end{bmatrix},$$

and corresponding individual variances for Fisher's Iris data:

$$d_1^2 = \begin{bmatrix} 48.4659 \\ 6.1221 \end{bmatrix}, \quad d_2^2 = \begin{bmatrix} 69.1947 \\ 10.4517 \end{bmatrix}, \quad d_3^2 = \begin{bmatrix} 14.9833 \\ 8.7936 \end{bmatrix},$$

which is comparable to the solution from the orthonormal INDSCAL. The variances of the first and the second CPCs are, respectively, 132.6438 and 25.3674 with total 158.0112.

The following general Algorithm 9.1 can find stepwise CPC, achieving either ML (Trendafilov, 2010, p. 3452) or LS fit. It does not work with fixed individual eigenvalues, as in the examples above. Instead, they are adjusted at each iteration step. For clarity, the indexes of the power iterations are given in parentheses:

For example, the LS-CPCs and their individual variances obtained using this sequential algorithm are almost identical to the ones obtained in Section 9.3 by the orthonormal INDSCAL. This is not surprising if you note that the (matrix) orthonormal INDSCAL (9.6) is in fact equivalent to

$$\max_{\substack{q, q^\top q = 1 \\ q \perp Q_{j-1}}} q^\top \left[\sum_{i=1}^k n_i (q^\top S_i q) S_i \right] q , \tag{9.32}$$

if one switches to vector notations.

Algorithm 9.1 CPC of $p \times p$ correlation matrices $S_1, \ldots, S_k$.

---

set $\Pi = I_p$
for $j = 1, 2, \ldots p$ **do**
  set $x_{(0)}$ with $x_{(0)}^\top x_{(0)} = 1$
  $x_{(0)} \leftarrow \Pi x_{(0)}$
  $\mu_{(0)}^i \leftarrow x_{(0)}^\top S_i x_{(0)}$ and $i = 1, 2, \ldots k$
  for $l = 1, \ldots, l_{max}$ **do**
    if ML case **then**
      $S \leftarrow \frac{n_1 S_1}{\mu_{(l-1)}^1} + \ldots + \frac{n_k S_k}{\mu_{(l-1)}^k}$
    else
      (LS case)
      $S \leftarrow n_1 \mu_{(l-1)}^1 S_1 + \ldots + n_k \mu_{(l-1)}^k S_k$
    end if
    $y \leftarrow \Pi S x_{(l-1)}$
    $x_{(l)} \leftarrow y / \sqrt{y^\top y}$
    $\mu_{(l)}^i \leftarrow x_{(l)}^\top S_i x_{(l)}$ and $i = 1, 2, \ldots k$
  end for
  $q_j \leftarrow x_{(l_{max})}$ and $\lambda_j^i \leftarrow \mu_{(l_{max})}^i$
  $\Pi \leftarrow \Pi - q_j q_j^\top$
end for

9.6 Proportional covariance matrices

9.6.1 ML and LS proportional principal components

Flury et al. (1994) demonstrated in a simulated study that even a simpler model than CPC with *proportional* covariance matrices (Flury, 1988, Ch 5) can provide quite competitive discrimination compared to other more complicated methods. For short, we call such PCs proportional (PPC). They are also interesting because they admit very simple and fast implementation suitable for large data.

As before, there are considered k normal populations with same mean vector and it is assumed that their $p \times p$ covariance matrices $\Sigma_i, i = 1, 2, \ldots, k$ are all PD and different. The hypothesis of the PPC model is that there exist $k - 1$ constants c_i such that

$$H_{\text{PPC}} : \Sigma_i = c_i \Sigma_1 , \ i = 2, \ldots, k , \tag{9.33}$$

where c_i are *specific* for each population. By substituting (9.33) in (9.1), the ML estimation of Σ_i, i.e. of Σ_1 and c_i, is formulated in (Flury, 1988, p. 103) as the following optimization problem (PPC-ML):

$$\min_{\Sigma_1,c} \sum_{i=1}^{k} n_i \left\{ \log\left[\det(c_i\Sigma_1)\right] + \text{trace}[(c_i\Sigma_1)^{-1}S_i] \right\} ,\qquad (9.34)$$

where S_i are given sample covariance matrices and $c = (c_1, c_2, ..., c_k) \in \mathbb{R}^k$. From now on, we assume that $c_1 = 1$.

Let $\Sigma_1 = QDQ^\top$ be its EVD and $Q = [q_1, ..., q_p]$. Then, after substitution, (9.34) becomes:

$$\min_{(Q,D,c)\in\mathcal{O}(p)\times\mathbb{R}^p\times\mathbb{R}^k} \sum_{i=1}^{k} n_i \left\{ \log\left[\det(c_iQDQ^\top)\right] + \text{trace}\left[(c_iQDQ^\top)^{-1}S_i\right] \right\} ,$$
$$(9.35)$$

which further simplifies to (Flury, 1988, p. 103, (2.2)):

$$\min_{(Q,D,c)\in\mathcal{O}(p)\times\mathbb{R}^p\times\mathbb{R}^k} \sum_{i=1}^{k} n_i \left\{ \sum_{j=1}^{p} \left[\log(c_id_j) + \frac{q_j^\top S_i q_j}{c_id_j} \right] \right\} ,\qquad (9.36)$$

where q_j and d_j are, respectively, the jth eigenvector and eigenvalue of Σ_1.

The ML estimations of q_j, d_j and c_i are derived from the first-order optimality conditions of (9.36). They are further used to construct a fast alternating iterative algorithm for their estimation (Flury, 1988, p. 211).

Similar to (9.6), the LS setting of the PPC problem (PPC-LS) is defined as

$$\min_{(Q,D,c)\in\mathcal{O}(p)\times\mathbb{R}^p\times\mathbb{R}^k} \sum_{i=1}^{k} n_i||S_i - c_iQDQ^\top||_F^2 .\qquad (9.37)$$

To find LS estimations of Q, D and $c_2, ..., c_k$ (assuming $c_1 = 1$), consider the objective function of (9.37) by letting $Y_i = S_i - c_iQDQ^\top$

$$f = \frac{1}{2}\sum_{i=1}^{k} n_i||Y_i||_F^2 = \frac{1}{2}\sum_{i=1}^{k} n_i\text{trace}(Y_i^\top Y_i) ,\qquad (9.38)$$

and its total derivative:

$$
\begin{aligned}
df &= \frac{1}{2}d\sum_{i=1}^{k} n_i \text{trace}(Y_i^\top Y_i) = -\sum_{i=1}^{k} n_i \text{trace}\left[Y_i d(c_i QDQ^\top)\right] \\
&= -\sum_{i=1}^{k} n_i \text{trace}\left\{Y_i\left[(dc_i)QDQ^\top + c_i Q(dD)Q^\top + 2c_i QD(dQ)^\top\right]\right\} .
\end{aligned}
$$

Then the partial gradients with respect to Q, D and $c_{i \in \{2,...,k\}}$ are

$$
\nabla_{c_i} f = -n_i \text{trace}(Y_i QDQ^\top) = n_i c_i \text{trace}(D^2) - n_i \text{trace}(Q^\top S_i QD) \quad (9.39)
$$

$$
\nabla_D f = -\sum_{i=1}^{k} n_i c_i Q^\top Y_i Q = \sum_{i=1}^{k} n_i c_i^2 D - \sum_{i=1}^{k} n_i c_i \text{diag}(Q^\top S_i Q) , \quad (9.40)
$$

$$
\nabla_Q f = -2\sum_{i=1}^{k} n_i c_i Y_i QD = 2\sum_{i=1}^{k} n_i c_i^2 QD^2 - 2\sum_{i=1}^{k} n_i c_i S_i QD . \quad (9.41)
$$

At the minimum of (9.38), the partial gradients (9.39) and (9.40) must be zero, which leads to the following LS estimations:

$$
c_i = \frac{\text{trace}(Q^\top S_i QD)}{\text{trace}(D^2)} = \frac{\sum_{j=1}^{p} q_j^\top S_i q_j d_j}{\sum_{j=1}^{p} d_j^2} , \quad i = 2, 3, ..., k , \quad (9.42)
$$

$$
D = \frac{\sum_{i=1}^{k} n_i c_i \text{diag}(Q^\top S_i Q)}{\sum_{i=1}^{k} n_i c_i^2} \quad \text{or} \quad d_j = q_j^\top \left(\frac{\sum_{i=1}^{k} n_i c_i S_i}{\sum_{i=1}^{k} n_i c_i^2}\right) q_j . \quad (9.43)
$$

The gradient (9.41) together with the constraint $Q^\top Q = I_p$ imply that at the minimum of (9.38) the matrix:

$$
Q^\top \left(\frac{\sum_{i=1}^{k} n_i c_i S_i}{\sum_{i=1}^{k} n_i c_i^2}\right) Q \quad (9.44)
$$

should be diagonal. This also indicates that PPCs Q can be found by consecutive EVD of $\frac{\sum_{i=1}^{k} n_i c_i S_i}{\sum_{i=1}^{k} n_i c_i^2}$, where updated values for c_i and d_i are found by (9.42) and (9.43). This is a very important feature which will be utilized in the next Section 9.6.2 for dimension reduction purposes.

Note, that as in the ML case, the equation for the proportionality constraints (9.42) holds also for $i = 1$, because $\sum_{j=1}^{p} q_j^\top S_1 q_j d_j = \sum_{j=1}^{p} d_j^2$.

For illustration, we solve the PPC-LS problem for Fisher's Iris data. The estimations are obtained by solving (9.37) making use of an alternating iterative algorithm similar to the one for the ML case (Flury, 1988, p. 211):

$$Q = \begin{bmatrix} .7307 & -.2061 & .5981 & -.2566 \\ .2583 & .8568 & .1586 & .4171 \\ .6127 & -.2209 & -.6816 & .3336 \\ .1547 & .4178 & -.3906 & -.8056 \end{bmatrix},$$

and, respectively:

$$d_1^2 = \begin{bmatrix} 48.4509 \\ 6.2894 \\ 6.3261 \\ 1.4160 \end{bmatrix}, \quad d_2^2 = \begin{bmatrix} 69.2709 \\ 10.5674 \\ 5.2504 \\ 3.7482 \end{bmatrix}, \quad d_3^2 = \begin{bmatrix} 14.7542 \\ 7.9960 \\ 6.3983 \\ 1.7719 \end{bmatrix}.$$

The proportionality constants are estimated as 1.0000, 1.4284 and .3343. For comparison with the ML solution obtained in (Flury, 1986, p. 32), we report here also the estimated population covariance matrices $(QD_i^2Q^\top, i = 1, 2, 3)$ for Fisher's Iris data:

$$\hat{\Sigma}_1 = \begin{bmatrix} 28.0935 & 8.0037 & 19.7198 & 4.1039 \\ 8.0037 & 9.3609 & 5.9809 & 3.5642 \\ 19.7198 & 5.9809 & 21.1279 & 4.5960 \\ 4.1039 & 3.5642 & 4.5960 & 4.7767 \end{bmatrix}$$

$$\hat{\Sigma}_2 = \begin{bmatrix} 40.1290 & 11.4325 & 28.1679 & 5.8621 \\ 11.4325 & 13.3711 & 8.5431 & 5.0911 \\ 28.1679 & 8.5431 & 30.1793 & 6.5650 \\ 5.8621 & 5.0911 & 6.5650 & 6.8231 \end{bmatrix}$$

$$\hat{\Sigma}_3 = \begin{bmatrix} 9.3914 & 2.6755 & 6.5921 & 1.3719 \\ 2.6755 & 3.1292 & 1.9993 & 1.1915 \\ 6.5921 & 1.9993 & 7.0629 & 1.5364 \\ 1.3719 & 1.1915 & 1.5364 & 1.5968 \end{bmatrix}.$$

The value of the PPC-LS objective function is 129.1579. The fit achieved by the LS-CPC solution produced in Section 9.3.2 is 93.3166.

9.6.2 Dimension reduction with PPC

For practical applications, and especially for problems with large number of variables p, it is more convenient to have versions of the ML- and LS-PPC algorithms producing only few PPCs, say $r \ll p$. In the ML case, the assumptions require PD population matrices $\Sigma_1, ..., \Sigma_k$. The ML-PPC equations (9.34) and (9.35) can still be used for some $r \ll p$, as they find low-dimensional approximations to $\Sigma_1, ..., \Sigma_k$ and the involved quantities are well defined.

For Fisher's Iris data, one can use a slightly modified algorithm from (Flury, 1988, p. 211) to obtain the following two ML-PPCs:

$$
Q = \begin{bmatrix} .7386 & .0277 \\ .3668 & -.8384 \\ .5423 & .5426 \\ .1608 & -.0449 \end{bmatrix},
$$

and corresponding individual variances:

$$
d_1^2 = \begin{bmatrix} 48.0952 \\ 6.0249 \end{bmatrix}, \quad d_2^2 = \begin{bmatrix} 67.4046 \\ 10.4143 \end{bmatrix}, \quad d_3^2 = \begin{bmatrix} 17.2391 \\ 9.5600 \end{bmatrix}.
$$

The variances of the first and the second CPCs are, respectively, 132.7389 and 25.9992 with total 158.738. The proportionality constants are estimated as 1.0000, 1.5211 and 0.8077. Here, we depict the estimated population covariance matrices for Fisher's Iris data:

$$
\hat{\Sigma}_1 = \begin{bmatrix} 20.6898 & 10.0804 & 15.3106 & 4.4926 \\ 10.0804 & 10.8892 & 3.7962 & 2.5461 \\ 15.3106 & 3.7962 & 13.5749 & 3.1056 \\ 4.4926 & 2.5461 & 3.1056 & 0.9969 \end{bmatrix}
$$

$$
\hat{\Sigma}_2 = \begin{bmatrix} 31.4716 & 15.3336 & 23.2892 & 6.8338 \\ 15.3336 & 16.5638 & 5.7745 & 3.8730 \\ 23.2892 & 5.7745 & 20.6490 & 4.7240 \\ 6.8338 & 3.8730 & 4.7240 & 1.5164 \end{bmatrix}
$$

$$\hat{\Sigma}_3 = \begin{bmatrix} 16.7120 & 8.1424 & 12.3670 & 3.6289 \\ 8.1424 & 8.7957 & 3.0664 & 2.0566 \\ 12.3670 & 3.0664 & 10.9650 & 2.5086 \\ 3.6289 & 2.0566 & 2.5086 & 0.8052 \end{bmatrix}.$$

The value of the ML-PPC objective function is 7.8811.

In the LS case, the LS-PPC problem can be readily defined to produce only r PPCs. For Fisher's Iris data, one obtains the following two LS-CPCs:

$$Q = \begin{bmatrix} .7306 & .2090 \\ .2574 & -.8551 \\ .6132 & .2164 \\ .1546 & -.4222 \end{bmatrix},$$

and corresponding individual variances:

$$d_1^2 = \begin{bmatrix} 48.4482 \\ 6.3010 \end{bmatrix}, \quad d_2^2 = \begin{bmatrix} 69.2773 \\ 10.5732 \end{bmatrix}, \quad d_3^2 = \begin{bmatrix} 14.7347 \\ 7.9355 \end{bmatrix}.$$

The variances of the first and the second CPCs are, respectively, 132.4601 and 24.8096 with total 157.2698. The proportionality constants are estimated as 1.0000, 1.4349 and 0.3233. Here, we depict the estimated population covariance matrices for Fisher's Iris data:

$$\hat{\Sigma}_1 = \begin{bmatrix} 26.0830 & 7.7158 & 21.9579 & 4.7790 \\ 7.7158 & 8.7541 & 6.2104 & 4.6634 \\ 21.9579 & 6.2104 & 18.4978 & 3.8799 \\ 4.7790 & 4.6634 & 3.8799 & 2.5075 \end{bmatrix}$$

$$\hat{\Sigma}_2 = \begin{bmatrix} 37.4266 & 11.0715 & 31.5076 & 6.8573 \\ 11.0715 & 12.5613 & 8.9113 & 6.6916 \\ 31.5076 & 8.9113 & 26.5426 & 5.5674 \\ 6.8573 & 6.6916 & 5.5674 & 3.5980 \end{bmatrix}$$

$$\hat{\Sigma}_3 = \begin{bmatrix} 8.4328 & 2.4946 & 7.0992 & 1.5451 \\ 2.4946 & 2.8303 & 2.0079 & 1.5077 \\ 7.0992 & 2.0079 & 5.9805 & 1.2544 \\ 1.5451 & 1.5077 & 1.2544 & 0.8107 \end{bmatrix}.$$

The value of the LS-PPC objective function is 161.727.

9.7 Some relations between CPC and ICA

CPC is a special case of a class of MDA problems requiring (approximate) simultaneous diagonalization of k symmetric (usually PD) matrices. We already mentioned its relation to methods for metric multidimensional scaling (MDS) of several proximity matrices (INDSCAL). In this section, we consider the relation of CPC to independent component analysis (ICA), which recently received considerable attention in the community of signal processing for blind separation of signals.

As ICA is mainly a tool for engineering applications and complex arithmetic is allowed. As a consequence, the ICA literature works with unitary matrices instead of orthogonal ones and the standard matrix transpose is replaced by its conjugate counterpart. To simplify and unify the notations and the text, we continue with real arithmetic: all vectors spaces and manifolds are real.

The ICA method is briefly formulated in Section 9.7.1. Then, two popular algorithms for ICA are presented in Section 9.7.2 as optimization problems on manifolds. Finally, ICA is considered in Section 9.7.3 as a kind of CPC problem, but defined on the manifold of all *oblique* rotations $\mathcal{OB}(p, r)$.

9.7.1 ICA formulations

In Section 4.5.4, we already considered a simplified version of the ICA problem. We remember that the main issue was the definition of some kind of measure for independence. In this section, a more formal definition of ICA is provided.

First, we briefly recollect some basic facts about ICA. Two centred random variables x_1 and x_2 are uncorrelated if

$$E(x_1 x_2) = E(x_1)E(x_2) , \qquad (9.45)$$

where $E()$ denotes the expectation operator. If x_1 and x_2 are independent, then the following identity holds for any measurable functions f_1 and f_2:

$$E(f_1(x_1)f_2(x_2)) = E(f_1(x_1))E(f_2(x_2)) . \qquad (9.46)$$

In particular, this is true for the power function x^k, with $k = 1, 2, \ldots$:

$$E(x_1^k x_2^k) = E(x_1^k)E(x_2^k) . \qquad (9.47)$$

Obviously, for x_1 and x_2 being independent is a much stronger relationship than just being uncorrelated. Independence implies uncorrelateness, but not the opposite. However, if x_1 and x_2 have joint Gaussian distribution, then independence and uncorrelateness are equivalent. Due to this property, the Gaussian distribution is the least interesting for ICA (Jones and Sibson, 1987). Like PCA, ICA is a linear technique for separating mixtures of signals into their sources (non-linear ICA versions also exist). However, ICA works with non-Gaussian distributions for which the covariance matrix does not contain the entire useful information. Thus, ICA is trying to reveal higher order information.

Here, the standard ICA notations are adopted. Let x be observed $p \times 1$ random vector. ICA of x consists of estimating the following model:

$$x = As + e ,\tag{9.48}$$

where the $r \times 1$ unknown vector (latent variable) s is assumed with independent components s_i. The $p \times r$ "mixing" matrix A is unknown but fixed (not random), and e is a $p \times 1$ random noise vector.

In this section, the more frequently used noise-free ICA version will be considered, where the noise vector e is omitted in (9.48), i.e.:

$$x = As ,\tag{9.49}$$

and the assumptions for A and s remain the same.

Clearly, the noise-free ICA definition (9.49) is a copy of the PCA one, while the general model (9.48) reminds for EFA from Section 5.2.

9.7.2   ICA by contrast functions

The considerations in this section are restricted to the noise-free ICA model (9.49). Thus, ICA is looking for a linear transformation $s = Wx$ so that the components of s are as independent as possible. The matrix W is called *demixing* matrix. In many situations, it is appropriate to do ICA with $p = r$, which simply leads to $W = A^{-1}$. It is argued that the general case of $p \geq r$ can be reduced to $p = r$ after initial dimension reduction of the data (from p to r), e.g. with PCA. For identification purposes, it is assumed additionally that all s_i are non-Gaussian, except possibly one (Comon, 1994).

In practice, we are given a $n \times p$ data matrix X. In ICA it is usually assumed that X is centred and already whitened, i.e. its covariance matrix is the identity matrix $X^\top X = I_p$. This assumption and the noise-free ICA equation (9.49) in matrix form $X = SA^\top$, imply that A is orthogonal (rotation) matrix. Thus, $W = A^{-1} = A^\top$ and $S = XA \in \mathcal{M}(n, p)$.

"The concept of lCA may actually be seen as an extension of the PCA, which can only impose independence up to the second order and, consequently, defines directions that are orthogonal" (Comon, 1994). Clearly, the tricky part that differentiates ICA from PCA is that those components s should be independent, which cannot be achieved by diagonalization of the sample covariance matrix. Instead, ICA needs some criterion which measures independence. In ICA, such criteria are called *contrast functions*. In addition, a contrast should be invariant under any sign change of its components. There exists a great number of contrast functions (Comon, 1994; Hyvärinen et al., 2001). This reminds for the jungle of simple structure criteria for PCA and EFA discussed in Section 4.5.2, and indicates that none of them is completely satisfactory.

The first contrast function to consider utilizes the likelihood of the noise-free ICA model (Hyvärinen et al., 2001, 9.1). The density functions f_i of the independent components s_i from (9.49) are assumed known. Their estimation is discussed in (Hyvärinen et al., 2001, 9.1.2). Also, the data are assumed already whitened and A is orthogonal $(A^{-1} = A^\top)$. This also implies that $|\det(A)| = 1$, i.e. the Jacobian of the linear transformation (9.49) is unity. Then, the density of the mixture $x(= As)$ is given by

$$f_x(x) = f_s(s) = \prod_{i=1}^{p} f_i(s_i) = \prod_{i=1}^{p} f_i(a_i^\top x) , \tag{9.50}$$

where a_i is the ith column of $A \in \mathcal{O}(p)$ and $s = A^\top x$.

In practice, we are given a $n \times p$ data matrix X. Then the likelihood of the noise-free ICA model can be obtained as a product of densities (9.50) evaluated at the n points. It is usually easier to work with its logarithm, which leads us to the following contrast function:

$$c_L(A) = \sum_{j=1}^{n} \sum_{i=1}^{p} \log f_i(a_i^\top x_j) , \tag{9.51}$$

where x_j is the jth observation, i.e. the jth row of the data matrix X.

Then, the noise-free ICA solution is found by maximization of (9.51) subject to A being orthogonal, i.e. $A \in \mathcal{O}(p)$, or:

$$\max_{A \in \mathcal{O}(p)} \sum_{j=1}^{n} \sum_{i=1}^{p} \log f_i(a_i^\top x_j) \ . \tag{9.52}$$

To solve (9.52), we need to find first the gradient of its cost function, which is given by

$$\nabla_A = X^\top G \ , \tag{9.53}$$

where $G = \left\{ \frac{f_i'(x_j a_i)}{f_i(x_j a_i)} \right\} \in \mathcal{M}(n, p)$ contains the derivatives of $\log f_i$. The first-order optimality condition for A to maximize (9.51) is that necessarily $A^\top X^\top G$ should be symmetric. Now, the noise-free ICA can be solved with Manopt on $\mathcal{O}(p)$ using either the Euclidean gradient ∇_A in (9.53), or its Riemannian version:

$$\nabla_A^R = \frac{\nabla_A - A \nabla_A^\top A}{2} \ . \tag{9.54}$$

Note, that most of the existing noise-free ICA gradient algorithms (Hyvärinen et al., 2001, 9.1) do not automatically produce orthogonal A. On the contrary, at each step they require some kind of projection or normalization of the update to make it orthogonal.

Arguably, the most popular contrast function is the *negentropy* of the random variable x with density f defined by

$$c_N(x) = h(z) - h(x) \ , \tag{9.55}$$

where z is a (zero-mean) Gaussian random variable with the same variance as x (Comon, 1994; Hyvärinen et al., 2001) and

$$h(x) = - \int f(x) \log f(x) dx \ , \tag{9.56}$$

is the *Boltzmann* entropy of x. Clearly, $c_N(x)$ measures the deviation of x_i from normality. The negentropy is statistically sound measure, but, unfortunately, it is difficult to apply for sample data, since (9.55) requires density estimation, possibly non-parametric. For this reason, the following simple approximation was proposed (Jones and Sibson, 1987):

$$c_N(x) \approx \frac{1}{12} c_3^2(x) + \frac{1}{48} c_4^2(x) \ , \tag{9.57}$$

where c_3 and c_4 are known as the *skewness* and the *kurtosis* of x. Unfortunately, this approximation is quite sensitive to outliers, and thus, of limited practical interest.

A better class of approximations is given by (Hyvärinen et al., 2001, (8.25))

$$c_N(x) \approx [E(g(x)) - E(g(z))]^2 , \tag{9.58}$$

where g is an arbitrary smooth function (non-quadratic and non-random) and z is a standard Gaussian random variable (zero mean and unit variance). Frequently suggested choices for g are

$$g(y) = \frac{1}{\alpha} \log \cosh(\alpha y), \alpha \in [1, 2] \quad \text{and} \quad g(y) = \exp(-y^2/2) . \tag{9.59}$$

Thus, the problem is to maximize the (approximate) negentropy as defined in (9.58). As the second term is constant, one is simply interested to solve the following problem (a noise-free ICA for one unit):

$$\max_{a^\top a = 1} E[g(a^\top x)] . \tag{9.60}$$

The FastICA algorithm (Hyvärinen et al., 2001, 8) is designed to solve (9.60). It makes a move in direction of the greatest gradient descent, followed by a normalization of a to unit length at every step.

An alternative algorithm is provided here that works directly on the p-dimensional unit sphere $\mathcal{S}^{p-1}$, coinciding with the oblique manifold $\mathcal{OB}(p, 1)$. We also need the Euclidean gradient of the cost function in (9.60) given by $\nabla_a = E[g'(a^\top x)x]$, where g' denotes the derivative of g from (9.59). In order to construct an algorithm working on $\mathcal{S}^{p-1}$, we need its Riemannian version, which is given by

$$\nabla_a^R = (I_p - aa^\top)\nabla_a . \tag{9.61}$$

Now, we can use Manopt to solve (9.60) by utilizing ∇_a^R on $\mathcal{S}^{p-1}$. The result will be an unit length vector $a = a_1$, which will be the first column of the unknown mixing matrix A.

To find the second column of A, say a_2, we cannot use (9.61) again, as we need to take into account that $a_2^\top a_1 = 0$. This can be achieved by correcting (9.61) a little bit as follows:

$$\nabla_a^R = (I_p - a_1 a_1^\top - aa^\top)\nabla_a . \tag{9.62}$$

In general, in order to find the kth column of A, say a_k, we use the following generalized gradient:

$$\nabla_a^R = (\Pi_k - aa^\top)\nabla_a \ . \tag{9.63}$$

where $\Pi_k = I_p - a_1 a_1^\top - a_2 a_2^\top - \ldots - a_{k-1} a_{k-1}^\top$ and $\Pi_1 = I_p$.

Thus, in order to find A one needs to construct Manopt code that sequentially finds its columns by utilizing (9.63) for $k = 1, 2, \ldots, p$.

It is probably better idea to find $A \in \mathcal{O}(p)$ in a single go. For this reason, let us rewrite (9.60) as follows:

$$\max_{A \in \mathcal{O}(p)} \sum_{j=1}^n \sum_{i=1}^p g(a_i^\top x_j) \ , \tag{9.64}$$

where g acts element-wise. The Euclidean gradient of (9.64) is given by

$$\nabla_A = X^\top G \ , \tag{9.65}$$

where $G = \{g'(x_j a_i)\} \in \mathcal{M}(n, p)$ is a matrix of derivatives of g. As with the likelihood ICA contrast, we see that the first-order optimality condition for A to maximize (9.64) is that necessarily $A^\top X^\top G$ is symmetric. The problem (9.64) can be solved with Manopt on $\mathcal{O}(p)$ using either the Euclidean gradient $\nabla_A = X^\top G$ in (9.65), or its Riemannian version (9.54).

9.7.3   ICA methods based on diagonalization

In the previous section, ICA methods based on the optimization of certain measures of independence were considered. Along with them, there exist another class of ICA methods based on simultaneous diagonalization of several matrices. As with CPC, some of those ICA methods work on $\mathcal{O}(p)$. However, they are methods that use oblique rotations, i.e. work on $\mathcal{OB}(p)$.

Belouchrani et al. (1997) proposed the second-order blind identification algorithm (SOBI) which is based on joint diagonalization of a set of k whitened sample covariance matrices C_k computed for k non-zero time lags. It is assumed that the unknown source s is stationary (p-variate) time series. The diagonalization is defined as minimization of the sum of the squared off-diagonal elements of $Q^\top C_k Q$ (Belouchrani et al., 1997, eq. (20)):

$$\min_{Q \in \mathcal{O}(p)} \sum_{i=1}^k \|QC_i Q^\top - \mathrm{diag}(QC_i Q^\top)\|_F^2 \ . \tag{9.66}$$

This problem is equivalent to LS-CPC (9.6). Belouchrani et al. (1997) proposed a generalization of the Jacobi method for several matrices, similar to the algorithm developed by Flury and Gautschi (1986), for solving the ML-CPC. The differences reflect the specifics of the least squares and the log-likelihood cost functions.

Another ICA method relying on joint diagonalization was proposed by Cardoso and Souloumiac (1993a). It is well known as the joint approximate diagonalization of eigenmatrices (JADE), and is based on the joint diagonalization of empirical fourth-order cumulant matrices of the observed (mixed) data. The standard application of JADE requires a pre-whitening phase, followed by orthogonal joint diagonalization of the transformed matrices using an generalized Jacobi algorithm (Cardoso and Souloumiac, 1993b). The idea behind this approach is that the eigenvectors of the cumulant matrices can form the demixing matrix for getting independent components. JADE also requires solution of a problem like (9.66) for some specially defined C_i (different from those in SOBI).

The data (pre-)whitening is usually done on the sample covariance/correlation matrix, which is usually C_1. However, Yeredor (2002) argued that "...*such a whitening phase...would attain exact diagonalization of the selected matrix at the possible cost of poor diagonalization of the others.*" As a remedy, Yeredor (2002) proposed to do joint diagonalization with a general non-singular $Q \in \mathcal{GL}(p)$, not necessarily orthogonal matrix. According to our agreement to work with real arithmetic, the problem considered in (Yeredor, 2002, eq (1)) becomes

$$\min_{Q \in \mathcal{GL}(p)} \sum_{i=1}^{k} ||C_i - QD^2Q^\top||_F^2 \, , \tag{9.67}$$

where C_i are symmetric PD matrices. The proposed AC–DC algorithm (Yeredor, 2002) outperformed JADE in terms of signals separation, however is computationally more demanding. Clearly, this approach seems promising, but its realization can definitely benefit from further development.

Afsari (2006) extended the research on the cost function (9.66) and showed that its minimization on $Q \in \mathcal{GL}(p)$ is not a good idea. This is because (9.66) is not invariant if Q is multiplied by a non-singular diagonal matrix D. To avoid this invariance problem with (9.66), Afsari (2006) proposed to solve the following modified problem:

$$\min_{Q \in \mathcal{GL}(p)} \sum_{i=1}^{k} ||C_i - Q^{-1}\text{diag}(QC_iQ^{\top})Q^{-\top}||_F^2 , \qquad (9.68)$$

which involves matrix inverses, and thus, not very practical.

Afsari (2006) also proposed an alternative way to avoid such invariance related problems by employing some convenient subset of $\mathcal{GL}(p)$, namely the
subset (group) of all non-singular matrices with *unit* determinant known as
the special linear group and denoted as $\mathcal{SL}(p)$. He proposed to minimize
(9.66) on $\mathcal{SL}(p)$, by utilizing the following iterative scheme:

$$Q_{i+1} = (I_p + \Delta)Q_i , \quad Q_0 = I_p , \qquad (9.69)$$

where Δ is triangular matrix with $\text{diag}(\Delta) = 0_p$. Then, $I_p + \Delta \in \mathcal{SL}(p)$,
and thus, $Q_{i+1} \in \mathcal{SL}(p)$ for any $i > 0$. Unfortunately, the limit point of
these iterations may not belong to $\mathcal{SL}(p)$, because it is also not compact.
However, the benefit from the iterations (9.69) on $\mathcal{SL}(p)$ is that for any i we
have $||Q_i||_F \geq 1$, i.e. the cost function (9.66) cannot be reduced by simply
reducing the norm of Q_i.

To avoid the above-listed difficulties, Absil and Gallivan (2006) proposed
to solve the problem (for simultaneous diagonalization) on the set of all
oblique rotations, i.e. as an optimization problem on $\mathcal{OB}(p,r)$. We know
that $\mathcal{OB}(p,r)$ is also not compact, but is a dense subset of the compact
product of r unit spheres in $\mathbb{R}^p$. As we mentioned before, Manopt treats
$\mathcal{OB}(p,r)$ as a product of unit spheres. Absil and Gallivan (2006) consider

$$\min_{Q \in \mathcal{OB}(p,r)} \sum_{i=1}^{k} ||C_i - QD^2Q^{\top}||_F^2 , \qquad (9.70)$$

and additionally proposed a super-linearly convergent algorithm, based on a
trust-region optimization method on $\mathcal{OB}(p,r)$. This is a considerable achieve-
ment, because the above-listed algorithms (and, in general, most of the ICA
ones) are based on steepest descent or direct-search ideas. Another important feature of (9.70) is that its solution can readily provide dimension reduction. In contrast, the rest of the ICA diagonalization methods work with
square matrices and need initial dimension reduction (see beginning of Section 9.7.2), which affects the resulting goodness-of-fit of the overall problem.

We only mention that diagonalization-based ICA, minimizing the original ML-CPC objective function (9.5) is considered in (Joho, 2008; Pham, 2001). They also work with orthogonal matrices on $\mathcal{O}(p)$, thus, one can reconsider the problem (9.4) on $\mathcal{OB}(p, r)$. However, it will involve matrix inverses, and can hardly be a reasonable alternative of its LS counterpart (9.70).

9.8 Conclusion

Section 9.1 reminds the origins of CPC and its relation to LDA aiming to overcome some of its limitations. The formal ML and LS definitions of CPC are given in Section 9.2 as constraint optimization problems on matrix manifolds, while Section 9.3 discusses their optimality conditions. Examples of CPC solutions as optimizers on matrix manifolds are compared to classical (FG) ones. Section 9.3.3 contains a MATLAB translation of the original FORTRAN codes of the FG algorithm.

Section 9.4 considers methods for CPCs estimation. First, the classical FG iterative algorithm is revisited in Section 9.4.1. Then, individual eigenvalues for the CPC are found in Section 9.4.2 as limit points of dynamical systems, starting from the (given) covariance matrices as initial points. With available individual eigenvalues, Section 9.4.3 shows that the CPC problem becomes a Procrustes problem.

Section 9.5 considers methods for finding only few CPCs explaining at least as much variance as the same number of standard CPCs (and usually more). Such methods are useful for simultaneous dimension reduction in a set of large covariance matrices.

Section 9.6 pays special attention to the particular case of proportional PCs. The general ML and LS definitions and estimation theory are considered in Section 9.6.1. Modifications related to finding only few PPCs for simultaneous dimension reduction are explored in Section 9.6.2.

Section 9.7 is dedicated to ICA and its relation to CPC. Section 9.7.1 briefly describes the ICA aims and its differences with PCA. Section 9.7.2 considers ICA methods that optimize certain criteria for "independence". The well-known FastICA algorithm is modified to work on the orthogonal group $\mathcal{O}()$. Section 9.7.3 is about methods modelling ICA as simultaneous diagonalization of certain data matrices which makes them similar to the CPC problem.

9.9   Exercises

1. Prove that (9.66) is equivalent to LS-CPC (9.6). Hint: make use of the optimality conditions for (9.6) and the unitary/orthogonally invariance of the Frobenius norm.

2. Prove that (9.68) does not change, if Q is replaced by DQ, for any non-singular diagonal matrix D. Hint: Prove $I \odot DAD = D(I \odot A)D$, for any diagonal D and an arbitrary square A (Horn and Johnson, 1994, p. 304).

3. Show that (9.68) is equivalent to (9.66) on $\mathcal{O}(p)$.

4. It is shown in Section 3.8, that if $\|A\|_F < 1$ for an arbitrary $A \in \mathcal{M}(n)$, then $I_p + A$ is non-singular. Find a more direct proof by assuming the opposite that $I_p + A$ is singular, and show that $\|A\|_F \geq 1$.

5. Give an example with $\mathcal{SL}(2)$ to show that the special linear group is not compact. Hint: Consider the matrix $\begin{bmatrix} 1 & \alpha \\ 0 & 1 \end{bmatrix}$, where $\alpha \in \mathbb{R}$.

6. Prove that if $Q \in \mathcal{SL}(p)$, then $\|Q\|_F \geq 1$. Hint: Assume that $\|Q\|_F < 1$, to reach contradiction.

7. Prove that $\mathcal{SL}(p)$ is connected. Hint: For $X \in \mathcal{SL}(p)$, make use of its polar decomposition to show that $\mathcal{SL}(p)$ can be expressed as a product of connected sets $\mathcal{O}(p) \times \mathcal{O}(p) \times \mathcal{D}_+(p)$.

Chapter 10

Metric multidimensional scaling (MDS) and related methods

© Springer Nature Switzerland AG 2021
N. Trendafilov and M. Gallo, *Multivariate Data Analysis on Matrix Manifolds*,
Springer Series in the Data Sciences, https://doi.org/10.1007/978-3-030-76974-1_10

10.1   Introduction

This section considers methods for analysing a special class of square data matrices collecting quantified proximities among subjects or variables. According to the Cambridge Dictionary, proximity means "the state of being near in space or time". For example, such a data matrix Δ can be obtained by taking the distances or journey times among ten cities. Clearly, Δ will be a 10×10 symmetric matrix with zeros on its main diagonal. Such tables are usually found in any road atlas. The purpose is to reconstruct the locations (coordinates) of the cities on the plane, such that the Euclidean distances among them approximate Δ. Another experiment, that leads to data of the same type is if we ask somebody to rate from, say, 1(different) to 5 (identical) the tastes of ten different beverages. In this case, the purpose would be to plot on paper a configuration of 10 points representing/visualizing those beverages, such that the distances among them approximate the rates or at least preserve correctly their magnitudes. The technique that is dealing with such problems is called multidimensional scaling (MDS). Particularly, problems of the former kind are classified as metric MDS, while the latter one as non-metric MDS (Cox and Cox, 2001; Mardia et al., 1979).

On many occasions, we are not given quantified proximities Δ, but the usual $n \times p$ (observations by variables) data matrix X. The interest is to plot the observations and/or the variables according to their proximities. To analyse such data with MDS, one needs first to *construct* the matrix of proximities among the observations and/or the variables of X. For this reason, there exists a great number of proximity measures which will be discussed briefly in the next Section 10.2.

Section 10.3 considers the metric MDS of a single proximity matrix. The basic metric MDS definitions and constructions are briefly summarized in Section 10.3.1. Reference to some modern problems related to metric MDS are given in Section 10.3.3. The exercises in Section 10.10 demonstrate some properties of the sample correlation matrix as a similarity matrix.

The main interest of this chapter is in fact on the MDS of three-way proximity data. Section 10.4 considers INDSCAL, which is a generalization of the metric MDS for analysing simultaneously several proximity matrices. Section 10.5 considers DINDSCAL, another version of the INDSCAL model, with improved fit. Section 10.6 considers the DEDICOM model for analysis of non-symmetric proximity matrices. Section 10.7 considers GIPSCAL, which is a special case of DEDICOM with better interpretation features.

10.2 Proximity measures

In specific applications, one speaks in terms of similarities or dissimilarities, rather than proximities. Usually, the similarities and the dissimilarities are considered non-negative. Naturally, the similarity and the dissimilarity of an object with itself is taken to be one and zero, respectively. Similarities are usually scaled to take values between 0 and 1, while the dissimilarities are simply non-negative.

For example, let X be a standardized $n \times p$ data matrix of continuous variables. The correlation coefficient is a well-known measure of the closeness between two variables. Then, the element-wise squared sample correlation matrix $(X^\top X) \odot (X^\top X)$ is an example of a *similarity* matrix of the involved p variables. In a similar way, $E_p - (X^\top X) \odot (X^\top X)$ is an example of a *dissimilarity* matrix, where E_p is a $p \times p$ matrix of ones, i.e. $1_p 1_p^\top$. Depending on the nature of the problem, one can consider the data matrix X as a collection of p columns $x_{.i} \in \mathbb{R}^n$ or a collection of n rows $x_{j.} \in \mathbb{R}^p$. Then, all known distances $(\ell_0, \ell_k, \ell_\infty)$ from Section 2.3 between vectors in the Euclidean spaces $\mathbb{R}^p$ and $\mathbb{R}^n$ can be employed to calculate dissimilarities between the columns/row of X. Note, that the created this way dissimilarities are in fact distances and satisfy all properties of a true distance function from Section 2.3. Dissimilarities created in another way may not posses these properties, i.e. may not be distances at all. A number of dissimilarity measures for quantitative data are listed in (Cox and Cox, 2001, Table 1.1).

Details on the features and the construction of proximity measures are given in (Cox and Cox, 2001, 1.3) and (Mardia et al., 1979, 13.4). Probably the most exhaustive lists of proximity measures covering all types of data (ratio, interval, ordinal and nominal variables) can be found in the User's Guides of statistical software as SAS and SPSS. However, the most difficult question remains: which measure is the most appropriate for particular dataset. Unfortunately, there is no universal answer and the choice should be made individually for every specific problem. As it can be expected, data involving mixture of different types of variables are more difficult to tackle. Some interesting examples can be found in (Hartigan, 1975, Ch 2).

10.3   Metric MDS

10.3.1   Basic identities and classic solution

We begin with several important definitions:

- A $n \times n$ matrix Δ is called a *dissimilarity matrix* if it is symmetric, has zeros on its main diagonal, and all off-diagonal entries are non-negative;

- A $n \times n$ matrix D is called a *distance matrix* if it is a dissimilarity matrix, and its entries satisfy the triangle inequality, i.e. $d_{ik} \leq d_{ij} + d_{jk}$ for every triple of distinct indices (i, j, k);

- A $n \times n$ distance matrix D is called *Euclidean* if there exists a natural number $r \in \mathbb{N}$ and n points $x_i \in \mathbb{R}^r$ such that

$$d_{ij}^2 = (x_i - x_j)^\top (x_i - x_j) , \qquad (10.1)$$

  i.e. D contains the Euclidean distances among those n points.

Apparently, not all distance matrices are Euclidean (Mardia et al., 1979, Exercise 14.2.7). Fortunately, there is a simple way to check that particular distance matrix D is Euclidean. Let X be an $n \times p$ data matrix, and $x_{i\cdot}$ denote its ith row. Then the Euclidean distance between the ith and the jth rows is given by

$$d_{ij}^2 = (x_{i\cdot} - x_{j\cdot})^\top (x_{i\cdot} - x_{j\cdot}) = x_{i\cdot} x_{i\cdot}^\top - 2 x_{i\cdot} x_{j\cdot}^\top + x_{j\cdot} x_{j\cdot}^\top , \qquad (10.2)$$

which was already utilized in (10.1). Let all possible distances d_{ij} be collected into an $n \times n$ matrix D. Apparently, its main diagonal is zero. With (10.2) in mind, one can easily check the following important matrix identity:

$$D \odot D = (I_n \odot X X^\top) E_n + E_n (I_n \odot X X^\top) - 2 X X^\top . \qquad (10.3)$$

where E_n denotes the $n \times n$ matrix of ones, i.e. $E_n = 1_n 1_n^\top$.

Now, consider the projector $J_n = I_n - E_n/n$, defined in Section 3.1.4 as the centring operator. Its name comes from the fact that $J_n X$ becomes a column-centred data matrix. Note, that $J_n 1_n = 1_n^\top J_n = 0_n$. After multiplying both sides of (10.3) by J_n, we find:

$$J_n (D \odot D) J_n = -2 J_n X X^\top J_n = -2 (J_n X)(J_n X)^\top . \qquad (10.4)$$

The matrix $B = -\frac{1}{2}J_n(D \odot D)J_n$ is called the double centred matrix of squared distances. In practice, we do not know X, but the identity (10.4) suggests that B is PSD, and X can be recovered exactly by EVD of B. This also implies that if B has negative eigenvalues, then, the initial distance matrix D *cannot* be Euclidean. However, if B is PSD, then, the number r of its positive eigenvalues (the rank of B), gives the dimension of the subspace $\mathbb{R}^r$, which is used in the definition of Euclidean matrix. Moreover, because B has zero column/row sums, its rank can be at most $n-1$. From $J_n 1_n = 0_{n \times 1}$, it follows that $B1_n = 0_{n \times 1} = 0 \times 1_n$, i.e. 0 is always eigenvalue of B and 1_n is its corresponding eigenvector. All the rest eigenvectors of B sum to 0, i.e. they are all *contrasts*.

By taking a truncated EVD of B, the metric MDS can additionally produce dimension reduction of the form:

$$-\frac{1}{2}J_n(D \odot D)J_n = Q\Lambda^2 Q^\top \approx Q_s\Lambda_s^2 Q_s^\top \,, \qquad (10.5)$$

where usually $s \ll r$ and Q_s contains only s eigenvectors, corresponding to the largest s eigenvalues of B collected in Λ_s^2.

The fundamental problem of MDS is to produce a low-dimensional $s \ll r(\leq n)$ representation of n high-dimensional observations based on their dissimilarities Δ. The metric MDS (10.5), finds low-dimensional coordinates the Euclidean distances among which approximate as close as possible the given dissimilarities among those high-dimensional observations. In matrix terms, the metric MDS finds an $n \times n$ Euclidean matrix $D_E(= Q_s\Lambda_s^2 Q_s^\top)$ of rank $s(\ll r)$, which gives the best LS fit (or rank s) to B, i.e.

$$\min_{\mathrm{rank}(D_E)=s} \left\| \frac{1}{2}J_n(\Delta \odot \Delta)J_n + D_E \right\|_F . \qquad (10.6)$$

Strictly speaking, the metric MDS admits fitting $f(\Delta)$ (not only Δ), for some given continuous monotone function f applied on Δ in a element-wise manner (Cox and Cox, 2001, 2.4). We restrain ourselves to the most frequently used case when f is the identity function. It is known as the classical MDS (Torgerson, 1957; Young and Householder, 1938). Hereafter, we refer to it as the metric MDS.

Nowadays, the metric MDS is related to a number of techniques for local and/or non-linear dimension reduction. Comprehensive introduction and taxonomy of these techniques can be found in (Izenman, 2008, Ch 16). There is also considerable interest in fast approximations to MDS applicable to large data (Platt, 2005).

10.3.2   MDS that fits distances directly

The metric MDS, based on the solution of (10.6), fits double-centred squared dissimilarities. Browne (1987) argued that better solutions can be achieved by fitting directly the original squared dissimilarities instead, i.e. using (10.4) rather than (10.3). Then, the MDS problem (10.6) becomes

$$\min_{\text{rank}(D_E)=s} \left\| \Delta \odot \Delta - (I_n \odot D_E)E_n - E_n(I_n \odot D_E) + 2D_E \right\|_F , \qquad (10.7)$$

which solution needs numerical optimization and will be discussed in Section 10.5, including Manopt implementation. Now, we briefly mention that we parametrize $D_E = Q\Lambda^2 Q^\top$ in (10.7) (or (10.6)) with its truncated EVD, and solve the following problem:

$$\min_{\substack{Q \in \mathcal{O}(n,s) \\ \Lambda \in \mathcal{D}(s)}} \left\| \Delta \odot \Delta - (I_n \odot Q\Lambda^2 Q^\top)E_n + E_n(I_n \odot Q\Lambda^2 Q^\top) - 2Q\Lambda^2 Q^\top \right\|_F ,$$

$$(10.8)$$

over the product of the Stiefel manifold $\mathcal{O}(n,s)$ and the linear sub-space $\mathcal{D}(s)$ of all $s \times s$ diagonal matrices.

We demonstrate the benefit of working with (10.7) over (10.6) on an artificial distance matrix considered in (Mardia et al., 1979, Exersise 14.2.7). For completeness, the upper triangular part of this distance matrix Δ_O is provided above the main (zero) diagonal of $\Delta_C \backslash \Delta_O$:

$$\Delta_C \backslash \Delta_O = \begin{bmatrix} 0 & 1 & 2 & 2 & 2 & 1 & 1 \\ 1 & 0 & 1 & 2 & 2 & 2 & 1 \\ 2 & 0 & 0 & 1 & 2 & 2 & 1 \\ 2 & 2 & 1 & 0 & 1 & 2 & 1 \\ 2 & 2 & 2 & 1 & 0 & 1 & 1 \\ 1 & 2 & 2 & 2 & 1 & 0 & 1 \\ 1 & 1 & 1 & 1 & 1 & 1 & 0 \end{bmatrix} . \qquad (10.9)$$

The entries of Δ_O do fulfil the triangle inequality, but the matrix is *not* Euclidean. Indeed, the eigenvalues of $B = -\frac{1}{2}(\Delta_O \odot \Delta_O)$ are -1.0, -0.1429, -0.0, 3.5, 3.5, 0.5, 0.5. Let $Q\Lambda^2 Q^\top$ be the EVD of B. The product $Q_2\Lambda_2$ of the eigenvectors Q_2 corresponding to the two largest eigenvalues with their square roots, i.e. $\Lambda_2 = \text{Diag}(\sqrt{3.5}, \sqrt{3.5})$ are given in the first pair of columns of Table 10.1. The second pair of columns of Table 10.1 contain the coordinates $(Q\Lambda)$ obtained by solving (10.8) with $s = 2$.

Table 10.1: Euclidean coordinates from two solutions.

	from EVD		from (10.8)	
	1	2	1	2
1	-1.0801	-0.0240	-0.5248	0.9394
2	-0.5401	0.9232	-1.0759	0.0153
3	0.5401	0.9472	-0.5512	-0.9242
4	1.0801	0.0240	0.5248	-0.9394
5	0.5401	-0.9232	1.0759	-0.0153
6	-0.5401	-0.9472	0.5512	0.9242
7	-0.0000	0.0000	-0.0000	0.0000

To compare the fit achieved by the solutions depicted in Table 10.1, one cannot simply look at the values of (10.6) and (10.8). They should be normalized by the norm of the corresponding data. The normalized fit of the classical MDS (EVD) is 0.2427, while the normalized fit of (10.8) is 0.1423, i.e. nearly two times better.

Then, it is interesting to see how well the Euclidean coordinates produced by the two approaches reconstruct the non-Euclidean distances Δ_O. The Euclidean distances D_E^{EVD} produced from the classical metric MDS (EVD) are given under the main diagonal of $D_E^{EVD} \backslash D_E^{(10.8)}$. Respectively, the reconstructed Euclidean distances obtained from the solution of (10.8) are given above its main diagonal.

$$D_E^{EVD} \backslash D_E^{(10.8)} = \begin{bmatrix} 0 & 1.08 & 1.86 & 2.15 & 1.86 & 1.08 & 1.08 \\ 1.09 & 0 & 1.08 & 1.86 & 2.15 & 1.86 & 1.08 \\ 1.89 & 1.08 & 0 & 1.08 & 1.86 & 2.15 & 1.08 \\ 2.16 & 1.85 & 1.07 & 0 & 1.08 & 1.86 & 1.08 \\ 1.85 & 2.14 & 1.87 & 1.09 & 0 & 1.08 & 1.08 \\ 1.07 & 1.87 & 2.18 & 1.89 & 1.08 & 0 & 1.08 \\ 1.08 & 1.07 & 1.09 & 1.08 & 1.07 & 1.09 & 0 \end{bmatrix} .$$

One notices, that the reproduced Euclidean distances from the two solutions are in fact reasonably similar. This is not what one could expect from the gap between the minima of their objective functions. To be precise, we check their fits to the non-Euclidean distances Δ_O. For the classical MDS (EVD), we find $\|\Delta_O - D_E^{EVD}\|_F = 0.7172$, while the solution obtained by (10.8) gives

$\|\Delta_O - D_E^{(10.8)}\|_F = 0.7073$, which is only slightly better. In other words, if the primary interest is in the reconstructed distances, then one could choose the simpler method.

10.3.3   Some related/adjacent MDS problems

In real applications, it is often the case that many dissimilarities are missing and/or are corrupted in some way. Then, it is advisable to fit the model by omitting them. Such type of problems are also know as matrix completion. As in the standard metric MDS, this approach approximates the (corrupted) dissimilarity matrix by a low rank one. As in MDS, the result is given by a set of points in a low-dimensional space and the distances among them recover the missing and/or corrupted dissimilarities (Fang and O'Leary, 2012; Mishra et al., 2011). The formulation of this completion problem by an Euclidean distance matrix is a slight modification of (10.6):

$$\min_{\text{rank}(D_E)=s} \left\| H \odot \left(\frac{1}{2} J_n (\Delta \odot \Delta) J_n + D_E \right) \right\|_F, \qquad (10.10)$$

where H is an $n \times n$ indicator matrix with 0's at the positions of the missing/corrupted entries, and 1's elsewhere. In other words, we restrain our efforts to the fit of the well-specified entries. As in the previous section, a better option is again to utilize (10.7), and reformulate (10.10) as follows:

$$\min_{\text{rank}(D_E)=s} \| H \odot [\Delta \odot \Delta - (I_n \odot D_E) E_n + E_n (I_n \odot D_E) - 2 D_E] \|_F .$$
$$(10.11)$$

To illustrate this completion-like metric MDS, we change the $(2, 3)$th entry of Δ_O in (10.9) from 1 to 0. The resulting "corrupted" matrix Δ_C is given below the main diagonal of $\Delta_C \backslash \Delta_O$ in (10.9).

First, we solve (10.8) with $\Delta_C \odot \Delta_C$. The Euclidean construction results in the distances given above the main diagonal of $D_{EH}^{(10.11)} \backslash D_E^{(10.8)}$. Then, we fit $\Delta_C \odot \Delta_C$ by solving (10.11), where $H(2, 3) = H(3, 2) = 0$ and 1 elsewhere. The estimated Euclidean distances are given below the main diagonal of

$D_{EH}^{(10.11)} \backslash D_E^{(10.8)}.$

$$D_{EH}^{(10.11)} \backslash D_E^{(10.8)} = \begin{bmatrix} 0 & 1.17 & 1.81 & 2.16 & 1.87 & 1.05 & 1.08 \\ 1.04 & 0 & 0.88 & 1.81 & 2.13 & 1.89 & 1.07 \\ 1.88 & 1.14 & 0 & 1.17 & 1.89 & 2.13 & 1.07 \\ 2.15 & 1.88 & 1.04 & 0 & 1.05 & 1.87 & 1.08 \\ 1.86 & 2.16 & 1.86 & 1.09 & 0 & 1.12 & 1.07 \\ 1.09 & 1.86 & 2.16 & 1.86 & 1.06 & 0 & 1.07 \\ 1.07 & 1.08 & 1.08 & 1.07 & 1.08 & 1.08 & 0 \end{bmatrix}.$$

Now, again one can compare the reconstructed distances produced from the solutions of (10.8) with the original and corrupted data, as well as the distances obtained from the solution of (10.11). As expected, the normalized fit of (10.11) is better than the one achieved from (10.8): 0.1417 against 0.1620. However, the estimated Euclidean distances from the solution of (10.11) are worse than the one from (10.8): $\|\Delta_C - D_{EH}^{(10.11)}\|_F = 1.7587$ against $\|\Delta_C - D_E^{(10.8)}\|_F = 1.4661$.

This example illustrates that the matrix completion approach is not very suitable if the primary interest is to find distances that fit well non-Euclidean distances. Recently, the so-called *metric nearness* problem for fitting dissimilarities Δ by a general, not necessarily Euclidean distance matrix D attracted attention. It can be stated as follows:

$$\min_{D \text{ metric}} \|\Delta - D\| , \tag{10.12}$$

for certain norm measuring the fit. Brickell et al. (2008) solve (10.12) for ℓ_0, ℓ_k ($1 < k < \infty$) and ℓ_∞ vector norms using linear and convex programming. The goal is for a given arbitrary dissimilarity matrix, possibly containing missing or corrupted entries, to find the closest distance matrix in some sense. An important difference with the metric MDS is that the metric nearness problem (10.12) is not interested in dimension reduction. This can be explained by the application area, e.g. clustering and graph theory, etc, where the main goal is to "repair" some given non-metric distances which improves the clustering (Gilbert and Jain, 2017). The optimal transport distance learning is another area where related problems arise (Cuturi and Avis, 2014).

It is interesting to explore the solution of (10.12) for the same corrupted non-Euclidean distances Δ_C considered above. Following (Brickell et al.,

2008), one can express the triangle inequality constraints in matrix form as follows. For each triangle ($n = 3$), one can write down exactly three different triangle inequalities. For each quadrilateral ($n = 4$), one can form four different triangles, which gives in total 12 triangle inequalities. Let δ be the vector containing the elements of the (strict) upper triangular part of Δ arranged row after row from left to right and denoted by $\delta = \text{vecu}(\Delta)$. For example, with $n = 4$ for some

$$\Delta = \begin{bmatrix} 0 & \delta_1 & \delta_2 & \delta_3 \\ \delta_1 & 0 & \delta_4 & \delta_5 \\ \delta_2 & \delta_4 & 0 & \delta_6 \\ \delta_3 & \delta_5 & \delta_6 & 0 \end{bmatrix}$$

we have $\delta = \text{vecu}(\Delta) = (\delta_1, \delta_2, \delta_3, \delta_4, \delta_5, \delta_6)^\top$. All possible 12 triangle inequalities between those six values δ_i can be expressed in a compact form as $T_4 \delta \geq 0_{12 \times 1}$ by making use of the following 12×6 matrix

$$T_4 = \begin{bmatrix} 1 & -1 & 0 & 1 & 0 & 0 \\ -1 & 1 & 0 & 1 & 0 & 0 \\ 1 & 1 & 0 & -1 & 0 & 0 \\ 1 & 0 & -1 & 0 & 1 & 0 \\ -1 & 0 & 1 & 0 & 1 & 0 \\ 1 & 0 & 1 & 0 & -1 & 0 \\ 0 & 1 & -1 & 0 & 0 & 1 \\ 0 & -1 & 1 & 0 & 0 & 1 \\ 0 & 1 & 1 & 0 & 0 & -1 \\ 0 & 0 & 0 & 1 & -1 & 1 \\ 0 & 0 & 0 & -1 & 1 & 1 \\ 0 & 0 & 0 & 1 & 1 & -1 \end{bmatrix}.$$

coding all possible triangle inequalities for $n = 4$.

In general, with n points, δ has $n(n-1)/2$ elements and one can get $\binom{n}{3}$ triangles, and thus, $3\binom{n}{3}$ triangle inequalities. One can collect them in a $n(n-1)(n-2)/2 \times n(n-1)/2$ matrix T_n. Then, the triangle inequalities constraints can be expressed as $T_n \delta \geq 0_{n(n-1)(n-2)/2 \times 1}$, where the inequalities are taken element-wise.

We consider here as illustration the corrupted non-Euclidean distances Δ_C from (10.9) and solve only the LS version of (10.12), which is rewritten as

$$\min_{T_n d \geq 0_{\frac{n(n-1)}{2} \times 1}} \|\delta_C - d\|_2 \, , \tag{10.13}$$

where $\delta_C = \text{vecu}(\Delta_C), d = \text{vecu}(D)$ and $n = 7$. This is a quadratic minimization problem with inequality constraints (polyhedral constraint set), which can be solved in a number of ways. We solve (10.13) as an interior point vector flow for d. Following (Helmke and Moore, 1994, p. 114-7), we first define the constraint set C of the problem:

$$C = \left\{ d \in \mathbb{R}^{\frac{n(n-1)}{2}} : Ad \geq 0_{\frac{n(n-1)(n-2)}{2} \times 1} \right\} , \qquad (10.14)$$

and its interior:

$$\mathring{C} = \left\{ d \in \mathbb{R}^{\frac{n(n-1)}{2}} : Ad > 0_{\frac{n(n-1)(n-2)}{2} \times 1} \right\} , \qquad (10.15)$$

by introducing the following block matrix:

$$A = \begin{bmatrix} I_{n(n-1)/2} \\ T_n \end{bmatrix} .$$

Now, for any $d \in \mathring{C}$, let $T_d(\mathring{C})$ denote the tangent space of $\mathring{C}$ at d. Then, one can define a Riemannian metric in the tangent space $T_d(\mathring{C})$ through the following inner product:

$$\ll h_1, h_2 \gg = h_1^\top A^\top \text{diag}(Ad)^{-1} Ah_2 ,$$

for any $h_1, h_2 \in T_d(\mathring{C}) = \mathbb{R}^{\frac{n(n-1)}{2}}$. The solution d of (10.13) is given by the following vector flow (Helmke and Moore, 1994, p. 115):

$$\dot{d} = [A^\top \text{diag}(Ad)^{-1}A]^{-1}(\delta_C - d) ,$$

which, after using the specific form of A, simplifies to

$$\dot{d} = [\Lambda_1^{-1} + T_n^\top \Lambda_2^{-1} T_n]^{-1}(\delta_C - d) , \qquad (10.16)$$

where $\Lambda_1 = \text{diag}(d)$ and $\Lambda_2 = \text{diag}(T_n d)$.

The numerical integration of (10.16) is performed with ode15s from MATLAB. First, a matrix of distances D is obtained using the original non-Euclidean distances Δ_O from (10.9) as an initial value. The resulting matrix is denoted as D_Δ and its upper triangular part is given in $D_{\Delta+U} \backslash D_\Delta$:

$$D_{\Delta+U} \backslash D_\Delta = \begin{bmatrix} 0 & 1.2500 & 1.7496 & 1.9998 & 1.9995 & 0.9997 & 0.9999 \\ 1.2511 & 0 & 0.4997 & 1.7499 & 1.9994 & 1.9997 & 0.9997 \\ 1.7537 & 0.5026 & 0 & 1.2503 & 1.9997 & 2.0001 & 1.0001 \\ 1.9933 & 1.7513 & 1.2487 & 0 & 1.0000 & 1.9999 & 0.9999 \\ 1.9934 & 1.9916 & 1.9977 & 1.0077 & 0 & 0.9999 & 0.9997 \\ 1.0086 & 1.9962 & 1.9978 & 1.9825 & 1.0153 & 0 & 1.0000 \\ 1.0008 & 1.0122 & 0.9990 & 1.0171 & 1.0093 & 1.0155 & 0 \end{bmatrix} ,$$

which LS fit to the corrupted data is $\|\Delta_C - D_\Delta\|_F = 1.0000$. Clearly, these non-Euclidean distances provide much better fit to the initial distances than any of the MDS-based procedures. This is understandable as such solutions are not restricted to be Euclidean.

Next, (10.16) is solved with random start $\Delta_O + U$, where U is an upper-triangular matrix with elements taken from uniform distribution in $(0, 0.2)$. The average fit of these solutions $D_{\Delta+U}$ to Δ_C over 15 runs is 1.2677 with standard deviation 0.2801. The lower triangular part of the solution $D_{\Delta+U}$ that achieves the best fit to Δ_C is given in $D_{\Delta+U} \backslash D_\Delta$. The achieved fit is $\|\Delta_C - D_{\Delta+U}\|_F = 1.0016$.

Finally, we note again that this approach is adopted here only for illustrative purposes. It is rather slow to be suitable/recommended for real large problems. Another issue is that $T_n d$ usually contains zero entries, and $\mathrm{diag}(T_n d)^{-1}$ in (10.16) should be replaced by some *generalized inverse*, e.g.:

$$
\begin{bmatrix} 2 & 0 & 0 \\ 0 & 1 & 0 \\ 0 & 0 & 0 \end{bmatrix}^{-1} = \begin{bmatrix} \frac{1}{2} & 0 & 0 \\ 0 & 1 & 0 \\ 0 & 0 & 0 \end{bmatrix} .
$$

A way to avoid this obstacle is by using the SMW formula (3.28):

$$
[\Lambda_1^{-1} + T_n^\top \Lambda_2^{-1} T_n]^{-1} = \Lambda_1 - \Lambda_1 T_n^\top (\Lambda_2 + T_n \Lambda_1 T_n^\top)^{-1} T_n \Lambda_1 ,
$$

which, however, in this case requires the inverse of a larger matrix.

10.4   INDSCAL – Individual Differences Scaling

INdividual **D**ifferences **SCAL**ing (INDSCAL) model is one of the first approaches for three-way data analysis (Carroll and Chang, 1970), and more generally to tensor data analysis. Its initial purpose is generalization of the metric MDS (Torgerson, 1957) in case of several double centred (dis)similarity matrices. Its interpretation is easier than other more complicated three-way models (Tucker, 1966). INDSCAL was originally aimed at psychometric and marketing research applications. More recent applications are in new areas as chemometrics (Courcoux et al., 2002), (multiple or tensor) face recognition (Turk and Pentland, 1991), signal processing (ICA) (De Lathauwer, 1997), and more generally to tensor decomposition and tensor data analysis (Kolda and Bader, 2009).

10.4.1 The classical INDSCAL solution and some problems

The original purpose of INDSCAL is simultaneous metric MDS of m symmetric $n \times n$ doubly centred dissimilarity matrices usually called slices S_i. INDSCAL applies the MDS decomposition (10.4) to each slice, i.e.:

$$S_i = -\frac{1}{2}J_n(D_i \odot D_i)J_n \approx Q\Xi_i Q^\top , \qquad (10.17)$$

where Q is an $n \times r$ matrix assumed of full column rank, and Ξ_i are diagonal matrices. In other words, all slices share a common loading matrix Q and differ each other only by the diagonal elements of Ξ_i called *idiosyncratic saliences*. Usually, they are assumed non-negative, i.e. Ξ_i is PSD.

We know from Section 10.3.1 that the doubly centred dissimilarity matrices are rank deficient, which explains why the INDSCAL model (10.17) cannot be written for square $n \times n$ matrices Q and Ξ_i.

Thus, the INSCAL problem is to find Q and Ξ_i such that

$$\min_{Q,\Xi_i} \sum_{i=1}^m \left\| S_i - Q\Xi_i Q^\top \right\|_F , \qquad (10.18)$$

i.e. to find Q and Ξ_i that provide the best LS fit to S_i.

This optimization problem has no analytical solution. Initially, it was proposed to solve (10.18) numerically as an unconstrained problem by an alternating LS algorithm, called CANDECOMP (Carroll and Chang, 1970). The two appearances of Q in (10.17) are represented by different matrices, say Q and P, and optimization is carried out on Q and P independently. In other words, instead of solving (10.18), they propose to solve

$$\min_{Q,\Xi_i,P} \sum_{i=1}^m \left\| S_i - Q\Xi_i P^\top \right\|_F , \qquad (10.19)$$

which is identical to the PARAFAC (**PARA**lel profiles **FAC**tor analysis) model proposed in (Harshman, 1970). For this reason, the algorithm is referred to as CANDECOMP/PRAFAC or simply CP for short.

The belief is that after convergence of CP eventually $Q = P$. This is known as the "symmetry requirement" and is frequently achieved in practice, though there is no general proof of the claim. On the contrary, ten Berge and Kiers

(1991) showed on a small example that CP algorithm can produce asymmetric INDSCAL solutions ($Q \neq P$) even for PSD S_i. However, for an orthonormality-constrained INDSCAL model, with $Q^\top Q = I_s$, they proved that CP yields solution with $Q = P$ for PSD S_i, with at least one of them strictly PD. Unfortunately, this result holds at a global CP minimum for $r = 1$ only. If the CP minimum is local, then $Q = P$ is not guaranteed.

Another serious complication with the INDSCAL problem (10.18) is that, in most applications, solutions with non-negative Ξ_i make sense only. Indeed, while Q represents the group stimulus space and $\Xi_i, i = 1, 2, \ldots, m$ represent the subject space, the individual stimulus space for ith individual is given by $Q\Xi_i^{1/2}$, i.e. the square root must be real. INDSCAL solutions with non-negative Ξ_i *cannot* be guaranteed by the original CP algorithm. Negative saliences can occur, even when S_i are PSD (ten Berge and Kiers, 1991). Moreover, ten Berge et al. (1993) provide an example with three symmetric matrices (two strictly PD and one PSD) where the CP solution has a negative salience. This non-negativity issue can be handled by ALSCAL for the non-metric version of INDSCAL (Takane et al., 1977), and by MULTISCALE, which provides maximum likelihood perspective of MDS (Ramsay, 1977). The problem with negative saliences for the metric (scalar-product) version of INDSCAL (10.18) is solved in (De Soete et al., 1993; ten Berge et al., 1993) for oblique Q, i.e. subject to $\mathrm{diag}(Q^\top Q) = I_r$.

Indeed, the INDSCAL problem (10.18) has a considerable weakness: the parameter set of all $n \times r$ matrices Q with full column rank is a *non-compact* Stiefel manifold. Optimization on a non-compact constraint set may be quite tricky. Indeed, Stegeman (2007) reports that in certain cases INDSCAL does not have a minimum, but only an *infimum*. In such situations, no optimal solution exists, and the INDSCAL algorithms are bound to produce sequence of parameters updates which are degenerate in the sense that some columns of Q are extremely highly correlated, and some of the diagonal elements of Ξ_i are arbitrarily large (Krijnen et al., 2008). A review and references related to a number of difficulties with CP for solving INDSCAL are given in (Takane et al., 2010).

10.4.2   Orthonormality-constrained INDSCAL

A natural remedy for the listed above problems with the INDSCAL definition is to assume that the $n \times r$ matrix Q in the INDSCAL model (10.17) is *orthonormal*, i.e. $Q \in \mathcal{O}(n, r)$, which is *compact* (Stiefel manifold). The complication with the non-negative entries in Ξ_i can be reduced by taking $\Xi_i \equiv \Lambda_i^2$. Then, the original INDSCAL problem (10.18) modifies to

$$\min_{Q, \, \Lambda_i} \sum_{i=1}^m \left\| S_i - Q\Lambda_i^2 Q^\top \right\|_F , \qquad (10.20)$$

where $(Q, \Lambda_1, \ldots, \Lambda_m) \in \mathcal{O}(n, r) \times \mathcal{D}(r)^m$. Here, the constraint set is a product of the Stiefel manifold $\mathcal{O}(n, r)$ and m copies of the subspace $\mathcal{D}(r)$ of all diagonal $r \times r$ matrices. The modified problem (10.20) is known as the *orthonormal* INDSCAL, or O-INDSCAL for short. A number of recent IND-SCAL applications actually employ O-INDSCAL (Courcoux et al., 2002; De Lathauwer, 1997).

This orthonormal model can also be validated by looking at INDSCAL as a kind of simultaneous diagonalization of several rank deficient dissimilarity matrices S_i. Indeed, the O-INDSCAL models each slice as

$$S_i = -\frac{1}{2} J_n (D_i \odot D_i) J_n \approx Q\Lambda_i^2 Q^\top , \qquad (10.21)$$

i.e. implicitly assumes that S_i form a commuting family (Horn and Johnson, 2013, p. 62). Thus, INDSCAL becomes a kind of LS version of CPC, a relation already mentioned in Chapter 9. The original INDSCAL model with full column rank or oblique Q provides a specific factorization of a family of symmetric matrices but cannot be interpreted as simultaneous diagonalization in the usual way.

Clearly, the parameters/unknowns of the O-INDSCAL model (10.20) are unique up to a permutation of columns in Q and Λ_i, and/or change of signs of entire column(s) in Q and Λ_i. Algorithms for solving (10.20) are discussed in (Takane et al., 2010; Trendafilov, 2004). The problem (10.20) can be readily solved by Manopt for optimization on matrix manifolds (Absil et al., 2008; Boumal et al., 2014).

The gradient ∇ of the objective function in (10.20) at $(Q, \Lambda_1, \ldots, \Lambda_m) \in \mathcal{O}(n, r) \times \mathcal{D}(r)^m$ with respect to the induced Frobenius inner product can be interpreted as $(m + 1)$-tuple of matrices:

$$\nabla_{(Q,\Lambda_1,\ldots,\Lambda_m)} = (\nabla_Q, \nabla_{\Lambda_1}, \ldots, \nabla_{\Lambda_m}) ,$$

where

$$\nabla_Q \;=\; \sum_{i=1}^{m} S_i Q \Lambda_i^2 \;, \tag{10.22}$$

$$\nabla_{\Lambda_i} \;=\; (Q^\top S_i Q) \odot \Lambda_i - \Lambda_i^3 \;. \tag{10.23}$$

With (10.22) and (10.23) in hand, we can write down the first-order optimality conditions for O-INDSCAL minimum as follows:

Theorem 10.4.1. *A necessary condition for* $(Q, \Lambda_1, \ldots, \Lambda_m) \in \mathcal{O}(n, r) \times \mathcal{D}(r)^m$ *to be a stationary point of the O-INDSCAL is that the following* $m + 2$ *conditions must hold simultaneously:*

- $\sum_{i=1}^{k} Q^\top S_i Q \Lambda_i^2$ *be symmetric;*
- $(I_n - QQ^\top) \sum_{i=1}^{m} S_i Q \Lambda_i^2 = O_{n \times n}$ *;*
- $(Q^\top S_i Q) \odot \Lambda_i = \Lambda_i^3.$

The middle condition reflects that $Q \in \mathcal{O}(n, r)$. It vanishes, for square *orthogonal* Q and the O-INDSCAL optimality conditions coincide with those for the LS-CPC from Theorem 9.3.2.

Similarly, one can derive conditions for INDSCAL with oblique Q, i.e. $Q \in \mathcal{OB}(n, r)$, as considered in (De Soete et al., 1993; ten Berge et al., 1993).

10.5   DINDSCAL – Direct INDSCAL

The INDSCAL model is designed for simultaneous MDS of several doubly centred matrices of squared dissimilarities. An alternative approach, called for short DINDSCAL (**D**irect **INDSCAL**), is proposed for analysing *directly* the input matrices of squared dissimilarities (Trendafilov, 2012). This is a straightforward generalization of the approach to the metric MDS introduced for the case of a single dissimilarity matrix (Browne, 1987). As we already know from Section 10.3.2, this is a way to achieve better fit to the dissimilarities than fitting their doubly centred counterparts. Another important benefit of this model is that missing values can be easily handled.

10.5.1 DINDSCAL model

The DINDSCAL model is based on the identity (10.3) and every slice here is simply $S_i = D_i \odot D_i$, i.e.:

$$S_i = D_i \odot D_i \approx (I_n \odot Q\Lambda_i^2 Q^\top)E_n + E_n(I_n \odot Q\Lambda_i^2 Q^\top) - 2Q\Lambda_i^2 Q^\top , \quad (10.24)$$

where $Q \in \mathcal{O}(n,r)$ and $E_n = 1_n 1_n^\top$. Thus, we do not need the re-phrase DINDSCAL as was the case with O-INDSCAL. The definition (10.24) generalizes the MDS model considered in (Browne, 1987) for the case $m = 1$. He additionally imposed an identification constraint on Q to have zero column sums, i.e. $1_n Q = 0_{n \times r}$. This is not really a constraint because it does not affect the fit and simply places the solution at the origin. This is also adopted in DINDSCAL and (10.24) is considered on the manifold $\mathcal{O}_0(n,r)$ of all $n \times r$ centred orthonormal matrices defined in Section 3.1.4.

To simplify the notations, let denote

$$Y_i = (I_n \odot Q\Lambda_i^2 Q^\top)E_n + E_n(I_n \odot Q\Lambda_i^2 Q^\top) - 2Q\Lambda_i^2 Q^\top ,$$

for $(Q, \Lambda_1, \Lambda_2, \ldots, \Lambda_m) \in \mathcal{O}_0(n,r) \times (\mathcal{D}(r))^m$. Then, the DINDSCAL fitting problem is concerned with the following constrained optimization problem: for given fixed $n \times n$ dissimilarity matrices $D_i, i = 1, 2, \ldots, m$ (Trendafilov, 2012):

$$\min_{\substack{Q \in \mathcal{O}_0(n,r), \\ \Lambda_i \in \mathcal{D}(r)}} \sum_{i=1}^m \|S_i - Y_i\|_F . \quad (10.25)$$

An important advantage of the DINDSCAL reformulation is that dissimilarity slices D_i with missing entries can also be analysed. Indeed, in (10.25) a direct fitting of the data D_i is sought and thus the missing entries of D_i can simply be excluded of consideration. The DINDSCAL problem (10.25) should simply be modified to fit the model to the available entries of the data D_i only. Suppose that there are missing values in the dissimilarity slices D_i. Let V_i are the m pattern matrices indicating the missing values in the corresponding D_i, i.e.: $V_i = \{v_{\iota_1 \iota_2, i}\}$ is defined as

$$v_{\iota_1 \iota_2, i} = \begin{cases} 0 & \text{if } d_{\iota_1 \iota_2, i} \text{ is missing} \\ 1 & \text{otherwise} \end{cases} .$$

Then, the DINDSCAL fitting problem (10.25) with missing values becomes

$$\min_{\substack{Q \in \mathcal{O}_0(n,r), \\ \Lambda_i \in \mathcal{D}(r)}} \sum_{i=1}^{m} \|V_i \odot (S_i - Y_i)\|_F \ . \tag{10.26}$$

10.5.2   DINDSCAL solution

The gradient ∇ of the objective function in (10.25) at $(Q, \Lambda_1, \ldots, \Lambda_m) \in \mathcal{O}(n,r) \times \mathcal{D}(r)^m$ can be interpreted as $(m+1)$-tuple of matrices (Trendafilov, 2012):

$$\nabla_{(Q, \Lambda_1, \ldots, \Lambda_m)} = (\nabla_Q, \nabla_{\Lambda_1}, \ldots, \nabla_{\Lambda_m}) \ ,$$

where

$$\nabla_Q = \sum_{i=1}^{m} W_i Q \Lambda_i^2 \ , \tag{10.27}$$

$$\nabla_{\Lambda_i} = -(Q^{\top} W_i Q) \odot \Lambda_i \ . \tag{10.28}$$

and

$$W_i = (S_i - Y_i) - [(S_i - Y_i)E_n] \odot I_n \ .$$

One can easily check, that $W_i E_n = E_n W_i = 0_n$. This implies that $J_n \nabla_Q = J_n W_i Q \Lambda_i^2 = W_i Q \Lambda_i^2$, i.e. the projection $J_n - QQ^{\top}$ can be replaced by the standard one $I_n - QQ^{\top}$.

With (10.27) and (10.28) in hand, we can write down the first order optimality conditions for DINDSCAL minimum as follows:

Theorem 10.5.1. *A necessary condition for* $(Q, \Lambda_1, \ldots, \Lambda_m) \in \mathcal{O}(n,r) \times \mathcal{D}(r)^m$ *to be a stationary point of the DINDSCAL is that the following* $m+2$ *conditions must hold simultaneously:*

- $\sum_{i=1}^{k} Q^{\top} W_i Q \Lambda_i^2$ *be symmetric;*
- $(I_n - QQ^{\top}) \sum_{i=1}^{m} W_i Q \Lambda_i^2 = O_n$;
- $(Q^{\top} W_i Q) \odot \Lambda_i = O_r$.

Everything remains valid for the DINDSCAL problem with missing values (10.26), after replacing Y_i by $Y_i^\star = V_i \odot Y_i$.

Particularly, for $m = 1$, these optimality conditions reduce to

- $Q^\top W Q \Lambda^2$ be *symmetric*,

- $Q Q^\top W Q \Lambda^2 = W Q \Lambda^2$,

- $(Q^\top W Q) \odot \Lambda = O_r$,

which are optimality conditions for the solution of the original metric MDS problem considered in (Browne, 1987).

The first optimality condition $Q^\top W Q \Lambda^2 = \Lambda^2 Q^\top W Q$ shows that $Q^\top W Q$ is diagonal if Λ^2 has different non-zero diagonal entries. Then, the third optimality condition additionally implies that $Q^\top W Q$ is identically zero at the solution of the problem. As Y is symmetric, $Y E_n \odot I_n = E_n Y \odot I_n$, and the third optimality condition becomes

$$Q^\top (S - Y) Q = Q^\top (D \odot D) Q + \Lambda^2 = Q^\top [(S - Y) E_n \odot I_n] Q .$$

10.5.3 Manopt code for DINDSCAL

As with the CoDa in Section 11.2, here we make use of the centred Stiefel manifold $\mathcal{O}_0(n, p)$.

```
%function [f S] = dindscal(X,p)

global X

% Insert the problem data.
% X(:,:,1) = [0 2 1 6 ; 2 0 3 5; 1 3 0 4; 6 5 4 0];
% X(:,:,2) = [0 3 1 5 ; 3 0 6 2; 1 6 0 4; 5 2 4 0];
% X(:,:,3) = [0 1 3 5 ; 1 0 6 4; 3 6 0 2; 5 4 2 0];
%X=X.^2;
p = 2;

n = size(X,1); mcas = size(X,3);

%=======random_start=======
Q = rand(n,p) -.5; Q = Q - repmat(sum(Q),n,1)/n;
[Q,R] = qr(Q,0); idx=find(diag(R)<=0);Q(:,idx)=-Q(:,idx); %sum(Q)
D = rand(p,mcas) - .5; %D
for i = 1:mcas; D(:,i) = diag(Q'*X(:,:,i)*Q); end
Q = Q';
```

```
20   %S.Q = Q,  S.D = D
21   %
22   %S
23
24   % Create the problem structure.
25   elements = struct();
26   elements.Q = stiefelcenteredfactory(n, p);
27   %elements.Q = stiefelfactory(n, p);
28   elements.D = euclideanfactory(p, mcas);
29   M = productmanifold(elements);
30
31   problem.M = M;
32
33   % Define the problem cost function and its gradient.
34   problem.costgrad = @(S) mycostgrad(S.Q,S.D);
35
36   % Numerically check gradient consistency.
37   %warning('off', 'manopt:stiefel:exp');
38   %checkgradient(problem);
39
40       % Solve.
41       tic
42       options.verbosity=1; % or 0 for no output
43       [S f info] = conjugategradient(problem,[],options)
44   %   [S f info] = trustregions(problem,[],options);
45       toc
46
47   %    f1 = 0; for i=1:mcas; f1 = f1 + norm(X(:,:,i),'fro'); end
48
49       % Display some statistics.
50       figure;
51       plot([info.iter], [info.gradnorm], '.-');
52       hold on;
53       semilogy([info.iter], [info.cost], 'ro-');
54       xlabel('Iteration #');
55       ylabel('Gradient norm');
56
57
58   % Calculating the objective function f and its gradient G
59   function [f G] = mycostgrad(Q,D)
60       global X
61       [n,p] = size(Q); mcas = size(D,2);
62
63       % objective function
64   %    f = 0;
65       Y = zeros(size(X)); W = zeros(n,p,mcas);
66       for j = 1:mcas
67           W(:,:,j) = Q*diag(D(:,j));
68           w = W(:,:,j)*W(:,:,j)';
69           E = repmat(diag(w),1,n);
70           Y(:,:,j) = X(:,:,j) - E - E' + 2*w;
71   %        f = f + norm(Y(:,:,j),'fro')^2;
72       end
73       f = sum(multisqnorm(Y))/4;
74   %    f = f/4;
75
```

```
    % gradient
    if nargout == 2
        s = zeros(n,p);
        for j=1:mcas
            w = 2*(Y(:,:,j) − diag(sum(Y(:,:,j))));
            s = s + w*W(:,:,j)*diag(D(:,j));
            G.D(:,j) = diag(Q'*w*Q).*D(:,j);
        end
        w = Q'*s;
        %G.Q = .5*Q*(w − w') + (eye(n) − Q*Q')*s;
        G.Q = s − .5*Q*(w + w');
    end
end
%end
```

10.6   DEDICOM

10.6.1   Introduction

A family of models, called DEDICOM (**DE**composition into **DI**rectional **COM**ponents) has been introduced for analysing *asymmetric* proximity data (Harshman, 1978). The simplest model of the family became the most popular one and nowadays is simply referred as the DEDICOM model. It decomposes a square (asymmetric) data matrix X of order n as

$$X \approx QRQ^{\top}, \tag{10.29}$$

where Q is an $n \times r(r < n)$ full rank matrix of weights for the n objects on r dimensions (aspects), R is a square (asymmetric) matrix of order r, representing the (asymmetric) relations among the r dimensions.

The DEDICOM model (10.29) can be readily generalized to analyse three-way data composed by several data matrices X_i collecting measurements on the same variables and observations, but in several occasions in an INDSCAL-like manner:

$$X_i \approx QR_iQ^{\top}. \tag{10.30}$$

Originally, as INDSCAL, the DEDICOM models targeted marketing and psychometric applications. Nowadays, such models are interesting for tensor data analysis/decomposition and applied to many different areas (Bader et al., 2007; Kolda and Bader, 2009).

In the hierarchy of DEDICOM models, (10.29) is referred to as the two-way DEDICOM. The simplest and most popular three-way DEDICOM model (for analysing three-way data) is defined as

$$X_i \approx AD_i RD_i A^\top ,$$ (10.31)

where A is a $n \times r(r < n)$ full rank loadings matrix, and D_i are $r \times r$ diagonal matrices specific for each X_i (Kiers, 1989a,b, 1993). The original algorithms are not suitable for large data. A faster version appropriate for modern applications is proposed in (Bader et al., 2007).

The DEDICOM model (10.31) looks too restrictive. Indeed, the variety among X_i's is expressed merely by the diagonal matrices D_i, which is rather simplistic. Another more enhanced model, GIPSCAL, is considered a better option for analysis of asymmetric data and is discussed in Section 10.7.1.

The section is organized as follows. First, the two-way DEDICOM problem is considered in Section 10.6.2. We refer to this problem as to the alternating DEDICOM, because one alternates between finding Q with (10.34) and updating R. Next, in Section 10.6.3, we derive a simultaneous solution of the two-way DEDICOM problem (10.33) with orthonormality constraints. Depending on the specific application one can find the symmetric and skew-symmetric parts of R in order to ease the interpretation and visualization of the data. Finally, we consider the three-way DEDICOM problem (10.30) with orthonormality constraints.

10.6.2   Alternating DEDICOM

For a given $n \times n$ data matrix X, the unknown DEDICOM parameters are found by fitting the model (10.29) to the data in LS sense, i.e.:

$$\min_{\substack{Q \in \mathcal{M}(n,r) \\ R \in \mathcal{M}(r)}} \left\| X - QRQ^\top \right\|_F^2 ,$$ (10.32)

where Q should be a full rank matrix. Several algorithms were developed for solving the DEDICOM problem (10.32). They are described and compared in (Harshman and Kiers, 1987). A number of other DINDSCAL algorithms are available in (Kiers, 1989a,b, 1993) that also generalize to the three-way problems (10.30) and (10.31).

The QR decomposition of Q suggests that the DEDICOM model (10.29) can be readily re-parametrized. Then, instead of solving (10.32), the DEDICOM problem boils down to

$$\min_{\substack{Q \in \mathcal{O}(n,r) \\ R \in \mathcal{M}(r)}} \left\| X - QRQ^\top \right\|_F^2 . \tag{10.33}$$

By noting that for fixed Q the minimum of (10.33) is simply $R = Q^\top X Q$, we transform (10.33) into the following problem:

$$\max_{Q \in \mathcal{O}(n,r)} \operatorname{trace}(QQ^\top X QQ^\top X^\top) . \tag{10.34}$$

Thus, we can solve (10.34) for Q, then update $R = Q^\top X Q$ and so on, until convergence. Kiers et al. (1990) proposed a monotonically convergent algorithm for solving (10.34). The unknown $Q \in \mathcal{O}(n,r)$ is found as a Gram–Schmidt orthonormalization of $(XQQ^\top X^\top Q + X^\top QQ^\top XQ + 2\alpha Q)$ for any α larger than the largest eigenvalue of the symmetric part of $(-X \otimes Q^\top XQ)$, where $\otimes$ is the standard Kronecker matrix product. Clearly, this approach will not be suitable for large data X.

The gradient ∇_Q of the objective function in (10.34) with respect to the Frobenius norm is

$$\nabla_Q = 2 \left(X^\top QQ^\top X + XQQ^\top X^\top \right) Q . \tag{10.35}$$

Theorem 10.6.1. *A first-order necessary condition for $Q \in \mathcal{O}(n,r)$ to be a stationary point of the DEDICOM problem (10.34) is that*

$$\left(I_n - QQ^\top \right) \left(X^\top QQ^\top X + XQQ^\top X^\top \right) Q = O_n .$$

Now, with the DEDICOM objective function (10.34) and its gradient (10.35) at hand, one can readily use Manopt to solve the alternating DEDICOM problem.

One can derive an explicit (projected) Hessian formula to further identify the stationary points of (10.34) (Chu and Driessel, 1990). Unfortunately, the second-order conditions are usually difficult to check and thus, of little practical interest. Another option to explore is to follow the formalism developed in (Edelman et al., 1998, 2.4).

10.6.3 Simultaneous DEDICOM

The gradient $\nabla_{(Q,R)}$ of the objective function in (10.33) at $(Q, R) \in \mathcal{O}(n,r) \times \mathcal{M}(r)$ can be interpreted as couple of matrices:

$$\nabla_{(Q,R)} = (\nabla_Q, \nabla_R) \ ,$$

where

$$
\begin{aligned}
\nabla_Q &= -(XQR^\top + X^\top QR) \ , & (10.36) \\
\nabla_R &= R - Q^\top XQ \ , & (10.37)
\end{aligned}
$$

and for convenience the objective function in (10.33) is divided by two.

With (10.36) and (10.37) in hand, we can write down the first-order optimality conditions for the minimum of the simultaneous DEDICOM:

Theorem 10.6.2. *A necessary condition for $(Q, R) \in \mathcal{O}(n,r) \times \mathcal{M}(r)$ to be a stationary point of the DEDICOM is that the following three conditions must hold simultaneously:*

- $Q^\top XQR^\top - R^\top Q^\top XQ$ *be symmetric;*

- $(I_n - QQ^\top)(XQR^\top + X^\top QR) = O_n$;

- $R = Q^\top XQ.$

10.7   GIPSCAL

10.7.1   GIPSCAL model

In data analysis, visualization of the results is of great importance. Usually, graphical presentation of the asymmetric relationships is given via separate displays for the symmetric and skew-symmetric parts. Probably the most direct way to obtain such results is analysing the symmetric and skew-symmetric parts of the data matrix X (i.e. $X_s = \frac{X+X^\top}{2}$ and $X_k = \frac{X-X^\top}{2}$) separately. Then the symmetric part X_s can be analysed by the standard techniques of the classical metric multidimensional scaling (MDS), and the skew-symmetric part X_k – by the technique proposed in (Constantine and

Gower, 1978; Gower, 1977), see also (Krzanowski, 2003, 3.4). As a result of such analysis two different representations of the same data are produced. For better interpretation of the data as a *whole* some further analysis is needed in order to link the graphical displays of X_s and X_k.

The DEDICOM model aims joint analysis of X and thus is an alternative approach to the problem for analysis and visualization of asymmetric data. The trouble is that DEDICOM does not produce results convenient for interpretation and direct visualization. For this reason, a simpler DEDICOM model (10.31) was considered more useful for practical data analysis. However, this model was criticized of being over-simplistic.

As a result, a model called GIPSCAL (**G**eneralized **I**nner **P**roduct **SCAL**ing), for both joint analysis of asymmetric data and graphical display of the results has been proposed in (Chino, 1978). Its disadvantage is the rather poor fit to the data. To improve this, a generalized GIPSCAL model has been proposed in (Kiers and Takane, 1994). The model has the form:

$$X \approx AA^\top + A\Psi A^\top + \gamma E_n, \tag{10.38}$$

where γ is a constant, A is an $n \times r(r < n)$ full rank matrix of weights, and E_n is a $n \times n$ matrix of ones. The $r \times r$ matrix Ψ is block-diagonal with skew-symmetric blocks $\begin{bmatrix} 0 & \psi_l \\ -\psi_l & 0 \end{bmatrix}$ for $l = 1, 2, ..., \lfloor \frac{s}{2} \rfloor$ and, if s is odd, a 0 in the last diagonal position. The original GIPSCAL (Chino, 1978) assumes *constant* Ψ which causes the poor data fit. Thus, the standard visualization known from MDS can be applied for the symmetric part $AA^\top + \gamma E_n$ and the Gower's technique (Constantine and Gower, 1978; Gower, 1977) can be applied to the skew-symmetric part $A\Psi A^\top$.

The GIPSCAL model (10.38) is fitted to the data in LS sense, which is, for a given, fixed $n \times n$ data matrix X, to

$$\min_{A,\Psi,\gamma} \left\| X - A(I_r + \Psi)A^\top - \gamma E_n \right\|_F . \tag{10.39}$$

Kiers and Takane (1994) show that if the symmetric part of R in the DEDICOM model (10.29) is PD then the DEDICOM is equivalent to the generalized GIPSCAL model for the case $\gamma = 0$ and vice versa. In contrast to the good visualization and fitting features of the model, the corresponding numerical algorithms proposed in (Chino, 1978; Kiers and Takane, 1994) suffer from some weaknesses. Simulated experiments show that Chino's iterative numerical procedure fails to decrease the objective function (dysmonotony) for certain datasets. The two algorithms proposed in (Kiers and Takane, 1994) are rather slow and not quite accurate, which is probably a consequence from the still rather restrictive form of the skew-symmetric part $A\Psi A^\top$, with Ψ having a pretty specific form.

To resolve these issues with GIPSCAL problem (10.39), let $A = QU$ be its QR decomposition, where $Q \in \mathcal{O}(n, r)$ and U is $r \times r$ upper triangular. Note that U should be non-singular because A is assumed full column rank. Then, $AA^\top = QUU^\top Q^\top = QPD^2P^\top Q^\top$, where $PD^2P^\top$ is the EVD of $UU^\top$. This implies that $A = QPD$. With this is mind, we can rewrite the GIPSCAL model (10.38) as follows:

$$AA^\top + A\Psi A^\top = (QP)D^2(QP)^\top + (QP)^\top D\Psi D(QP)^\top ,$$

where $D\Psi D$ is again an $r \times r$ block-diagonal matrix with skew-symmetric blocks $\begin{bmatrix} 0 & \psi_l d_{2l-1}d_{2l} \\ -\psi_l d_{2l-1}d_{2l} & 0 \end{bmatrix}$, where d_i are the diagonal entries of D and l comes from the definition of (10.38). This shows that without loss of generality the GIPSCAL model (10.38) can be considered with orthonormal loadings A.

To improve the GIPSCAL model fit we can generalize (10.38) by replacing $D\Psi D$ with a general $r \times r$ skew-symmetric matrix K. Thus, the GIPSCAL problem (10.39) becomes (Trendafilov, 2002):

$$\min_{Q,D,K} \left\| X - Q(D^2 + K)Q^\top \right\|^2 , \tag{10.40}$$

where $(Q, D, K) \in \mathcal{O}(n, r) \times \mathcal{D}(r) \times \mathcal{K}(r)$. The constraint set is a product of the Stiefel manifold $\mathcal{O}(n, r)$, the subspace of all diagonal $r \times r$ matrices is $\mathcal{D}(r)$ and the subspace of all $r \times r$ skew-symmetric matrices $\mathcal{K}(r)$.

The reformulation (10.40) of the GIPSCAL model reveals its explicit link to the original DEDICOM model (10.33). It becomes obvious that the GIPSCAL is a special case of the DEDICOM when the symmetric part of R in (10.29) is restricted to be non-negative diagonal. From this reformulation it follows also that GIPSCAL can be seen as an extended version of INDSCAL for asymmetric data (Carroll and Chang, 1970; Krzanowski, 2003).

The GIPSCAL model can be generalized for analyzing three-way array consisting of m asymmetric $n \times n$ slices X_i. Similarly to the INDSCAL model the three-way GIPSCAL model decomposes each asymmetric slice as

$$X_i \approx Q(D_i^2 + K_i)Q^\top. \tag{10.41}$$

This means that all slices share a common loading matrix Q and differ each other only by the diagonal elements of D_i and the skew-symmetric component K_i. Thus, the three-way GIPSCAL problem seeks for $(Q, D_1, \ldots, D_m, K_1, \ldots, K_m)$ such that the model fits the data in LS sense, i.e. for given m asymmetric $n \times n$ data matrices X_i (Trendafilov, 2002):

$$\min_{Q, D_i, K_i} \sum_{i=1}^{m} \left\| X_i - Q(D_i^2 + K_i)Q^\top \right\|_F , \tag{10.42}$$

where $(Q, D_1, \ldots, D_m, K_1, \ldots, K_m) \in \mathcal{O}(n,r) \times \mathcal{D}(r)^m \times \mathcal{K}(r)^m$.

10.7.2   GISPSCAL solution

The gradient $\nabla_{(Q,D,K)}$ of the objective function in (10.40) at $(Q, R, K) \in \mathcal{O}(n,r) \times \mathcal{D}(r) \times \mathcal{K}(r)$ can be interpreted as a triple of matrices:

$$\nabla_{(Q,R,K)} = (\nabla_Q, \nabla_D, \nabla_K) ,$$

where

$$
\begin{aligned}
\nabla_Q &= -2X_s Q D^2 + 2X_k Q S , &\tag{10.43}\\
\nabla_D &= 2(D^2 - Q^\top X_s Q) \odot D , &\tag{10.44}\\
\nabla_K &= S - Q^\top X_k Q , &\tag{10.45}
\end{aligned}
$$

and for convenience the objective function in (10.40) is divided by 2.

With these gradients in hand, we can write down the first-order optimality conditions for the GIPSCAL minimum:

Theorem 10.7.1. *A necessary condition for $(Q, R, K) \in \mathcal{O}(n,r) \times \mathcal{K}(r)$ to be a stationary point of the GIPSCAL is that the following four conditions must hold simultaneously:*

- $Q^\top X_k Q K - Q^\top X_s Q D^2$ *be symmetric;*
- $(I_n - QQ^\top)(X_s Q D^2 - X_k Q S) = O_n$;

- $D^2 = I_s \odot (Q^\top X_s Q)$ *(assuming D is invertible);*
- $K = Q^\top X_k Q.$

From the first and the fourth conditions, it follows that

$$D^2 Q^\top X_s Q - Q^\top X_s Q D^2 = K Q^\top X_k Q - Q^\top X_k Q K = O_r \ ,$$

which, in turn, implies that at the stationary point of GIPSCAL:

$$D^2 Q^\top X_s Q = Q^\top X_s Q D^2 \ . \tag{10.46}$$

For an arbitrary D with non-zero diagonal entries, (10.46) can be true only if $Q^\top X_s Q$ is diagonal, and thus, $D^2 = Q^\top X_s Q$. In other words, GIPSCAL diagonalizes the symmetric part of the data X.

Second-order optimality conditions are derived in (Trendafilov, 2002, p. 140).

Example: Consider the following artificial 7×7 data matrix X which elements are random numbers taken from uniform distribution in $[-0.5, 0.5]$:

$$X = \begin{bmatrix}
.0129 & -.4219 & .1617 & -.1635 & .4339 & -.3418 & -.2280 \\
-.0395 & -.1307 & -.0681 & -.3267 & -.2732 & .1012 & .4280 \\
-.1496 & -.4664 & -.0540 & -.4139 & .2859 & -.3824 & .4213 \\
-.4050 & -.3078 & .0083 & -.1067 & -.0893 & .1261 & .0420 \\
-.0663 & -.0286 & .0281 & .3044 & -.3806 & .3351 & .3129 \\
.2092 & -.3551 & .0729 & -.4889 & .1344 & .4404 & -.3336 \\
-.3840 & .2178 & -.1392 & -.2669 & .3624 & -.0844 & -.1796
\end{bmatrix} \ .$$

The GIPSCAL solution with $s = 3$ is

$$Q^\top = \begin{bmatrix}
-.2325 & -.4497 & -.3373 & .1304 & .7290 & -.2838 & -.0309 \\
.5363 & -.3858 & .5326 & -.3486 & .0910 & -.3874 & .0009 \\
-.3013 & .3625 & .0633 & .0563 & -.1376 & -.7732 & .3923
\end{bmatrix} \ ,$$

$$D^2 = \begin{bmatrix} .0000 \\ .3486 \\ .6487 \end{bmatrix} \ , \quad K = \begin{bmatrix}
-.0000 & -.5611 & .0346 \\
.5611 & -.0000 & .3860 \\
-.0346 & -.3860 & .0000
\end{bmatrix} \ .$$

One can check how the GIPSCAL optimality conditions from Theorem 10.7.1 are fulfilled.

10.7.3 Three-way GIPSCAL

The gradient $\nabla_{(Q,D_1,\ldots,D_m,K_1,\ldots,K_m)}$ of the objective function in (10.41) at $(Q, D_1, \ldots, D_m, K_1, \ldots, K_m) \in \mathcal{O}(n,r) \times \mathcal{D}(r) \times \mathcal{K}(r)$ can be interpreted as a $m+2$-ple of matrices:

$$\nabla_{(Q,D_1,\ldots,D_m,K_1,\ldots,K_m)} = (\nabla_Q, \nabla_{D_1}, \ldots, \nabla_{D_m}, \nabla_{K_1} \cdots, \nabla_{K_m}) \, ,$$

where

$$\begin{aligned}
\nabla_Q &= -2X_{i,s}QD_i^2 + 2X_{i,k}QS_i \, , && (10.47) \\
\nabla_{D_i} &= 2(D_i^2 - Q^\top X_{i,s}Q) \odot D_i \, , && (10.48) \\
\nabla_{K_i} &= K_i - Q^\top X_{i,k}Q \, , && (10.49)
\end{aligned}$$

and for convenience the objective function in (10.40) is divided by 2.

With these gradients in hand, we can write down the first-order optimality conditions for the three-way GIPSCAL minimum:

Theorem 10.7.2. *A necessary condition for $(Q, R, K) \in \mathcal{O}(n,r) \times \mathcal{K}(r)$ to be a stationary point of the three-way GIPSCAL is that the following four conditions must hold simultaneously:*

- $Q^\top \sum_{i=1}^m (X_{i,k}QK_i - X_{i,s}QD_i^2)$ *be symmetric;*

- $(I_n - QQ^\top) \sum_{i=1}^m (X_{i,s}QD_i^2 - X_kQK_i) = O_n$;

- $D_i^2 = I_s \odot (Q^\top X_{i,s}Q)$ *(assuming D is invertible);*

- $K_i = Q^\top X_{i,k}Q$.

10.8 Tensor data analysis

Nowadays, there is a trend to call any m-dimensional data array a tensor, which is not completely correct. The definition (2.15) states that the tensor is a multi-linear functional, which already suggests that it is an extension of certain mathematical objects and the related to them operations and transformations. We keep the tradition to identify the linear operators with their matrices, and use the same notations for tensors and their "multidimensional" matrices also called *hypermatrices* (Lim, 2013). Thus, we call $X \in \mathbb{R}^{n_1 \times \cdots \times n_m}$ a (real) tensor of *order* m, or m-tensor, with entries $x_{i_1,\ldots,i_m} \in \mathbb{R}$. We also write $X \in \mathcal{M}(n_1, \ldots, n_m)$.

Traditionally tensors are associated with differential geometry (Levi-Civita, 1927) and their applications to gravitation and relativity theory (Wald, 1984), continuum mechanics (Truesdell, 1991), etc. The topic of tensor decomposition/factorization into "simpler" tensors was initiated in (Hitchcock, 1927a,b), but was left untouched in the main stream theory for years.

In the last couple of decades, the interest in adopting tensor methods for analysis of multi-way data arrays increased considerably (Kolda and Bader, 2009). Probably, the first attempt to model and analyse higher than two-way data (observations × variables) is made in (Tucker, 1966). This pioneering work aims at extending the standard PCA/SVD to three-way data. Such an attempt seems quite natural because PCA and its dimension reduction (low-rank approximation) purpose is one of the most fundamental tasks in data analysis.

The development of the three-way PCA was triggered by applied problems is psychometrics and marketing research. Unfortunately, it turned out that the results from such a three-way PCA are difficult to interpret. As a result, there were introduced a great number of simpler models for analysis of three-way data. We already met several of them in the previous sections. Their common feature is that they are meant for three-way data in which the third mode is not really another dimension but some kind of replication of the measurements made in the other two. For example, the third mode of such data represents different occasions, or "judges", etc., when the measurements are made/evaluated on the same observations and variables.

However, in many modern applications, e.g. face and image recognition, the third, forth and so on modes are genuine further dimensions. In this sense, such applications naturally require true multi-way methods, and the introduction of tensor methods becomes essential. An excellent review of the history and the theory and methods for tensor decomposition can be found in (Kolda and Bader, 2009). Further technical details are given in (Eldén, 2007, Ch 8) and (Golub and Van Loan, 2013, 12.4). There exists plenty of MATLAB-based software for tensor manipulations and decompositions:

- *Tensorlab* https://www.tensorlab.com (some procedures utilize optimization methods on manifolds)

- *Tensor Toolbox* https://www.sandia.gov/~{}tgkolda/TensorToolbox/index-2.6.html

- *TT-Toolbox* http://github.com/oseledets/TT-Toolbox

10.8.1 Basic notations and definitions

Consider the m-tensor $X \in \mathcal{M}(n_1, \ldots, n_m)$. Its different dimensions are called modes. Thus, every matrix is a two-mode tensor, while the vectors are one-mode tensors. According to this new terminology, the classical expressions as "three-way data" used in the previous sections, will be translated in the sequel into "three-mode data".

Like with the PCA of data matrices, the main purpose when analysing high-dimensional data arrays is to find some convenient decomposition such that the given tensor is presented as or approximated by a sum of simpler tensors which can be easily interpreted and/or make its storage more compact. For the classical PCA of a $n \times p$ data matrix X, we use its SVD

$$X = UDV^\top = \sum_1^r d_i u_i v_i^\top = \sum_1^r d_i u_i \otimes v_i = \sum_1^r d_i u_i \circ v_i \ ,$$

where r is the rank of X (equal to the number of non-zero singular values), and u_i, v_i and d_i are, respectively, the singular vectors and values of X (collected in U, V and D). The symbol $\circ$ denotes the *outer product* of two vectors resulting in a matrix containing the products of their elements for all possible combination of indexes.

The main idea is to produce similar kind of decomposition for tensors. For this reason, a *rank-one tensor* U is defined as an outer product of m vectors $u_i \in \mathbb{R}^{n_i}$, $i = 1, \ldots, m$. This is denoted by $U = u_1 \circ \ldots \circ u_m$ and the entries of U are given by $U_{i_1 \ldots i_m} = u_{i_1} \ldots u_{i_m}$ for all index combinations. On the contrary, a m-tensor U is said to have rank 1, if there exist m vectors $u_i \in \mathbb{R}^{n_i}$, $i = 1, \ldots, m$, such that $U = u_1 \circ \ldots \circ u_m$. More generally, the purpose of the **C**anonical **P**olyadic **D**ecomposition (CPD) is to present a tensor X as a sum of finite number rank-one tensors U_i (Hitchcock, 1927b):

$$X \approx \sum_{i=1}^r U_i = \sum_{i=1}^r u_{1,i} \circ \ldots \circ u_{m,i} \ , \tag{10.50}$$

or in terms of Kronecker products as

$$\text{vec}(X) \approx \sum_{i=1}^{r} \text{vec}(U_i) = \sum_{i=1}^{r} u_{m,i} \otimes \ldots \otimes u_{1,i} . \tag{10.51}$$

By definition, the minimal number r for which (10.50) or (10.51) becomes equalitie is called the *rank* of X. Otherwise, the closeness is measured with respect to some norm. In this relation, we also speak about the best rank-r approximation to a m-tensor $X \in \mathcal{M}(n_1, \ldots, n_m)$, if we can find r rank-1 tensors $U_i = u_{1,i} \circ \ldots \circ u_{m,i}$ ($u_{1,i} \in \mathbb{R}^{n_1}, \ldots, u_{m,i} \in \mathbb{R}^{n_m}$), such that

$$\min_{U_i} \left\| X - \sum_{i=1}^{r} U_i \right\| , \tag{10.52}$$

exists. It turns out that this is not always possible (de Silva and Lim, 2008).

Let us use (10.50) to create a rank-one tensor of order three (3-tensor) X by the outer product of the following three vectors:

$$u_1 = \begin{bmatrix} u_{1,1} \\ u_{2,1} \\ u_{3,1} \end{bmatrix} , \quad u_2 = \begin{bmatrix} u_{1,2} \\ u_{2,2} \end{bmatrix} , \quad u_3 = \begin{bmatrix} u_{1,3} \\ u_{2,3} \end{bmatrix} .$$

Then, $u_1 \circ u_2 \circ u_3$ gives a $3 \times 2 \times 2$ rank-one 3-tensor (data array). It is accepted to form its elements $x_{i_1 i_2 i_3}$ by cycling multiplication of the vectors' elements over each index starting from the first, as shown in Algorithm 10.1.

Algorithm 10.1 Outer product of u_1, u_2 and u_3.

---

  for $i = 1 : 3$ **do**
    for $j = 1 : 2$ **do**
      for $k = 1 : 2$ **do**
        $x(i, j, k) := u_1(i)u_2(j)u_3(k)$
      end for
    end for
  end for

We are unable to reproduce on paper higher than two-dimensional data arrays. Instead, we produce matrix "slices", by keeping all but two indexes/modes fixed. In our example, the resulting $3 \times 2 \times 2$ tensor X is composed by the following two *vertical* slices:

$$X(:,:,1) = \begin{bmatrix} x_{111} & x_{121} \\ x_{211} & x_{221} \\ x_{311} & x_{321} \end{bmatrix} = \begin{bmatrix} u_{1,1}u_{1,2}u_{1,3} & u_{1,1}u_{2,2}u_{1,3} \\ u_{2,1}u_{1,2}u_{1,3} & u_{2,1}u_{2,2}u_{1,3} \\ u_{3,1}u_{1,2}u_{1,3} & u_{3,1}u_{2,2}u_{1,3} \end{bmatrix} ,$$

and

$$X(:,:,2) = \begin{bmatrix} x_{112} & x_{122} \\ x_{212} & x_{222} \\ x_{312} & x_{322} \end{bmatrix} = \begin{bmatrix} u_{1,1}u_{1,2}u_{2,3} & u_{1,1}u_{2,2}u_{2,3} \\ u_{2,1}u_{1,2}u_{2,3} & u_{2,1}u_{2,2}u_{2,3} \\ u_{3,1}u_{1,2}u_{2,3} & u_{3,1}u_{2,2}u_{2,3} \end{bmatrix} .$$

For every m-tensor, the sub-tensors defined by fixing all indexes except *two* are called *slices*. The slices are 2-tensors (matrices). The sub-tensors defined by fixing all indexes except *one* are called *fibres*. The fibres are 1-tensors (vectors).

Particularly, for a three-mode data array, the elements $x_{i_1 i_2 i_3}$ of X for a fixed i_1 form a matrix $X_{(i_1)} = \{x_{i_1,:,:}\}$ called horizontal slice. In the same way, the matrices $X_{(i_2)} = \{x_{:,i_2,:}\}$ and $X_{(i_3)} = \{x_{:,:,i_3}\}$ are, respectively, called lateral and frontal slices. Of course, the three-mode array can also be thought as composed by vectors (one-mode matrices), usually called fibres. In particular, keeping two indexes of $x_{i_1 i_2 i_3}$ fixed gives respectively columns, e.g. $X_{(i_2 i_3)} = \{x_{:,i_2,i_3}\}$, rows, e.g. $X_{(i_1 i_3)} = \{x_{i_1,:,i_3}\}$ and tubes, e.g.: $X_{(i_1 i_2)} = \{x_{i_1,i_2,:}\}$.

Another useful experiment is to arrange the elements of the 2-tensor (matrix) $Y = u_1 \circ u_2$ in a 6×1 column-vector, in order they are produced by Algorithm 10.1:

$$\begin{aligned} Y &= u_1 \circ u_2 \\ &\Leftrightarrow \begin{bmatrix} u_{1,1}u_{1,2} & u_{1,1}u_{2,2} & u_{2,1}u_{1,2} & u_{2,1}u_{2,2} & u_{3,1}u_{1,2} & u_{3,1}u_{2,2} \end{bmatrix}^\top \\ &= \begin{bmatrix} y_{11} & y_{12} & y_{21} & y_{22} & y_{31} & y_{32} \end{bmatrix}^\top \\ &= u_2 \otimes u_1 , \end{aligned}$$

where $\otimes$ denotes the usual Kronecker product. However, note that:

$$\begin{aligned} \mathrm{vec}(Y) &= \begin{bmatrix} y_{11} & y_{21} & y_{31} & y_{12} & y_{22} & y_{32} \end{bmatrix}^\top \\ &= \begin{bmatrix} u_{1,1}u_{1,2} & u_{2,1}u_{1,2} & u_{3,1}u_{1,2} & u_{1,1}u_{2,2} & u_{2,1}u_{2,2} & u_{3,1}u_{2,2} \end{bmatrix}^\top \\ &= u_1 \otimes u_2 . \end{aligned}$$

We can further use the same $3 \times 2 \times 2$ tensor X from the previous pages to illustrate the *tensor matricization or unfolding*. This is an operation that arranges the tensor's entries into a matrix. It is very important because it makes it possible to deal with tensors by working with well-known matrix operations. The most frequently used unfoldings are the modal ones, when a tensor is unfolded along one of its modes. In general, for a tensor $X \in \mathcal{M}(n_1, \ldots, n_m)$ with $n = \prod_{i=1}^{m} n_i$, its mode-i_0 unfolding is an $n_{i_0} \times n/n_{i_0}$ matrix whose columns are the mode-i_0 fibers. In the example above, the mode-1, -2 and -3 unfoldings of X are, respectively, $3 \times 4, 2 \times 6$ and 2×6 matrices, as follows:

$$X_{(1)} = \begin{bmatrix} x_{111} & x_{121} & x_{112} & x_{122} \\ x_{211} & x_{221} & x_{212} & x_{222} \\ x_{311} & x_{321} & x_{312} & x_{322} \end{bmatrix} ,$$

$$X_{(2)} = \begin{bmatrix} x_{111} & x_{211} & x_{311} & x_{112} & x_{211} & x_{212} \\ x_{121} & x_{221} & x_{321} & x_{122} & x_{221} & x_{322} \end{bmatrix} ,$$

$$X_{(3)} = \begin{bmatrix} x_{111} & x_{211} & x_{311} & x_{121} & x_{221} & x_{321} \\ x_{112} & x_{212} & x_{312} & x_{122} & x_{222} & x_{322} \end{bmatrix} .$$

One can check that: $X_{(1)} = u_1 \otimes (u_3^\top \otimes u_2^\top), X_{(2)} = u_2 \otimes (u_3^\top \otimes u_1^\top)$ and $X_{(3)} = u_3 \otimes (u_2^\top \otimes u_1^\top)$. In general, if U is a rank-one tensor defined as an outer product of m vectors $u_i \in \mathbb{R}^{n_i}$, $i = 1, \ldots, m$, then its modal unfoldings are rank-one matrices given by

$$U_{(i)} = u_i \otimes (u_m^\top \otimes \ldots u_{i+1}^\top \otimes u_{i-1}^\top \ldots \otimes u_1^\top) .$$

10.8.2   CANDECOMP/PARAFAC (CP)

The main purpose of MDA is as in (10.50) to approximate a given m-dimensional data array (m-tensor) X by a sum of small number rank-one tensors $U_i = u_{1,i} \circ \ldots \circ u_{m,i}$, such that $\sum_i U_i$ is as close as possible to X.

The CPD (10.50) was the first attempt in this direction. Later, it was re-discovered/readopted and applied to real data analysis problems by (Carroll and Chang, 1970) and (Harshman, 1970) as the CP model, already mentioned in Section 10.4.1. Thus, CPD/CP is a generalization of SVD, but not all of the SVD properties are preserved.

In reality, neither r nor U_i in (10.50) are known. Thus, the CPD/CP problem is to find them. In particular, for 3-tensors $X \in \mathcal{M}(n_1, n_2, n_3)$, the problem (10.50) reduces to

$$X \approx \sum_{i=1}^{r} X_i = \sum_{i=1}^{r} u_{1,i} \circ u_{2,i} \circ u_{3,i} . \tag{10.53}$$

We know from Section 10.4.1 to approach (10.53) by re-writing the data tensor X in terms of its frontal slices $X(:,:,j), j = 1, 2, 3$. Then, in terms of the three matrices $U_j = [u_{j,1}, \ldots, u_{j,r}] \in \mathcal{M}(n_j, r)$ for $j = 1, 2, 3$, collecting those unknown vectors, each slice is approximated by $U_1 D_j U_2^\top$, where $D_j = \operatorname{diag}(u_{j,3})$. Fixing the unknown r (a common choice in MDA is $r = 2$), the CP solves

$$\min_{U_1, U_2, D_i} \sum_{j=1}^{3} \left\| X(:,:,j) - U_1 D_j U_2^\top \right\|_F , \tag{10.54}$$

which is another form of the INDSCAL problem (10.20).

For $m > 3$, this way of CP matricization becomes inconvenient. Instead, one can fit the mode-j unfording of X as follows:

$$X_{(j)} \approx U_j \Lambda (U_m \circledast \ldots \circledast U_{j+1} \circledast U_{j-1} \circledast \ldots U_1) ,$$

or directly

$$\operatorname{vec}(X) \approx (U_m \circledast \ldots \circledast U_1) \operatorname{vec}(\Lambda) ,$$

where Λ is diagonal matrix of λ_is and $\circledast$ denotes the *Khatri–Rao matrix product* between $U = [u_1, \ldots, u_r] \in \mathcal{M}(n_1, r)$ and $V = [v_1, \ldots, v_r] \in \mathcal{M}(n_2, r)$ defined by

$$[u_1, \ldots, u_r] \circledast [v_1, \ldots, v_r] = [u_1 \otimes v_1, \ldots, u_r \otimes v_r] ,$$

and known also as the *vectorized* Kronecker product.

If an exact CP decomposition (10.53) exists, then it is called the *rank decomposition* of X and r is its rank. However, there is no algorithm to find such an exact fit. In practice, the rank of a tensor is determined numerically by fitting CP models (10.53) with different values of r, e.g. making use of Tensorlab. Although the tensor rank looks as a straightforward generalization of the matrix rank, it has many peculiar features (Kolda and Bader, 2009; Lim, 2013). For example, tensors of order three or higher may not have best rank-r approximation (de Silva and Lim, 2008). Moreover, the probability that a randomly chosen tensor does not admit a best rank-r approximation is *positive*. For example, in the CP problem (10.54) we can consider U_1 and U_2 evolving on compact sets, e.g. $U_1 \in \mathcal{O}(n_1, r)$ and $U_2 \in \mathcal{O}(n_2, r)$. However, we need $D_j > 0$ for all j, which is not guaranteed because $\mathcal{D}_+(n_3)$ is not compact, it is an open cone.

What makes the CP decomposition attractive is the fact that it can be unique (up to a permutation and/or scaling with constant product) under quite mild conditions. A sufficient condition for the uniqueness of the CP decomposition (10.53) of m-tensor X is given as (Sidiropoulos and Bro, 2000)

$$\sum_1^m k_{U_i} \geq 2r + (m - 1) \ ,$$

where k_{U_i} denotes the *Kruskal's rank* of U_i. By definition, k_{U_i} is the maximum value k, such that *any* k columns of U_i are linearly independent (Kruskal, 1977, 1993). In particular, for $m = 3$, if $\mathrm{rank}(U_1) = \mathrm{rank}(U_2) = r$ and $k_{U_3} \geq 2$, then the rank of X is r and its CP is unique (Harshman, 1972).

Another attractive aspect of CP for large X is that it requires storage space of only $m \sum_j^m n_j$, compared to the storage space $\prod_j^m n_j$ for the original X.

10.8.3   Three-mode PCA (TUCKER3)

Let $X \in \mathcal{M}(n_1, n_2, n_3)$ be a three-mode data array. For example, $x_{i_1 i_2 i_3}$ denotes the i_1th observation on the i_2th variable under the i_3th condition. Tucker (1966) introduced three-mode PCA model (TUCKER3) to seek for a PCA-like representation of X in the form:

$$x_{i_1 i_2 i_3} = \sum_{j_1=1}^{r_1} \sum_{j_2=1}^{r_2} \sum_{j_3=1}^{r_3} c_{j_1 j_2 j_3} u_{i_1 j_1} u_{i_2 j_2} u_{i_3 j_3} \ , \tag{10.55}$$

where

- $U_1 = \{u_{i_1 j_1}\} \in \mathcal{M}(n_1, r_1)$, $U_2 = \{u_{i_2 j_2}\} \in \mathcal{M}(n_2, r_2)$ and $U_3 = \{u_{i_3 j_3}\} \in \mathcal{M}(n_3, r_3)$ are factor-loading matrices of the "idealized individuals", "idealized variables" and "idealized conditions", respectively;

- $C = \{c_{j_1 j_2 j_3}\} \in \mathcal{M}(r_1, r_2, r_3)$ is a three-mode array called "core" and its elements can be seen as an interaction indicators of the three modes.

Tucker (1966) also proposed a matrix formulation of the three-way model (10.55). Three equivalent representations are possible, which can be obtained from each other by cycling change of the involved parameters:

$$
\begin{aligned}
X_{(1)} &= U_1 C_{(1)} (U_2^\top \otimes U_3^\top), \\
X_{(2)} &= U_2 C_{(2)} (U_3^\top \otimes U_1^\top), \\
X_{(3)} &= U_3 C_{(3)} (U_1^\top \otimes U_2^\top),
\end{aligned}
\tag{10.56}
$$

where $\otimes$ denotes the Kronecker matrix product, and $X_{(i)}$ and $C_{(i)}$ denote the mode-i unforldings of X and C, respectively, for $i = 1, 2, 3$.

Reformulation of TUCKER3 as a LS problem for the estimation of its parameters, U_1, U_2, U_3 and C, is given in (Kroonenberg and De Leeuw, 1980). Consider the function:

$$
F(U_1, U_2, U_3, C) = \| X_{(1)} - U_1 C_{(1)} (U_2^\top \otimes U_3^\top) \|_F^2 .
\tag{10.57}
$$

It is shown in (Kroonenberg and De Leeuw, 1980) that, for fixed U_1, U_2 and U_3, there exists an unique minimizer of (10.57) given by

$$
\hat{C}_{(1)} = U_1^\top X_{(1)} (U_2 \otimes U_3) .
\tag{10.58}
$$

After substituting $\hat{C}_{(1)}$ in (10.57), the function to be minimized becomes

$$
f(U_1, U_2, U_3) = \| X_{(1)} - U_1 U_1^\top X_{(1)} (U_2 U_2^\top \otimes U_3 U_3^\top) \|_F^2 ,
$$

i.e. it depends on U_1, U_2 and U_3 only.

In the initial model, U_1, U_2 and U_3 are (almost) not constrained; they are simply required to be full column rank matrices (Tucker, 1966). Later, for computational convenience, orthonormality constrains are imposed on U_1, U_2 and U_3 (Kroonenberg and De Leeuw, 1980). Thus, TUCKER3 transforms into the following constrained optimization problem:

$$
\min_{\substack{U_1 \in \mathcal{O}(n_1, r_1) \\ U_2 \in \mathcal{O}(n_2, r_2) \\ U_3 \in \mathcal{O}(n_3, r_3)}} \| X_{(1)} - U_1 U_1^\top X_{(1)} (U_2 U_2^\top \otimes U_3 U_3^\top) \|_F^2 ,
\tag{10.59}
$$

on the product of three Stiefel manifolds $\mathcal{O}(n_1, r_1) \times \mathcal{O}(n_2, r_2) \times \mathcal{O}(n_3, r_3)$. Clearly, (10.59) is equivalent to the following maximization problem:

$$\max_{\substack{U_1 \in \mathcal{O}(n_1, r_1) \\ U_2 \in \mathcal{O}(n_2, r_2) \\ U_3 \in \mathcal{O}(n_3, r_3)}} \text{trace} \left(U_1^\top X_{(1)} (U_2 U_2^\top \otimes U_3 U_3^\top) X_{(1)}^\top U_1 \right) , \tag{10.60}$$

which is solved in (Kroonenberg and De Leeuw, 1980) for U_1 as EVD of $X_{(1)}(U_2 U_2^\top \otimes U_3 U_3^\top) X_{(1)}^\top$, with U_2 and U_3 fixed. Note, that the objective function in (10.60) does not change if U_1, U_2 and U_3 are cyclically replaced. The algorithm solves (10.60) by EVDs in turn for U_1, U_2 and U_3, and then, updating the remaining parameters.

The gradients of the objective function (10.60) with respect to the Frobenius norm are given by

$$\nabla_{U_1} = X_{(1)}(U_2 U_2^\top \otimes U_3 U_3^\top) X_{(1)}^\top U_1 ,$$
$$\nabla_{U_2} = X_{(2)}(U_3 U_3^\top \otimes U_1 U_1^\top) X_{(2)}^\top U_2 ,$$
$$\nabla_{U_3} = X_{(3)}(U_1 U_1^\top \otimes U_1 U_1^\top) X_{(3)}^\top U_3 ,$$

and the corresponding Riemannian gradients are

$$\nabla_{U_1} = (I_{n_1} - U_1 U_1^\top) X_{(1)} (U_2 U_2^\top \otimes U_3 U_3^\top) X_{(1)}^\top U_1 ,$$
$$\nabla_{U_2} = (I_{n_2} - U_2 U_2^\top) X_{(2)} (U_3 U_3^\top \otimes U_1 U_1^\top) X_{(2)}^\top U_2 ,$$
$$\nabla_{U_3} = (I_{n_3} - U_3 U_3^\top) X_{(3)} (U_1 U_1^\top \otimes U_1 U_1^\top) X_{(3)}^\top U_3 .$$

They can be used for Manopt codes to find a simultaneous solution (U_1, U_2, U_3) of TUCKER3 on the product manifold $\mathcal{O}(n_1, r_1) \times \mathcal{O}(n_2, r_2) \times \mathcal{O}(n_n, r_3)$. Thus, for $(U_1, U_2, U_3) \in \mathcal{O}(n_1, r_1) \times \mathcal{O}(n_2, r_2) \times \mathcal{O}(n_3, r_3)$ to be a stationary point of TUCKER3, the following three conditions must hold simultaneously:

- $(I_{n_1} - U_1 U_1^\top) X_{(1)} (U_2 U_2^\top \otimes U_3 U_3^\top) = O_{n_1 \times n_2 n_3};$

- $(I_{n_2} - U_2 U_2^\top) X_{(2)} (U_3 U_3^\top \otimes U_1 U_1^\top) = O_{n_2 \times n_3 n_1};$

- $(I_{n_3} - U_3 U_3^\top) X_{(3)} (U_1 U_1^\top \otimes U_2 U_2^\top) = O_{n_3 \times n_1 n_2}.$

It is easy to check that the TUCKER3 solution obtained in (Kroonenberg and De Leeuw, 1980), also fulfil these conditions.

Before moving on, we mention two special cases of the TUCKER3 model. The TUCKER2 decomposition of a 3-tensor (data array) sets one of the loadings matrices U_1, U_2 or U_3 to be the identity matrix. For example, one can set $n_3 = r_3$ and $U_3 = I_{n_3}$. In a similar fashion, the TUCKER1 decomposition sets two loadings matrices to identity, say, $U_2 = I_{n_2}$ and $U_3 = I_{n_3}$.

In contrast to the CP decomposition, the TUCKER3 solution is not unique: U_1, U_2 and U_3 can be multiplied by orthogonal matrix without affecting the model fit. Then, as in the standard PCA and EFA this rotational freedom is used to simplify the loadings U_1, U_2 and U_3, and/or the core C (Kiers, 1997, 1998). Procrustes types of such rotations were discussed in Section 6.5.2. Note, that the core C is considered simple for interpretation if it is *diagonal*, i.e. when $c_{ijk} \neq 0$ only for $i = j = k$, or if it is sparse (Ikemoto and Adachi, 2016).

Note, that TRCKER3 with $r_1 = r_2 = r_3$ and diagonal C reduces to the CP model. However, TUCKER3 requires the storage of $r_1 r_2 r_3$ core elements, which is potentially a weakness for large and/or higher mode tensor data.

10.8.4   Multi-mode PCA

A straightforward extension of the standard (two-mode) PCA and the three-mode Tuckers's PCA for analysing m-dimensional data arrays/tensors is studied in (Magnus and Neudecker, 1988, p. 363-6). It is called the *multi-mode* PCA and was already mentioned in Section 6.5.2 in relation to Procrustes analysis.

Making use of the notations adopted in Section 6.5.2 the standard (two-mode) PCA can be written as $X = U_1 C U_2^\top$, where the matrix C containing the singular values of X is renamed to "core" matrix. After applying the vec-operator, the PCA model becomes

$$\text{vec}(X) = (U_2 \otimes U_1)\text{vec}(C) .$$

Then, the multi-mode PCA model is defined as extrapolation of the two-mode case and is expressed by

$$x = (U_m \otimes \ldots \otimes U_1)c , \tag{10.61}$$

where $U_i \in \mathcal{O}(n_i \times r_i)$ for $i = 1, 2, \ldots, m$, and the vectors x and c are the column vectors of the m-tensors X and C, respectively, i.e. $x = \text{vec}(X)$ and $c = \text{vec}(C)$. Then, for some initial C, the loadings U_i are found by alternating algorithm that fits the data in LS sense, i.e. by solving

$$\max_{U_i \in \mathcal{O}(n_i \times p_i)} \|x - (U_m \otimes \ldots \otimes U_1)c\|_F \ . \tag{10.62}$$

It is shown in (Magnus and Neudecker, 1988, p. 363-6) that the optimal U_i, minimizing (10.62) is composed by the r_i largest eigenvectors of $X_{(i)}^\top U^i U^{i\top} X_{(i)}$. Here $U^i = U_m \otimes \ldots \otimes U_{i+1} \otimes U_{i-1} \otimes \ldots \otimes U_1$ and $X_{(i)}$ is the mode-i unfolding of X. Eventually, one finds $c = (U_m \otimes \ldots \otimes U_1)^\top x$.

As with the TUCKER3 model, one can set any subset of loadings matrices U_i to identity and work with a simplified model.

10.8.5   Higher order SVD (HOSVD)

Nowadays, we know that the TUCKER3 model (10.55), as well as the multimode PCA, can be thought as special cases of a generalized SVD version for tensors known as the higher order SVD (HOSVD) (De Lathauwer et al., 2000).

The HOSVD idea is to directly attack (10.56) by applying SVD to the modal unfoldings of X, rather than eliminating the core and solving the EVD problems (10.60). The HOSVD solution results again in orthonormal loadings U_1, U_2 and U_3. After they are found, one can calculate/update the core by (10.58). The slices of the core C for every mode are orthogonal in the following sense: for the first mode $\text{trace}[C(i_1, :, :)^\top C(i_2, :, :)] = 0$, where $1 \leq i_1 < i_2 \leq r_1$; for the second mode $\text{trace}[C(:, i_1, :)^\top C(:, i_2, :)] = 0$, where $1 \leq i_1 < i_2 \leq r_2$, and etc. The singular values $\sigma_i^{(1)}$ of the first mode are defined by

$$\sigma_i^{(1)} = \sqrt{\text{trace}[C(i, :, :)^\top C(i, :, :)]} \ , \text{ for } i = 1, \ldots, r_1 \ .$$

The singular values of the other modes are defined similarly. Also, the singular values of every mode of the core are ordered in decreasing order: $\sigma_{i_1}^{(1)} \geq \sigma_{i_2}^{(1)}$ for $i_1 > i_2$, and the same for the other modes. These properties follow from

$$X_{(i)} = U_i C_{(i)} (U_m \otimes \ldots \otimes U_{i+1} \otimes U_{i-1} \otimes \ldots \otimes U_1)^\top \ , \tag{10.63}$$

which, in turn, comes from the model equation (10.61).

Now, it makes sense to consider a truncated HOSVD tensor which is supposed to provide a kind of "low-rank" approximation to X. *Multilinear rank* of a m-tensor is the vector of ranks of its modal unfoldings. Then, one can consider some $s_i \leq \mathrm{rank}(X_{(i)})$, and truncate the HOSVD loadings to $\tilde{U}_i \in \mathcal{O}(n_i, s_i)$. A frequent choice in MDA is $s_1 = \ldots = s_m = 2$. The corresponding core is calculated as

$$\tilde{c} := (\tilde{U}_m \otimes \ldots \otimes \tilde{U}_1)x ,$$

and the low-rank approximation $\tilde{X}$ to X is given by

$$\tilde{x} := (\tilde{U}_m \otimes \ldots \otimes \tilde{U}_1)\tilde{c} .$$

In contrast to the matrix case, when the truncated SVD provides the best low-rank LS approximation, the truncated tensor $\tilde{X}$ is usually *not* optimal.

For large tensors and/or cores, the multi-mode PCA becomes impractical because of increasing memory requirements. Several new approaches addressing this issue are reviewed in (Grasedyck et al., 2013), notably the *tensor train* (TT) decomposition (Oseledets, 2011) and the *hierarchical Tucker* (HT) decomposition (Hackbusch, 2012).

For example, the TT approach defines TT rank of m-tensor X as the vector:

$$\mathrm{rank}_{TT}(X) = (r_0, r_1, \ldots, r_m) = (1, \mathrm{rank}(X_{(1)}), \ldots, \mathrm{rank}(X_{(m-1)}), 1) ,$$

making use of its modal unforlings. By definition, the tensor A is said to be in TT-format, if each of its terms is presented as

$$a_{i_1 \ldots i_m} = \sum_{k_1=1}^{r_1} \ldots \sum_{k_{m-1}=1}^{r_{m-1}} G_1(1, i_1, k_1)G_2(k_1, i_2, k_2) \ldots G_m(k_{m-1}, i_n, 1) ,$$

$$(10.64)$$

where the involved 3-tensors $G_j \in \mathcal{M}(r_{j-1} \times n_j \times r_j)$ are called TT cores or tensor "carriages". They are calculated from the SVDs of the corresponding modal unfoldings of A, making use of (Oseledets and Tyrtyshnikov, 2010, Algorithm 1). The elements of A can be further expressed compactly as products of m matrices:

$$a_{i_1 \ldots i_m} = U_1(i_1) \ldots U_m(i_m) ,$$

where $U_j \in \mathcal{M}(r_{j-1}, r_j)$ is formed from the corresponding carriage G_j, for $j = 1, \ldots, m$. For clarity and convenience, this process of turning a standard tensor into a TT-format is summarized in Algorithm 10.2, reproducing the algorithm from (Oseledets and Tyrtyshnikov, 2010; Oseledets, 2011).

Algorithm 10.2 Full2TT tensor compression (Oseledets, 2011, Oseledets and Tyrtyshnikov, 2010).

{Initialization: tensor $X \in \mathcal{M}(n_1, \ldots, n_m)$ and accuracy bound ϵ}
$\delta \Leftarrow \frac{\epsilon \|X\|_F}{\sqrt{m-1}}$, where $\|X\|_F^2 = \sum_{i_1 \ldots i_m} x_{i_1 \ldots i_m}^2$ is the Frobenius norm of X
$r_0 \Leftarrow 1$
$W \Leftarrow X$
for $i = 1 : m - 1$ **do**
  $W \Leftarrow \mathtt{reshape}\left(W, r_{i-1}n_i, \frac{\mathtt{numel}(W)}{r_{i-1}n_i}\right)$
  Compute δ-truncated SVD of $W = UDV^\top + E$, such that $\|E\|_F \leq \delta$
  $r_i \Leftarrow \mathrm{rank}(D)$
  $G_k \Leftarrow \mathtt{reshape}(U, r_{i-1}, n_i, r_i)$ (reshape U into tensor)
  $W \Leftarrow DV^\top$
end for
$G_m \Leftarrow W$
form A using (10.64), for which $\|X - A\|_F \leq \epsilon \|X\|_F$

Thus, in case of $n_1 = \ldots = n_m = n$ and $r_1 = \ldots = r_m = r$, the storage space for X is reduced from order n^m to order mnr^2 for the TT cores. The approach is really efficient if the ranks are small, i.e. for $r \ll n$.

10.9   Conclusion

Section 10.2 briefly reminds the main properties of proximity (similarity and dissimilarity) measures widely used in MDS. The metric MDS of a single proximity matrix is considered in Subsection 10.3. The classical MDS definition is given in Subsection 10.3.1 as EVD of a certain matrix of squared proximities, while Subsection 10.3.2 improves it by *directly* fitting the proximities. Subsection 10.3.3 treats problems with missing proximities, as well as finding distances that fit well non-Euclidean distances, known as metric nearness problems.

Subsection 10.4 is about the INDSCAL model for simultaneous MDS of several matrices of squared distances. Subsection 10.4.1 outlines the classical INDSCAL and approaches for its solution, Subsection 10.4.2 considers INDSCAL with orthonormal (and oblique) loadings, i.e. as problems on matrix manifolds.

Subsection 10.5 is about a INDSCAL version (DINDSCAL) that directly fits the set of proximity matrices. The basic model (including tacking missing proximities) is described in Subsection 10.5.1, while its solution as optimization on a product of Stiefel manifold and spaces of diagonal matrices is described in Subsection 10.5.2. Manopt codes for solving DINDSCAL are presented in Subsection 10.5.3.

Subsection 10.6 discusses the DEDICOM model(s) generalizing the IND-SCAL in case of asymmetric proximity matrices. The classical alternating approach to DEDICOM is presented in Subsection 10.6.2, while the simultaneous procedure on a product of matrix manifolds is discussed in Subsection 10.6.3.

The results from DEDICOM are usually difficult for interpretation and direct visualization. Subsection 10.7 is about a special sub-model GIPSCAL to resolve those practical matters. Subsection 10.7.1 defines the GIPSCAL model and explains how it can be interpreted. It solution on a product of manifolds is discussed in Subsection 10.7.2, while Subsection 10.7.3 treats the case of several proximity matrices.

Subsection 10.8 provides some preliminary facts related to the initial attempts analyse multidimensional data arrays. Subsection 10.8.1 introduces relevant basic notations, the concept of rank-one tensor and the idea to present a tensor as a sum of finite number rank-one tensors. The tensor matricization/unfolding is a very important manipulation helping to present it as a set of standard matrices. Subsection 10.8.2 considers the problem of approximation a given arbitrary m dimensional data array (tensor) by a sum of small number rank-one tensors and the CP algorithm for its solution. Subsection 10.8.3 concentrates on the three-mode PCA model known as TUCKER3 seeking for a PCA-like representation of three-dimensional data array. It is shown that it can be solved as optimization problem on a product of three Stiefel manifolds. These considerations are generalized to arbitrary m-dimensional data array in Subsection 10.8.4. Further, Subsection 10.8.5 shows that the TUCKER3 model, as well as the multi-mode PCA, can be thought as special cases of the generalized SVD version for tensors known as the higher order SVD (HOSVD). For large tensors, some other types of decompositions are regarded as more suitable and are briefly sketched.

10.10   Exercises

1. Let C be a correlation $p \times p$ matrix, i.e. a PSD symmetric matrix with unit main diagonal. Define a matrix D by letting $d_{ij}^2 = 1 - |c_{ij}|$.

   (a) Show that D is a dissimilarity matrix, i.e. $d_{ij} \geq 0$ for every $i \neq j$. Hint: Show that every principal 2×2 minor of C is non-negative.

   (b) Show that D is a dissimilarity matrix by constructing $d_{ij}^2 = e_{ij}^\top C e_{ij}$ for some special e_{ij} depending on i and j only. Hint: As C is PSD, then $x^\top C x \geq 0$ for any $x \in \mathbb{R}^p$ and $x \neq 0$.

   (c) Show that D is a distance matrix, i.e. $d_{ij} \leq d_{ik} + d_{kj}$ for every triple of different indexes. Hint: Consider vectors e_{ij} with 1 at its i position, -1 at its jth position, and zeros elsewhere, and make use of $(x^\top C y)^2 \leq (x^\top C x)(y^\top C y)$ valid for any PSD C.

   (d) Show that D is Euclidean distance matrix by proving that B is PSD, because C is PSD and $B = J_p C J_p$. Hint: Rewrite (10.3) and (10.4) for distances among the columns of standardized X.

2. Check that the same results hold true for an arbitrary PSD matrix C, where the dissimilarity matrix D is defined by $d_{ij}^2 = c_{ii} + c_{jj} - 2|c_{ij}|$.

3. Consider an *orthonormal* version of CP, by taking $Q, P \in \mathcal{O}(n, s)$ in (10.19), and construct a Procrustes-based algorithm for its solution.

4. Consider a simplified version of O-INDSCAL (10.21), called the *poor man's* INDSCAL (Gower and Dijksterhuis, 2004, 13.2), where Q is found from the EVD of

$$\frac{1}{m} \sum_{i=1}^{m} S_i \ ,$$

and Λ_i are constrained to

$$\frac{1}{m} \sum_{i=1}^{m} \Lambda_i^2 = I_s \ .$$

5. Let $V_s D_s^2 V_s^\top$ be the truncated EVD of S_i, the ith given slice in (10.21). Then, form "raw" data $X_i = V_r D_r$ for every i, and consider the following Procrustes version of O-INDSCAL (Gower and Dijksterhuis, 2004, 13.2):

$$\min_{\substack{Q \in \mathcal{O}(n,s), \\ P_i \in \mathcal{O}(s), \Lambda_i \in \mathcal{D}(s)}} \sum_{i=1}^{m} \|X_i P_i - Q \Lambda_i\|_F \ .$$

6. Consider the following iterative algorithm (suggested by Bart De Moor) for solving the alternating DEDICOM (10.34). Let Q_k denote the problem unknown at the kth iteration step. Form:

$$W_{k+1} = X Q_k Q_k^T X^T Q_k + X^T Q_k Q_k^T X Q_k \ ,$$

and then orthogonalize its columns using the QR decomposition:

$$W_{k+1} = Q_{k+1} R_{k+1},$$

where $Q_{k+1}^T Q_{k+1} = I_s$. Repeat until convergence. Can you prove its convergence?

7. Column-wise alternating DEDICOM. Consider the case of $s = 1$. Then, (10.34) becomes:

$$\max_{u^T u=1} (u^T X u)^2 \ .$$

Introducing a Lagrange multiplier λ, one finds

$$2(X + X^T)u(u^T X u) = 2u\lambda \ ,$$

which, after left multiplying with u^T leads to

$$u^T (X + X^T)u(u^T X u) = \lambda = 2(u^T A u)^2 \ .$$

So, this DEDICOM problem requires the maximum eigenvalue of the symmetric part $X + X^T$ of the data. Construct a column-wise DEDICOM algorithm for arbitrary $s(> 1)$.

8. Check that the Kronecker product of two orthonormal matrices is also orthonormal, i.e. if $Q_1 \in \mathcal{O}(m,n)$ and $Q_2 \in \mathcal{O}(p,q)$, then $Q = Q_1 \otimes Q_2 \in \mathcal{O}(mp, nq)$. Hint: see (6.48) and (6.85).

9. For the matrix

$$A = \begin{bmatrix} 0 & 1 & 11 \\ 0 & 2 & 22 \\ 0 & 3 & 33 \\ 0 & 4 & 44 \end{bmatrix},$$

find $\mathrm{rank}(A)$ and the Kruskal's rank k_A. Do they differ, and why?

10. Check that the objective function in (10.60) can be rewritten as

$$\mathrm{trace}\left(G^\top X_{(1)}(HH^\top \otimes EE^\top)X_{(1)}^\top G\right) =$$
$$\mathrm{vec}\left(X_{(1)}^\top\right)^\top (GG^\top \otimes HH^\top \otimes EE^\top)\mathrm{vec}\left(X_{(1)}^\top\right),$$

where vec() is the vec-operator of its argument. Hint:

$$\mathrm{trace}\left(G^\top X_{(1)}(HH^\top \otimes EE^\top)X_{(1)}^\top G\right) = \|(H^\top \otimes E^\top)X_{(1)}^\top G\|_F^2 =$$
$$\|vec[(H^\top \otimes E^\top)X_{(1)}^\top G]\|_2^2 = \|(G^\top \otimes H^\top \otimes E^\top)vec(X_{(1)}^\top)\|_2^2 .$$

11. Find the gradients of the objective function in (10.60). It might be help-ful to use the following special permutation $mn \times mn$ matrix K_{mn}, also known as the *commutation matrix* (Magnus and Neudecker, 1988, p. 46). For every $A \in \mathcal{M}(m,n)$ and $B \in \mathcal{M}(p,q)$, K_{mn} relates $A \otimes B$ and $B \otimes A$ through $K_{pm}(A \otimes B) = (B \otimes A)K_{qn}$. See also (Golub and Van Loan, 2013, 12.3.5) and (Lancaster and Tismenetsky, 1985, p. 408-9). The commuta-tion matrix K_{mn} is a block matrix composed by mn blocks with sizes $n \times m$ each. For example, for $m = 3$ and $n = 2$, we have:

$$K_{3,2} = \begin{bmatrix} 1 & 0 & 0 & : & 0 & 0 & 0 \\ 0 & 0 & 0 & : & 1 & 0 & 0 \\ \cdots & \cdots & \cdots & : & \cdots & \cdots & \cdots \\ 0 & 1 & 0 & : & 0 & 0 & 0 \\ 0 & 0 & 0 & : & 0 & 1 & 0 \\ \cdots & \cdots & \cdots & : & \cdots & \cdots & \cdots \\ 0 & 0 & 1 & : & 0 & 0 & 0 \\ 0 & 0 & 0 & : & 0 & 0 & 1 \end{bmatrix}.$$

The commutation matrix has also the following properties:

(a) $K_{mn}\mathrm{vec}(A) = \mathrm{vec}(A^\top)$, for every $A \in \mathcal{M}(m,n)$;

(b) $K_{mn}^\top = K_{mn}^{-1} = K_{nm}$;

(c) $K_{mn}K_{nm}\mathrm{vec}(A) = \mathrm{vec}(A)$;

(d) $K_{mn}K_{nm} = K_{nm}K_{mn} = I_{mn}$.

2. Check that for $A \in \mathcal{M}(m,n)$ and $B \in \mathcal{M}(p,q)$ the following identity

$$\mathrm{vec}(A \otimes B) = (I_n \otimes K_{qm} \otimes I_m)[\mathrm{vec}(A) \otimes \mathrm{vec}(B)] \ ,$$

holds (Magnus and Neudecker, 1988, p. 47). Hint: Use that $A = \sum_{i=1}^{n} a_i e_i^\top$, where $A = [a_1, \ldots, a_n]$ and $I_n = [e_1, \ldots, e_n]$.

3. Write Manopt codes to find simultaneous solution (G, H, E) of TUCKER3 on the product manifold $\mathcal{O}(n_1, r_1) \times \mathcal{O}(n_2, r_2) \times \mathcal{O}(n_n, r_3)$.

Chapter 11

Data analysis on simplexes

© Springer Nature Switzerland AG 2021
N. Trendafilov and M. Gallo, *Multivariate Data Analysis on Matrix Manifolds*,
Springer Series in the Data Sciences, https://doi.org/10.1007/978-3-030-76974-1_11

11.1 Archetypal analysis (AA)

11.1.1 Introduction

Cutler and Breiman (1994) introduced the archetypal analysis (AA) in attempt to produce several "representative" observations for a particular dataset, called "archetypes", such that the data as a whole can be well approximated by a *convex* combination of them. The archetypes, by themselves, are also considered being *convex* combinations of the original observations.

First, note that the linear combination x of r vectors y_i:

$$x = \alpha_1 y_1 + \ldots + \alpha_r y_r$$

is called convex when the weights are all non-negative and sum to one, i.e. $\alpha_i \geq 0$, $i = 1, \ldots, r$ and $\sum_{i=1}^{r} \alpha_i = 1$. Now, AA can be stated more formally. Let X be an $n \times p$ standardized data matrix. AA looks for n convex linear combinations of r archetypes Y, such that they approximate X:

$$X \approx AY^\top , \tag{11.1}$$

and the weights $A \in \mathcal{M}(n, r)$ are non-negative and the rows sum to one. The next AA goal is to find those archetypes Y as convex linear combinations of the original data X, i.e.:

$$Y = X^\top B , \tag{11.2}$$

where the weights $B \in \mathcal{M}(n, r)$ are again required non-negative and with *columns summing to one*. Thus, the AA problem is to find the matrices of weights A and B for the convex linear combinations (11.1)–(11.2).

For illustration, AA is applied to Five Socio-Economic Variables, (Harman, 1976, p. 14). Solutions with two, three, and four archetypes are produced. Figure 11.1 depicts three two-dimensional plots of the observations (circles) and the corresponding archetypes (stars).

The chapter is organized as follows. The next Section 11.1.2 defines formally the AA problem. In Section 11.1.3, AA is stated as a LS optimization on the product of interiors of two simplexes. In Section 11.1.4, AA is alternatively presented as a LS optimization on product of two oblique manifolds. Codes for solving this problem are also provided. Finally, Section 11.1.5 treats the AA solution as a limit point of two matrix interior point (gradient) flows.

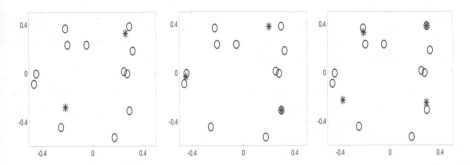

Figure 11.1: AA of Five Socio-Economic Variables, (Harman, 1976, p. 14) with two, three and four archetypes (from left to right).

11.1.2 Definition of the AA problem

By combining both equations (11.1) and (11.2) into one, we can see that AA approximates the data as follows:

$$X \approx AB^\top X \, , \tag{11.3}$$

where the right-hand side matrix has rank at most r (provided $r < \mathrm{rank}(X)$). This resembles (4.6) in PCA, which gives the best approximation to X of rank r. The considerable difference from PCA is that AA looks for convex linear combinations. In order to find the unknown weights $A, B \in \mathcal{M}(n, r)$, AA solves the following problem:

$$\min_{A,B} \|X - AB^\top X\|_F^2 \, , \tag{11.4}$$

where it is assumed that A and B have only non-negative entries and satisfy $A 1_r = 1_n$ and $B^\top 1_n = 1_r$, where 1_n denotes a column-vector of n ones. The LS problem (11.4) uses the standard Frobenius matrix norm.

Note, that the AA objective function (11.4) depends on the data only through the $n \times n$ product $XX^\top$, i.e. the number of variables p is immaterial for AA. This is extremely important for modern applications with huge p.

Cutler and Breiman (1994) solve the AA problem (11.4) using an alternating minimization algorithm, alternating over A and B. They modify the classical non-negative LS (NNLS) algorithm (Lawson and Hanson, 1974) by adding penalty terms to enforce solutions with $A 1_r = 1_n$ and $B^\top 1_n = 1_r$. A weakness of their method is that (as any other alternating method for optimization) the process can get stuck while optimizing one of the unknowns, and thus, to destroy the whole solution.

The remaining of the chapter discusses methods that find A and B simultaneously. Moreover, they tackle the AA constraints in a very natural way, making use of their geometry.

11.1.3 AA solution on multinomial manifold

To see that the AA problem (11.4) is another LS optimization problem on manifolds, consider the *standard* $(n-1)$-*dimensional simplex* or simply $(n-1)$-simplex in $\mathbb{R}^n$:

$$\mathcal{V}_{n-1} = \{v = (v_1, ..., v_n) \in \mathbb{R}^n \mid v_i \geq 0, \sum_{i=1}^{n} v_i = 1\}. \tag{11.5}$$

The simplex (11.5) is a compact convex set in $\mathbb{R}^n$. Moreover, it is the smallest convex set containing the canonical basis $\{e_1, \ldots, e_n\}$ of $\mathbb{R}^n$, i.e. it is its convex hull. In this notations, the 0-simplex $\mathcal{V}_0$ is a single point, say, $\{1\}$; the one-dimensional simplex $\mathcal{V}_1$ is the line segment connecting the points $(0, 1)$ and $(1, 0)$ on the plane $\mathbb{R}^2$; the 2-simplex is the (filled-in) triangle connecting $(0, 0, 1), (0, 1, 0)$ and $(1, 0, 0)$; the 3-simplex is a solid tetrahedron, etc. The points $\{e_1, \ldots, e_n\}$ are called *vertices* of $\mathcal{V}_{n-1}$. The simplexes formed by subsets of $\{e_1, \ldots, e_n\}$ are called *faces* of $\mathcal{V}_{n-1}$. The n simplexes of dimension $(n-2)$, formed by omitting each of the vertices $\{e_1, \ldots, e_n\}$ in turn, are called *boundary* faces of $\mathcal{V}_{n-1}$. The collection/union of the boundary faces of $\mathcal{V}_{n-1}$ is called its boundary. The numbers v_i are called *barycentric coordinates*. The arithmetic mean of the simplex vertices is its *barycenter*, or the centre of gravity.

Every compact convex set $\mathcal{C} \subset \mathbb{R}^n$ (with non-empty interior) is homeomorphic to the closed unit ball $\overline{\mathcal{B}_n(0, 1)} = \{x \in \mathbb{R}^n \mid \|x\|_2 \leq 1\}$ by a homeomorphism preserving their boundaries (Lee, 2000, 4.26). In particular, the n-simplex $\mathcal{V}_n$ and $\overline{\mathcal{B}_n(0, 1)}$ are homeomorphic, and their boundaries $\partial\mathcal{V}_n = \mathcal{V}_{n-1}$ and $\mathcal{S}^{n-1}$ are also homeomorphic. In other words, the n-simplex is a n-dimensional topological manifold with boundary.

Clearly, each row of $A \in \mathcal{M}(n, r)$ from (11.4) is a point in $\mathcal{V}_{r-1} \subset \mathbb{R}^r$. Similarly, each column of B is a point in $\mathcal{V}_{n-1} \subset \mathbb{R}^n$. Thus, the matrices A

and B can be considered as elements of the following products of simplexes:
$A \in \underbrace{\mathcal{V}_{r-1} \times \ldots \times \mathcal{V}_{r-1}}_{n}$ and $B \in \underbrace{\mathcal{V}_{n-1} \times \ldots \times \mathcal{V}_{n-1}}_{r}$.

These products can also be rewritten as the following (constraint) sets:

$$\mathcal{C}_A = \{A \in \mathbb{R}^{n \times r} \mid A \geq 0, \ A1_r = 1_n\}, \tag{11.6}$$

and

$$\mathcal{C}_B = \{B \in \mathbb{R}^{n \times r} \mid B \geq 0, \ B^\top 1_n = 1_r\}, \tag{11.7}$$

where $\geq$ in (11.6) and (11.7) are taken element-wise.

The simplex (11.5) itself has "edges" and thus, is not smooth. For this reason its interior, also known as the *open* simplex:

$$\mathring{\mathcal{V}}_{n-1} = \{v = (v_1, \ldots, v_n) \in \mathbb{R}^n \mid v_i > 0, \sum_{i=1}^{n} v_i = 1\}, \tag{11.8}$$

is more interesting and convenient for practical use. Moreover, it forms a smooth sub-manifold in $\mathbb{R}^n$. The structure of a smooth (Riemannian) can be given to $\mathring{\mathcal{V}}_{n-1}$ by the *Fisher information* inner product (metric):

$$\langle h, g \rangle_v = \sum_{i=1}^{n} \frac{h_i g_i}{v_i}, \tag{11.9}$$

where $h, g \in T_v \mathring{\mathcal{V}}_{n-1}$ are tangent vectors to $\mathring{\mathcal{V}}_{n-1}$ at v. Note, that in general $\mathring{\mathcal{V}}_{n-1}$ is not an open subset of $\mathbb{R}^n$, i.e. it is not open in the $\mathbb{R}^n$ topology.

The manifold $\mathring{\mathcal{V}}_{n-1}$ is known in the field of *information geometry* as the *multinomial manifold* (Kass, 1989). The reason behind this is that it gives the parameter space of the *multinomial distribution*, which generalizes the binomial distribution in case of $n(> 2)$ outcomes (Cramér, 1945, p. 418). A sequence of independent experiments is performed and each experiment can result in any one of n possible outcomes with probabilities v_i, $i = 1, \ldots, n$, such that $\sum_1^n v_i = 1$. The probability mass function of the multinomial distribution is $f(x|v) \sim v_1^{x_1}, \ldots, v_n^{x_n}$, where x_i denote the number of occurrences of outcome i and $\sum_{i=1}^n x_i$ equals the number of the independent experiments. The corresponding log-likelihood function is $l(x|v) = \sum_{i=1}^n x_i \ln v_i$.

In this relation, we consider the interiors of $\mathcal{C}_A$ and $\mathcal{C}_B$ from (11.6) and (11.7), which form smooth manifolds and are given as:

$$\mathring{\mathcal{C}}_A = \{A \in \mathbb{R}^{n \times r} \mid A > 0, \ A1_r = 1_n\},$$

and

$$\mathring{\mathcal{C}}_B = \{B \in \mathbb{R}^{n \times r} \mid B > 0, \ B^\top 1_n = 1_r\} \ .$$

Sun et al. (2016) consider the multinomial manifold as a Riemannian submanifold embedded in Euclidean space and equipped with (11.9). They show how to find Riemannian gradient and Hessian of the objective function of an optimization problem defined on multinomial manifold. They also develop Riemannian trust-region optimization algorithm which is implemented in Manopt.

The application of this approach to the AA problem (11.4) is straightforward. The Euclidean gradients of the AA objective function are as follows:

$$\nabla_A = (AB^\top - I_n)XX^\top B \ , \tag{11.10}$$
$$\nabla_B = XX^\top(BA^\top - I_n)A \ , \tag{11.11}$$

and can be readily used to write Manopt codes for solving (11.4).

11.1.4   AA solution on oblique manifolds

Formulation and gradients

Unaware of the work of Sun et al. (2016), Hannachi and Trendafilov (2017) proposed an alternative solution of the AA problem (11.4), considered as a optimization problem on the manifold of oblique matrices $\mathcal{OB}(n, r)$. This is based on the surjective mapping $f : \mathcal{S}^{n-1} \to \mathcal{V}^{n-1}$, defined by

$$f(x_1, \ldots, x_n) = (x_1^2, \ldots, x_n^2) \ .$$

This is a quadratic map from the unit sphere $\mathcal{S}^{n-1} \subset \mathbb{R}^n$ to the simplex $\mathcal{V}^{n-1}$. It becomes bijection (and diffeomorphism), if restricted to the positive orthant of the unit sphere $\mathcal{S}^{n-1}$.

Now, optimization problems on $\mathcal{V}^{n-1}$, can be considered on $\mathcal{S}^{n-1}$, which is endowed with the (Riemannian) inner product

$$\langle h, g \rangle_x = 2h^\top g \ , \tag{11.12}$$

where $h, g \in \mathcal{T}_x\mathcal{S}^{n-1}$ are tangent vectors to $\mathcal{S}^{n-1}$ at x (Helmke and Moore, 1994, 4.1). Alternatively, we can use (11.12) and the geodesic distance between $x, y \in \mathcal{S}^{n-1}$, which is the distance from x to y along their great circle also called *arc length*, to define inner product between $v = f(x)$ and $w = f(y)$ in $\mathcal{V}^{n-1}$ by

$$\langle v, w \rangle_G = \arccos \sqrt{v^\top w} \,. \tag{11.13}$$

It was discussed in Section 4.8.2 that Manopt, in fact, redefines $\mathcal{OB}(n, r)$ as the set of all $n \times r$ matrices with unit length columns, which is a product of r unit spheres in $\mathbb{R}^n$, i.e. $\prod_r \mathcal{S}^{n-1}$. Now, observe that for every $A \in \mathcal{M}(n, r)$, such that $A^\top \in \prod_n \mathcal{S}^{r-1}$, the element-wise matrix product $A \odot A$ has only non-negative entries and each of its rows sum to 1. In the same way, for every $B \in \mathcal{M}(n, r)$ such that $B \in \prod_r \mathcal{S}^{n-1}$ the element-wise matrix product $B \odot B$ has only non-negative entries and each of its columns sum to 1.

With this in mind, the AA problem (11.4) can be reformulated as follows:

$$\min_{A,B} \| X - (A \odot A)(B \odot B)^\top X \|^2 \,, \tag{11.14}$$

where $A \in \mathcal{OB}(r, n)$ and $B \in \mathcal{OB}(n, r)$.

The gradients of the AA objective function in (11.14) with respect to the Frobenius norm are as follows:

$$\nabla_A = \left\{ [(A \odot A)(B \odot B)^\top - I_n] X X^\top (B \odot B) \right\} \odot A \,, \tag{11.15}$$

$$\nabla_B = \left\{ X X^\top [(B \odot B)(A \odot A)^\top - I_n](A \odot A) \right\} \odot B \,. \tag{11.16}$$

They are used in the next Section 11.1.4 to construct Manopt codes for solving the AA problem (11.14).

Manopt code for AA with oblique matrices

The following codes are used in (Hannachi and Trendafilov, 2017) to analyse large climate data matrix (1740×24501). The readers are encouraged to write codes for solving the AA problem on the multinomial manifold by utilizing the gradients (11.10)–(11.11) and `multinomialfactory` in Manopt. Our experiments show that such a solver is slightly faster than the one working on oblique manifolds.

```
1   function [f S] = aaM(XX,r)
2   % solves min ||X - AB'X||^2 over A abd B
3   %          s.t. sum(A,2)=1_n and sum(B)=1_r
4   % XX = X*X'
5   %=======random_start=======
6   n = size(XX,1);
7   A = rand(r,n)-.5; A = A./repmat(sqrt(sum(A.*A)),r,1);
8   B = rand(n,r)-.5; B = B./repmat(sqrt(sum(B.*B)),n,1);
9
10  S.A = A; S.B = B;
11  %=========================
12  %S
13
14  % Create the problem structure.
15  elements = struct();
16  elements.A = obliquefactory(r, n);
17  elements.B = obliquefactory(n, r);
18  M = productmanifold(elements);
19
20  problem.M = M;
21
22  % Define the problem cost function and its gradient.
23  problem.costgrad = @(S) mycostgrad(S.A,S.B);
24
25  % Numerically check gradient consistency.
26  % warning('off', 'manopt:stiefel:exp');
27  % checkgradient(problem);
28
29      % Solve
30
31      options.verbosity=1; % or 0 for no output
32  %    [S f info] = steepestdescent(problem,[],options);
33      [S f info] = conjugategradient(problem,[],options);
34  %    [S f info] = trustregions(problem,[],options);
35
36      S.A; S.B;
37
38      % Display some statistics.
39      plot_ok = 'Y' % 'N' %
40      if plot_ok == 'Y'
41          figure(2);
42          %semilogy([info.iter], [info.gradnorm], 'k.-');
43          plot([info.iter], [info.gradnorm], 'k.-');
44          hold on;
45          %semilogy([info.iter], [info.cost], 'ro-');
46          plot([info.iter], [info.cost], 'ro-');
47          xlabel('Iteration #'); ylabel('Gradient norm');
48      end
49
50      %Subroutine for the objective function f and its gradient
51  function [ f G] = mycostgrad (A,B)
52
53  [n , r] = size(B) ;
54  A = A./repmat(sqrt(sum(A.*A)),r,1);
55  B = B./repmat(sqrt(sum(B.*B)),n,1);
56
```

```
AA = (A.*A)' ; BB = B.*B;
W = BB'*XX; WB = W*BB; WA = AA'*AA;

% objective function
 f = trace(WB*WA) - 2*trace(W*AA);
 f = f/2;

% gradient
 if nargout == 2
     gradA = (AA*WB' - W).*A';
     gradB = (W*WA - XX*AA).*B;

     G.A = gradA' - A.*repmat(sum(gradA'.*A),r,1);
     G.B = gradB - B.*repmat(sum(gradB.*B),n,1);
 end
end

end
```

11.1.5 AA as interior point flows

We already know that the interior of $\mathcal{V}^{n-1}$ is more interesting for practical purposes. The algorithms to deal with optimization on simplex $\mathcal{V}^{n-1}$ are *interior point algorithms*, which suggests that they evolve in the interior of the simplex $\overset{\circ}{\mathcal{V}}{}^{n-1}$ anyway (Karmarkar, 1984, 1990).

Traditionally, interior point flows are used for optimizing a linear functional on a simplex. Faybusovich (1991, 2003) extended the theory by considering interior point gradient flows for an arbitrary cost function defined on the interior of a simplex (as well as on an arbitrary *polytope*, a compact convex subset of $\mathbb{R}^n$).

Here, we develop interior point flows to solve the AA problem (11.4). We follow the comprehensive exposition in (Helmke and Moore, 1994, Ch 4).

Let $a_i \in \mathbb{R}^r$ denote the ith row of A and $b_j \in \mathbb{R}^n$ denote the jth column of B. Then, the constraint sets (11.6)–(11.7) of the AA problem (11.4) can be rewritten as:

$$\mathcal{C}_A = \mathcal{C}_{a_1} \times \mathcal{C}_{a_2} \times \ldots \times \mathcal{C}_{a_n}$$

and

$$\mathcal{C}_B = \mathcal{C}_{b_1} \times \mathcal{C}_{b_2} \times \ldots \times \mathcal{C}_{b_r} ,$$

where

$$\mathcal{C}_{a_i} = \{a_i \in \mathbb{R}^r \mid a_i \geq 0, \ a_i^\top 1_r = 1\} , \tag{11.17}$$

and
$$\mathcal{C}_{b_j} = \{b_j^\top \in \mathbb{R}^n \mid b_j \geq 0, \ b_j^\top 1_n = 1\} , \tag{11.18}$$
and $\geq$ are taken element-wise in (11.17) and (11.18).

The interiors of $\mathcal{C}_{a_i}$ and $\mathcal{C}_{b_j}$ form smooth manifolds and are given by
$$\mathring{\mathcal{C}}_{a_i} = \{a_i \in \mathbb{R}^r \mid a_i > 0, \ a_i^\top 1_r = 1\} ,$$
and
$$\mathring{\mathcal{C}}_{b_j} = \{b_j \in \mathbb{R}^n \mid b_j > 0, \ b_j^\top 1_n = 1\} .$$

To make use of the Fisher information inner product (11.9), form the following two types of diagonal matrices:
$$D(a_i) = \operatorname{diag}(a_{i,1}, a_{i,2}, \ldots, a_{i,r},) \in \mathcal{D}(r) , \tag{11.19}$$
and
$$D(b_j) = \operatorname{diag}(b_{1,j}, b_{2,j}, \ldots, b_{n,j},) \in \mathcal{D}(n) . \tag{11.20}$$

Clearly, $D(a_i)$ is PD for any $i = 1, \ldots, r$, and $D(b_j)$ is PD for any $j = 1, \ldots, n$.

For any $y \in \mathring{\mathcal{C}}_{a_i}$, let $T_y(\mathring{\mathcal{C}}_{a_i})$ denote the tangent space of $\mathring{\mathcal{C}}_{a_i}$ at y. Thus, $T_y(\mathring{\mathcal{C}}_{a_i})$ coincides with the tangent space of the affine subspace of $\{y \in \mathbb{R}^r \mid y^\top 1_r = 1\}$ at y:
$$T_y(\mathring{\mathcal{C}}_{a_i}) = \{h \in \mathbb{R}^r \mid h^\top 1_r = 0\} .$$
For any $y \in \mathring{\mathcal{C}}_{a_i}$, the diagonal matrix $D_r(y)$ in (11.19) is PD and thus, for any $h_1, h_2 \in T_y(\mathring{\mathcal{C}}_{a_i})$:
$$\ll h_1, h_2 \gg = h_1^\top D(y)^{-1} h_2 , \tag{11.21}$$
defines a positive definite inner product on $T_y(\mathring{\mathcal{C}}_{a_i})$, and in fact, a Riemannian metric on $\mathring{\mathcal{C}}_{a_i}$.

In a similar fashion, the tangent space of $\mathring{\mathcal{C}}_{b_j}$ at any y coincides with the tangent space of the affine subspace of $\{y \in \mathbb{R}^n \mid y^\top 1_n = 1\}$ at y:
$$T_y(\mathring{\mathcal{C}}_{b_j}) = \{h \in \mathbb{R}^n \mid h^\top 1_n = 0\} .$$

For any $y \in \mathring{\mathcal{C}}_{b_j}$, the diagonal matrix $D(y)$ in (11.20) is PD and thus, for any $h_1, h_2 \in T_y(\mathring{\mathcal{C}}_{b_j})$:
$$\ll h_1, h_2 \gg = h_1^\top D(y)^{-1} h_2 , \tag{11.22}$$

defines a positive definite inner product on $T_y(\mathring{\mathcal{C}}_{b_j})$, which is a Riemannian metric on $\mathring{\mathcal{C}}_{b_j}$.

For fixed B and A except its ith row a_i, the gradient $\mathrm{grad}_{a_i} f$ of $f(*, B)$: $\mathring{\mathcal{C}}_{a_i} \to \mathbb{R}$ with respect to the Riemannian metric (11.21), at $a_i \in \mathring{\mathcal{C}}_{a_i}$ is characterized by the properties:

(i) $\mathrm{grad}_{a_i} f \in T_{a_i}(\mathring{\mathcal{C}}_{a_i})$

(ii) $\ll \mathrm{grad}_{a_i} f, h \gg = \nabla_{a_i} h$,

for all $h \in T_{a_i}(\mathring{\mathcal{C}}_{a_i})$, where ∇_{a_i} is the ith row of the Euclidean gradient (11.10) of f with respect to A.

Here (i) simply means that $\mathrm{grad}_{a_i}^\top f 1_r = 0$. The other item (ii) is equivalent to
$$[\mathrm{grad}_{a_i} f^\top D(a_i)^{-1} - \nabla_{a_i}] h = h^\top [D(a_i)^{-1} \mathrm{grad}_{a_i} f - \nabla_{a_i}^\top] = 0 ,$$
for any $h \in \mathbb{R}^r$ that satisfies $h^\top 1_r = 0$. This suggests that for some constant $\alpha \in \mathbb{R}$, the vector $D(a_i)^{-1} \mathrm{grad}_{a_i} f - \nabla_{a_i}^\top$ should have the form $\alpha 1_r$. Then, we can write $\mathrm{grad}_{a_i} f = \alpha 1_r^\top D(a_i) + \nabla_{a_i} D(a_i)$, and after right multiplication by 1_r, we obtain:
$$\alpha = -\frac{\nabla_{a_i} D(a_i) 1_r}{1_r^\top D(a_i) 1_r} .$$

After substitution, one finally finds
$$\mathrm{grad}_A f = \nabla_{a_i} D(a_i) - \frac{\nabla_{a_i}^\top D(a_i) 1_r}{1_r^\top D(a_i) 1_r} 1_r^\top D(a_i) = \nabla_{a_i} D(a_i) \left(I_r - \frac{E_r D(a_i)}{1_r^\top D(a_i) 1_r} \right) .$$
$$(11.23)$$

Similarly, for fixed A and B except its jth column b_j, the gradient $\mathrm{grad}_{b_j} f$ of $f(A, *)$: $\mathring{\mathcal{C}}_{b_j} \to \mathbb{R}$ with respect to the Riemannian metric (11.22), at $b_j \in \mathring{\mathcal{C}}_{b_j}$ is characterized by the properties:

(i) $\mathrm{grad}_{b_j} f \in T_{b_j}(\mathring{\mathcal{C}}_{b_j})$

(ii) $\ll \mathrm{grad}_{b_j} f, h \gg = \nabla_{b_j}^\top h$,

for all $h \in T_{b_j}(\mathring{\mathcal{C}}_{b_j})$, where ∇_{b_j} is the jth column of the Euclidean gradient (11.11) of f with respect to B.

Here, (i) simply means that $\text{grad}_{b_j} f^\top 1_n = 0$. The other item (ii) is equivalent to

$$[\text{grad}_{b_j}^\top f D(b_j)^{-1} - \nabla_{b_j}^\top]h = h^\top[D(b_j)^{-1}\text{grad}_{b_j} f - \nabla_{b_j}] = 0 ,$$

for any $h \in \mathbb{R}^n$ that satisfies $h^\top 1_n = 0$. This suggests that for some scalar $\beta \in \mathbb{R}$, the matrix $D(b_j)^{-1}\text{grad}_{b_j} f - \nabla_{b_j}$ should have the form $\beta 1_n$. Then, we can write $\text{grad}_B f = D(b_j)\nabla_{b_j} + \beta D(b_j)1_n$, and after transposing it and right multiplication by 1_n, we obtain:

$$\beta = -\frac{\nabla_{b_j}^\top D(b_j)1_n}{1_n^\top D(b_j)1_n} .$$

After substitution, one finally finds:

$$\text{grad}_{b_j} f = D(b_j)\nabla_{b_j} - D(b_j)1_n \frac{1_n^\top D(b_j)\nabla_{b_j}}{1_n^\top D(b_j)1_n} = \left(I_n - \frac{D(b_j)E_n}{1_n D(b_j)1_n}\right)D(b_j)\nabla_{b_j} .$$
$$(11.24)$$

Thus, with (11.23) and (11.24) in hand, we have proved the following:

Theorem 11.1.1 *The gradient flow $[\dot{A}, \dot{B}] = \text{grad } f(A, B)$ of $f(A, B) : \mathring{\mathcal{C}}_A \times \mathring{\mathcal{C}}_B \longrightarrow \mathbb{R}$, with respect to the product of the Riemannian metrics (11.21) and (11.22) on the interiors of the constraint sets, is given by*

$$\dot{a}_i = \nabla_{a_i} D(a_i)\left(I_r - \frac{E_r D(a_i)}{1_r^\top D(a_i)1_r}\right) , \quad i = 1, \dots, r \qquad (11.25)$$

$$\dot{b}_j = \left(I_n - \frac{D(b_j)E_n}{1_n^\top D(b_j)1_n}\right)D(b_j)\nabla_{b_j} , \quad j = 1, \dots, n, \qquad (11.26)$$

where ∇_{a_i} and ∇_{b_j} are respectively the ith row and the jth column of the gradients ∇_A and ∇_B available from (11.10)–(11.11).

11.1.6   Conclusion

The AA problem is explained and formally defined in Section 11.1.1 and Section 11.1.2, together with its initial alternating iterative solution. Section 11.1.3 redefines and solves the AA problem on a product of two multinomial manifolds (interior of simplex), while Section 11.1.4 treats it as optimization problem on a product of two oblique manifolds and provides Manopt codes for its solution. Section 11.1.5 derives a interior point gradient dynamical system to express the AA problem on a product of the interiors of two simplexes.

11.1.7   Exercises

1. Show that $\mathcal{V}_{n-1}$ is the convex hull of $\{e_1, \ldots, e_n\}$.

2. Let $\mathcal{C} \in \mathbb{R}^n$ be a compact convex set containing the origin, i.e. $0_n \in \mathring{\mathcal{C}}$. Then, any ray from 0_n intersects with its boundary $\partial\mathcal{C}$ in exactly one point. Hint: Consider the distance between 0_n and another point on the ray as a continuous function and find its maximum.

3. The embedding of $\mathcal{S}^{n-1}$ in $\mathbb{R}^n$ induces a (locally flat) metric $\sum_i dx_i dx_i$ on the sphere inherited from the Euclidean metric in $\mathbb{R}^n$. Show that it becomes the Fisher information metric on the simplex $\mathring{\mathcal{V}}^{n-1}$. Hint: Change the variables $x_i = \sqrt{v_i}$. to see that

$$\sum_{i=1}^n dx_i dx_i = \sum_{i=1}^n d\sqrt{v_i}\,d\sqrt{v_i} = \frac{1}{4}\sum_{i=1}^n \frac{dv_i dv_i}{v_i} = \frac{1}{4}\sum_{i=1}^n v_i d(\ln v_i) d(\ln v_i) \ .$$

4. Let z be a draw/sample from the multinomial distribution, i.e. $z_i \in \{0, 1\}$ and $\sum_i z_i = 1$. Show that the Fisher information inner product (metric) at $v \in \mathring{\mathcal{V}}_{n-1}$ is given by (11.9). Hint: Use the following general definition of Fisher information inner product (metric):

$$\langle x, y\rangle_v = \sum_{i=1}^n \sum_{j=1}^n x_i y_j g_{ij}(v) = -\sum_{i=1}^n \sum_{j=1}^n x_i y_j E\left[\frac{\partial^2 l(z|v)}{\partial v_i \partial v_j}\right] \ ,$$

where g_{ij} are the elements of the Fisher information matrix and E is the expectation operator.

5. The *Hellinger distance* between $v, w \in \mathring{\mathcal{V}}_{n-1}$ is expressed by the Euclidean distance between their images on the sphere (instead of the arc length):

$$d_H(v, w) = \|\sqrt{v} - \sqrt{w}\|_2 = \sqrt{\sum_1^n (\sqrt{v_i} - \sqrt{w_i})^2} \ .$$

Show that d_H is related to d_G in (11.13) by

$$d_H = 2\sin\left(\frac{d_G(v, w)}{2}\right) \ .$$

6. Let $v, w \in \mathring{\mathcal{V}}_{n-1}$ and consider the two intersection points a and b of the line vw with $\mathcal{V}_{n-1}$. Order the points on the line as $avwb$. Then the Hilbert metric/distance between $v, w \in \mathring{\mathcal{V}}_{n-1}$ is defined as follows:

$$
d_H(v, w) = \begin{cases} \left| \ln \left(\frac{|bv|}{|bw|} : \frac{|av|}{|aw|} \right) \right| & , \ a \neq b \\ 0 & , \ a = b \end{cases} ,
$$

where $|ab|$ denotes the length between the points a and b. There are a number of ways to express $d_H(v, w)$. For example, (Nielsen and Sun, 2019) propose the following. Let $l(t) = (1-t)p + tq$ be the line passing through p and q, and $t_0 (\leq 0)$ and $t_1 (\geq 0)$ be the two points for which $l(t_0) = a$ and $l(t_1) = b$ i.e. at which the line l intersects with $\mathcal{V}_{n-1}$. Then, $d_H(v, w)$ is expressed as

$$
d_H(v, w) = \left| \ln \left(1 - \frac{1}{t_0} \right) - \ln \left(1 - \frac{1}{t_1} \right) \right| ,
$$

and is used for clustering in a simplex.

(a) If $v \neq w$, then $d_H(v, w) \neq 0$.

(b) Let w lie between u, v on the same line defined by them. Check that $d_H(u, v) = d_H(u, w) + d_H(w, v)$.

(c) Can you prove the general triangle inequality for d_H? Hint: You may need to use the *cross-ratio invariance* under projective lines. The original Hilbert's proof is reproduced in (Troyanov, 2014). See also (de la Harpe, 1993).

7. Consider the following vector subspace (hyperplane):

$$
\mathcal{H}_2 = \{ x \in \mathbb{R}^3 \, | \, x_1 + x_2 + x_3 = 0 \} .
$$

(a) (de la Harpe, 1993) Show that the map $f : \mathring{\mathcal{V}}_2 \to \mathcal{L}_2$ defined by

$$
f(v) = \frac{1}{3} \begin{pmatrix} 2 \ln v_1 - \ln v_2 - \ln v_3 \\ - \ln v_1 + 2 \ln v_2 - \ln v_3 \\ - \ln v_1 - \ln v_2 + 2 \ln v_3 \end{pmatrix} ,
$$

is a bijection whose inverse is:

$$
f^{-1}(x) = \frac{1}{e^{x_1} + e^{x_2} + e^{x_3}} \begin{pmatrix} e^{x_1} \\ e^{x_2} \\ e^{x_3} \end{pmatrix} .
$$

(b) (Vernicos, 2014) Show that the map $f : \overset{\circ}{\mathcal{V}}_2 \to \mathcal{L}_2$ defined by

$$f(v) = \left(\ln \frac{v_1}{v_2}, \ln \frac{v_2}{v_3}, \ln \frac{v_3}{v_1} \right)^{\mathsf{T}},$$

is a bijection whose inverse is:

$$f^{-1}(x) = \frac{1}{e^{x_1} + e^{x_1+x_2} + 1} \begin{pmatrix} e^{x_1} \\ e^{x_1+x_2} \\ 1 \end{pmatrix}.$$

11.2   Analysis of compositional data (CoDa)

11.2.1   Introduction

The compositional data (CoDa) are very different from the usual data. Every observation on, say, p variables is a vector which components are nonnegative and must sum to unity, or in general, to some fixed constant $\kappa > 0$. Most frequently, CoDa are composed by vectors of proportions describing the relative contributions of each of p variables or categories to the whole.

These restrictions make the CoDa difficult to handle because the standard statistical techniques need to be properly adjusted. The major progress in the analysis of CoDa is due to the contributions (Aitchison and Shen, 1980; Aitchison, 1982, 1983) and culminated in the monograph (Aitchison, 1986).

Aitchison introduced several logratio transformations and demonstrated that the logratios are easier to handle than the original CoDa. In addition they provide one-to-one mapping of CoDa to real Euclidean space. This is very important because one can apply the standard MDA methods developed for unconstrained data matrices. These useful features led to the wide acceptance of the methodology involving a great variety of logratio transformations (Egozcue et al., 2003; Hron et al., 2017a,b; Greenacre, 2018a,b).

Nowadays, the compositional analysis is widely used in many experimental fields. The usual CoDa high dimensionality requires MDA methods to find an adequate low-dimensional description of the compositional variability (Bergeron-Boucher et al., 2018; Engle et al., 2014; Simonacci and Gallo, 2017). In particular, we discuss later some distinct features of the classical PCA when applied to compositions. In addition, we demonstrate how sparse PCA is tailored to analyse CoDa. Its application is very effective because it suggests constant logcontrasts without prior knowledge on the sampled phenomenon.

We already mentioned that the logratio transformations provide one-to-one correspondence between compositional vectors and associated logratio vectors. They help to transform the space of compositions into specific Euclidean spaces with respect to the operations of perturbation and powering (Billheimer et al., 2001; Pawlowsky-Glahn and Egozcue, 2001). This implies that compositional problems can be investigated directly on these Euclidean spaces.

The chapter is organized as follows. The next Section 11.2.2 defines the main properties of compositions and the most popular logratio transformations involved in the traditional way to approach CoDa. Section 11.2.3 concentrates on the clr transformation of CoDa and how it transforms the data simplex into a specific Euclidean space. Section 11.2.4 considers the specific features of PCA when applied to CoDa together with its sparse version. An illustration of sparse PCA applied to a small real data example is given in Section 11.2.5.

11.2.2 Definition and main properties

Suppose that n compositions are formed by p non-negative parts or components that sum to κ. For simplicity, we assume that $\kappa = 1$.

Let the compositions be collected in an $n \times p$ data matrix V. Thus, the sample space for the generic row v of V is the simplex $\mathcal{V}^{p-1}$, already defined in (11.5). Traditionally, it is considered more convenient to exclude from consideration the compositions with zero components. In other words, the CoDa sample space is the interior of the simplex $\overset{\circ}{\mathcal{V}}^{p-1} \left(\subset \mathbb{R}^p_+ \right)$ defined as

$$\overset{\circ}{\mathcal{V}}^{p-1} = \{v = (v_1, \ldots, v_p)^\top \mid v_i > 0, \sum_{i=1}^p v_i = 1\} .$$

This corresponds to what we already know from Section 11.1.3: working with the smooth manifold $\overset{\circ}{\mathcal{V}}^{p-1}$ is more convenient than with the simplex $\mathcal{V}^{p-1}$ itself.

There are a number of problems to apply standard MDA techniques to CoDa. For example, one basic obstacle is that the correlations are not free any more to range over $[-1, 1]$, as the usual correlations. To overcome such kind of difficulties, the suggested strategy is to move from the constrained sample space (manifold) $\overset{\circ}{\mathcal{V}}^{p-1}$ to an unconstrained space, e.g. the Euclidean space $\mathbb{R}^p$. For this reason, a number of *logratio transformations* are proposed. The presumption is that the standard MDA techniques can be then applied to the transformed CoDa. However, we will see in Section 11.2.4, that the standard MDA techniques exhibit some distinctive features triggered by the CoDa.

There is a number of logratio transformations proposed in the CoDa literature. Here is a list the most popular ones:

- The *pairwise logratio* (plr) converts the generic compositional vector $v \in \mathring{\mathcal{V}}^{p-1}$ into a new vector $x \in \mathbb{R}^{\frac{p(p-1)}{2}}$ of the logarithms of the ratios between all possible pairs of elements of v. In other words, the compositional matrix $V \in \mathcal{M}(n, p)$ becomes a matrix of plr coordinates $X_{plr} \in \mathcal{M}\left(n, \frac{p(p-1)}{2}\right)$ with generic element $\log(v_{ij}/v_{ij'})$, $j' > j$. The disadvantage of this transformation is that the dimensionality of the new data matrix is considerably increased.

- The *additive logratio* (alr) converts the generic composition vector $v \in \mathring{\mathcal{V}}^{p-1}$ into a new vector $x \in \mathbb{R}^{p-1}$ composed by the logarithms of the ratios v_-/v_i, where $v_- = [v_1, v_2, \ldots, v_{i-1}, v_{i+1}, \ldots, v_p]$ and $1 \leq i \leq p$. For convenience, the general choice is $i = p$. Thus, the generic element of the new matrix of alr coordinates $X_{alr} \in \mathcal{M}(n, p-1)$ is $\log(v_{ij}/v_{ip})$ for $i = 1, \ldots, n$, $j = 1, \ldots, p-1$. This transformation is isomorphic, asymmetric (in treating all p parts of the composition) and with major disadvantage of being non-isometric.

- The *centred logratio* (clr) considers the logarithms of the ratios between each element of $v \in \mathring{\mathcal{V}}^{p-1}$ and its geometric mean $g(v) = \sqrt[p]{v_1 \ldots v_p}$. It yields a matrix of clr coordinates $X_{clr} \in \mathcal{M}(n, p)$ with generic element $v_{ij}/g(v_i)$, where v_i is the ith row of V. This transformation has the advantage of being both symmetric (in treating all p parts) and isometric. However, the covariance matrix obtained from clr coordinates is always singular.

- The *isometric logratio* (ilr) yields an orthonormal basis in the simplex with respect to the Aitchison inner product (Egozcue et al., 2003). The compositional matrix $V \in \mathcal{M}(n, p)$ becomes a matrix of ilr coordinates $X_{ilr} \in \mathcal{M}(n, p-1)$ with generic element $\sqrt{\frac{p-j}{p-j+1}} \log\left(\frac{v_{ij}}{(p-i)\sqrt{v_{i,j+1} \ldots v_{i,p}}}\right)$, where v_i is the ith row of V and $j = 1, \ldots, p-1$. The transformed compositions are orthonormal, but the results are not easy to interpret.

- The *summated logratio* (slr) is an amalgamation logratio. The slr converts the generic compositional vector v into a new vector given by the logratios of two subsets of components of v, i.e. $\log\left(\frac{\sum_{i \in I} v_i}{\sum_{j \in J} v_j}\right)$, where I and J contain the indexes of the partition of v, and can be chosen in a number of ways. This transformation has no particular mathematical properties but yields simple and easily interpretable results.

All existing transformations have advantages and disadvantages. It looks that clr is the most frequently used logratio in real applications. Many find it intuitive and straightforward to interpret. However, others insist that alr is the only one providing clear practical interpretation. Which transformation of the original CoDa is most appropriate (at least for the specific application) is still open question and subject to active research and comparisons.

In the sequel, it is assumed that clr transformation of CoDa is considered, i.e. $X = X_{clr}$, which main properties are given in the next Section 11.2.3.

11.2.3   Geometric clr structure of the data simplex

Consider the following two operations defined in $\mathring{\mathcal{V}}^{p-1}$. The *perturbation operation* between two compositions $v, w \in \mathring{\mathcal{V}}^{p-1}$ is defined as

$$v \boxplus w = c(v \odot w) = c(v_1 w_1, \ldots, v_p w_p) ,$$

and the *powering operation* defined for any $\alpha > 0$ and a composition v as

$$\alpha \boxtimes v = c(v_1^\alpha, v_2^\alpha, \ldots, v_p^\alpha) ,$$

which is equivalent to the multiplication of a vector by a scalar. In both definitions $c()$ is the *closure operation* in which the elements of a positive vector are divided by their sum, i.e. $c : \mathbb{R}_+^p \to \mathring{\mathcal{V}}^{p-1}$:

$$c(v) = \left[\frac{v_1}{\sum_{i=1}^p v_i}, \ldots, \frac{v_p}{\sum_{i=1}^p v_i} \right] .$$

The data simplex furnished with these two operations becomes a vector space. They play the role of the addition and multiplication by scalar in $\mathbb{R}^p$.

It is convenient to define an operation clr : $\mathring{\mathcal{V}}^{p-1} \to \mathbb{R}^p$ as follows:

$$\mathrm{clr}(v) = \left[\ln \frac{v_1}{g(v)}, \ldots, \ln \frac{v_p}{g(v)} \right] , \tag{11.27}$$

where $g(v)$ denotes the geometric mean of $v \in \mathring{\mathcal{V}}^{p-1}$, i.e. $g(v) = \sqrt[p]{v_1 \ldots v_p}$. Let $x \in \mathbb{R}^p$ be such that $x = \mathrm{clr}(v)$. Clearly, the coordinates of such x sum to 0, i.e. $1_p^\top x = 0$. In other words, the clr transformation maps $\mathring{\mathcal{V}}^{p-1}$ into the hyperplane $\mathbb{H}^{p-1} = \{x \in \mathbb{R}^p \mid 1_p^\top x = 0\} \subset \mathbb{R}^p$, which is a vector subspace of $\mathbb{R}^p$ orthogonal to 1_p (in $\mathbb{R}^p$). Thus, it has dimension $p - 1$ and is subset of $\mathbb{R}^{p-1}$. As a finite-dimensional vector space has no proper subspace of the same dimension, we conclude that $\mathbb{H}^{p-1}$ and $\mathbb{R}^{p-1}$ are identical. This, in turn, demonstrates that $\mathring{\mathcal{V}}^{p-1}$ is isomorphic to $\mathbb{R}^{p-1}$.

Let $\langle,\rangle_2$ be the standard inner product in $\mathbb{R}^p$. Then, one can introduce an inner product in $\mathring{\mathcal{V}}^{p-1}$ defined as follows (Egozcue et al., 2011):

$$\langle v, w \rangle_{clr} = \langle \mathrm{clr}(v), \mathrm{clr}(w) \rangle_2 = \mathrm{clr}(v)^\top \mathrm{clr}(w) = \sum_{i=1}^{p} \ln \frac{v_i}{g(v)} \ln \frac{w_i}{g(w)} \ , \quad (11.28)$$

which by definition is an isometry. According to the general theory in Section 2.3, the inner product (11.28) induces the following norm in $\mathring{\mathcal{V}}^{p-1}$:

$$\|v\|_{clr} = \sqrt{\langle v, v \rangle_{clr}} = \sqrt{\mathrm{clr}(v)^\top \mathrm{clr}(v)} = \left[\sum_{i=1}^{p} \left(\ln \frac{v_i}{g(v)} \right)^2 \right]^{1/2} \ .$$

As every finite-dimensional normed space is complete, we conclude that $\mathring{\mathcal{V}}^{p-1}$ is a real Euclidean space. Some authors call $\mathring{\mathcal{V}}^{p-1}$ a Hilbert space (Billheimer et al., 2001, Theorem A.3.), however we prefer to keep the term for infinite-dimensional spaces. Several less known aspects and features of the clr transformation are considered in (de la Harpe, 1993).

Section 2.3 also suggests that one can define the following induced metric in $\mathring{\mathcal{V}}^{p-1}$:

$$d_{clr}(v, w) = \|v \boxminus w\|_{clr} \ , \quad (11.29)$$

where $v \boxminus w = c \left(\frac{v_1}{w_1}, \dots, \frac{v_p}{w_p} \right)$ is the modified perturbation operation playing the role of subtraction in $\mathring{\mathcal{V}}^{p-1}$.

In addition, the distance (11.29) posses the following useful properties:

- Permutation invariance: $d_{clr}(Pv, Pw) = d_{clr}(v, w)$, for any permutation matrix P

- Perturbation invariance: $d_{clr}(v \boxplus u, w \boxplus u) = d_{clr}(v, w)$, where u is the perturbation vector

- Sub-compositional dominance: $d_{clr}(v^*, w^*) \leq d_{clr}(v, w)$, where v^* is a sub-composition formed by certain subset of elements of v, and then subject to closure operation.

Conversely, one can check that for every $x \in \mathbb{H}^{p-1}$ we have

$$c(\exp(x)) = v \in \overset{\circ}{\mathcal{V}}{}^{p-1} ,$$

which can be expressed as $\mathrm{clr}^{-1} : \mathbb{H}^{p-1} \to \overset{\circ}{\mathcal{V}}{}^{p-1}$. This relation can be utilized to define an orthonormal, with respect to (11.28), basis of $\overset{\circ}{\mathcal{V}}{}^{p-1}$ by simply mapping an orthonormal basis from $\mathbb{R}^{p-1}$. If $x_1, \ldots, x_{p-1}$ is a basis in $\mathbb{R}^{p-1}$, then, every $v \in \overset{\circ}{\mathcal{V}}{}^{p-1}$ can be identified by its, so-called, ilr coordinates:

$$
\begin{aligned}
\mathrm{ilr}(v) &= \left(\langle v, \mathrm{clr}^{-1}(x_1) \rangle_{clr}, \ldots, \langle v, \mathrm{clr}^{-1}(x_{p-1}) \rangle_{clr} \right) \\
&= \left(\langle \mathrm{clr}(v), x_1 \rangle_2, \ldots, \langle \mathrm{clr}(v), x_{p-1} \rangle_2 \right) .
\end{aligned}
$$

To find orthonormal (ilr) coordinates of $v \in \overset{\circ}{\mathcal{V}}{}^{p-1}$, one can use Algorithm 11.1 (adopting some Matlab notations for matrix operations). Note, that they depend (and differ) on the choice of the orthonormal basis X of $\mathbb{R}^p$.

Algorithm 11.1 comp2ilr

{Initialization: orthonormal basis X of $\mathbb{R}^p$ and a composition $v \in \overset{\circ}{\mathcal{V}}{}^{p-1}$}
{Center X in H to have zero columns sums}
$J \Leftarrow \mathrm{eye}(p) - \mathrm{ones}(p)/p$
$H \Leftarrow J * X$
{Centred orthonormal basis for $\mathbb{R}^{p-1}$ (economy QR decomposition)}
$Q \Leftarrow \mathrm{qr}(H(:, 1 : p - 1), 0)$
{Exponent and closure to get basis for $\overset{\circ}{\mathcal{V}}{}^{p-1}$}
$V \Leftarrow \exp(Q)$
$c \Leftarrow \mathrm{sum}(V)$
$V \Leftarrow V./(1_p * c)$
$V \Leftarrow [V \; v]$
{clr coordinates for V}
$g \Leftarrow \sqrt[p]{\mathrm{prod}(V)}$
$\mathrm{clr} \Leftarrow V./(1_p * g)$
$\mathrm{clr} \Leftarrow \log(\mathrm{clr})$
{ilr coordinates for v}
$\mathrm{ilr} \Leftarrow \mathrm{clr}(:, 1 : p - 1)^{\top} * \mathrm{clr}(:, p)$

For example, if we decide to choose the canonical basis of $\mathbb{R}^4$, we have to start Algorithm 11.1 with $X = I_4$. Then, the orthonormal basis in $\overset{\circ}{\mathcal{V}}{}^3$ is given by the columns of:

$$Y = \begin{bmatrix} .0951 & .2247 & .2212 \\ .3016 & .0993 & .2212 \\ .3016 & .3380 & .1091 \\ .3016 & .3380 & .4486 \end{bmatrix} .$$

However, if we start with another (random) orthonormal basis in $\mathbb{R}^4$, say:

$$X = \begin{bmatrix} -.2100 & -.4791 & -.1397 & -.8407 \\ .6260 & -.5247 & .5750 & .0471 \\ -.4707 & -.6849 & -.1562 & .5338 \\ -.5852 & .1615 & .7909 & -.0773 \end{bmatrix} ,$$

then the corresponding orthonormal basis in $\mathring{\mathcal{V}}^3$ is given by:

$$Y = \begin{bmatrix} .2062 & .1828 & .5046 \\ .4984 & .2486 & .1745 \\ .1566 & .1146 & .1367 \\ .1388 & .4540 & .1842 \end{bmatrix} .$$

An interesting aspect of the inner product (11.28) is that it makes the log-contrasts play the role of compositional "linear combinations" required in PCA for dimension reduction of high-dimensional compositional data. By definition, a *logcontrast* of a composition v is any loglinear combination $a^\top \ln v = \sum_{i=1}^p a_i \ln v_i$ with $\sum_{i=1}^p a_i = 0$ (Aitchison, 1986, p. 84).

In case of clr transformation, one can see that:

$$a^\top \mathrm{clr}(v) = \sum_{i=1}^p a_i \ln \frac{v_i}{g(v)} = \sum_{i=1}^p a_i \ln v_i - g(v) \sum_{i=1}^p a_i ,$$

which implies $a_1 + \cdots + a_p = 0$.

11.2.4   PCA and sparse PCA for compositions

Suppose that $L \in \mathcal{M}(n, p)$ contains the element-wise logarithmic transformations of the elements of V, i.e. $L = \ln(V)$. Then, the matrix of the clr-coordinates of V can be expressed as $X = LJ_p$, where $J_p = I_p - p^{-1}1_p1_p^\top$

is the usual $p \times p$ centring matrix. In other words, the clr data matrix X is row-wise centred, which implies that it is always rank deficient with rank at most $\min\{n, p\} - 1$.

The corresponding sample covariance matrix has to form $X^{\top}X = J_p L^{\top} L J_p$. This explains why the sum of its columns/rows is 0, as well as the sums of the elements of its eigenvectors. Clearly, such a covariance matrix is always singular: it has a zero eigenvalue, and the corresponding eigenvector is arbitrary.

Then, the definition of the standard PCA problem (4.1) can be adjusted to suit such a data matrix $X(= LJ_p)$ as follows:

$$\max_{a \in \mathcal{O}_0(p,r)} a^{\top} X^{\top} X a , \tag{11.30}$$

where $\mathcal{O}_0(p, r)$ is the Stiefel sub-manifold of all centred orthonormal $p \times r$ matrices (with zero column sums) considered in Section 3.3.4.

As we know from Section 4.2, the solution a of (11.30) contains the component loadings, which magnitudes express the importance of each variable for the PCs. A loading of large magnitude shows the strong positive or negative contribution of a certain part (variable) to a particular PC. The common practice for interpreting component loadings is to ignore the ones with small magnitude, or set to zero loadings smaller than certain threshold values. This makes the loading matrix *sparse* artificially and subjectively. The sparse PCA (Section 4.8) eliminates this subjective thresholding of the component loadings. In case of compositional data, sparse PCA can be used to make logcontrasts easier to interpret. In this way it is possible to distinguish and isolate subcompositions in dataset with huge number of parts.

Thus, to analyse compositions by sparse PCA we simply need to solve some sparsified version of (11.30) on $\mathcal{O}_0(p, r)$. Here we make use of the sparse PCA, approximating the PCs' variances and defined in (4.33) as:

$$\min_{A \in \mathcal{O}_0(p,r)} \|A\|_{\ell_1} + \tau \|\mathrm{diag}(A^{\top} X^{\top} X A) - \mathrm{diag}(D)\|_2^2 , \tag{11.31}$$

where $\|A\|_{\ell_1} = \sum_{i,j} |a_{ij}|$.

As in the standard sparse PCA, the tuning parameter τ controls the importance of the two terms of the objective function (11.31). With the increase of τ, the second term in (11.31) becomes more important, and the solution of the problem becomes closer/identical to the original PCA (with dense component loadings). Respectively, smaller values of τ make $\|A\|_{\ell_1}$ more important, which results in sparser component loadings A.

11.2.5   Case study

Here we consider a small example involving data for the category of hotels preferred by foreign tourists that visit the island of Ischia (in the bay of Naples - Italy) in 2012. The six parts of compositions (or parts) considered are: one star (1S), two stars (2S), three stars (3S), four stars (4S), five stars (5S) and others (Oth - bed and breakfast, private apartments etc.). We solve (11.31) by taking the matrix $X^\top X$ of the clr-transformed data:

$$X^\top X = \begin{bmatrix} 3.9730 & .2838 & -.9194 & -.9789 & -1.4162 & -.9423 \\ .2838 & 1.8167 & -.5479 & -.5164 & -.9983 & -.0379 \\ -.9194 & -.5479 & .5030 & .4172 & .4182 & .1290 \\ -.9789 & -.5164 & .4172 & .4720 & .4244 & .1817 \\ -1.4162 & -.9983 & .4182 & .4244 & 1.8201 & -.2482 \\ -.9423 & -.0379 & .1290 & .1817 & -.2482 & .9176 \end{bmatrix},$$

and consider $r = 2$ sparse components such that $D = \mathrm{Diag}(5.5329, 2.3748)$.

The variance $V_o = \mathrm{trace}(D)$ of the first two principal components of $X^\top X$ is 7.9077, and the percentage of the explained by them variance is 83.22% ($=100V_o/\mathrm{trace}(X^\top X)$). The sparse loadings A are orthonormal, but the components that they form are *not* uncorrelated as the original principal components. Thus, $V_s = \mathrm{trace}(A^\top X^\top X A)$ gives an over-estimation of the variance of the sparse components. The adjusted variance introduced by Zou et al. (2006) and discussed in Section 4.6.1 is a more reliable choice. It is given by $V_a = \mathrm{trace}(C \odot C)$, where C is the Cholesky factor from $A^\top X^\top X A = CC^\top$.

The choice of the tuning parameter τ is of great importance. For small applications, as the considered here, the optimal tuning parameter τ can be easily located by solving the problem for several values of τ and compromising between sparseness and fit. Another option is to solve (11.31) for a range of values of τ and choose the most appropriate of them based on some index of sparseness. Here we use the one already introduced in (4.29)

$$\mathrm{IS} = \frac{V_a V_s}{V_o^2} \times \frac{\#0}{pr},$$

where V_a, V_s and V_o are defined above, and $\#_0$ is the number of zeros among all *pr* loadings of A. IS increases with the goodness-of-fit (V_s/V_o), the higher adjusted variance (V_a/V_o) and the sparseness.

We solve (11.31) for several values of τ and the results are summarized in Table 5.6. The second and third columns give the percentage of explained and explained adjusted variances, respectively TV$= 100V_s/\text{trace}(X^\top X)$ and TVA$= 100V_a/\text{trace}(X^\top X)$.

The highest IS is for $\tau = 0.5$. The corresponding sparse loadings A_μ are:

$$
A_{0.5} = \begin{bmatrix} .855 & & \\ & .707 & \\ -.187 & & \\ -.261 & & \\ -.001 & -.707 \\ -.407 & & \end{bmatrix}, \; A_{8.0} = \begin{bmatrix} -.883 & -.227 \\ & .797 \\ .190 & \\ .203 & \\ .353 & -.566 \\ .138 & \end{bmatrix}, \; A_{0.0} = \begin{bmatrix} -.707 & \\ .707 & \\ & \\ & .707 \\ & -.707 \end{bmatrix}.
$$

For comparison, we reproduce the sparse loadings for $\tau = 8.0$. They are less sparse, but the explained and the adjusted explained variances are quite high and pretty close to those of the first two principal components. Finally, the most extreme case is when $\tau = 0.0$ and the second term of (11.31) is switched off. The sparsest possible loadings are $A_{0.0} \in \mathcal{O}_0(6,2)$. They are not related to $X^\top X$ and the locations of the non-zero entries depend only on the initial value for solving (11.31).

11.2.6 Manopt code for sparse PCA of CoDa

The Manopt code for sparse PCA of CoDa can be produced easily by modifying, say, the one given in Section 4.6.3 for SPARSIMAX. The objective function is different, and can be easily adjusted with the proper gradient of (11.31). Here, more important is that the CoDa problem is defined on the Stiefel sub-manifold $\mathcal{O}_0(p,r)$ of all $p \times r$ orthonormal matrices with zero column sums and discussed in Section 3.1.4. Thus, one has to replace

```
manifold = stiefelfactory(p, r)
```

with something that produces centred orthonormal matrices, e.g.:

```
manifold = stiefelcenteredfactory(p, r) ,
```

Table 11.1: Solutions of (11.31) for a range of τ values for comparing their merits

τ	TV	TVA	# 0s	IS
10.0	80.5870	79.4583	3	.2312
9.5	80.5615	79.4092	3	.2309
9.0	80.5126	79.3290	3	.2306
8.5	80.5112	79.3270	3	.2306
8.0	80.4960	79.3127	4	.3073
7.5	80.4916	79.3099	4	.3073
7.0	80.4882	79.3079	4	.3073
6.5	80.4798	79.3004	4	.3072
6.0	80.4615	79.2851	4	.3071
5.5	80.4486	79.2759	4	.3070
5.0	80.4196	79.2514	4	.3068
4.5	80.3991	79.2389	4	.3067
4.0	80.3569	79.2087	4	.3064
3.5	80.2549	79.1288	4	.3057
3.0	80.0848	79.0023	4	.3045
2.5	79.6932	78.7040	4	.3019
2.0	78.8883	78.0568	4	.2964
1.5	77.5811	76.9538	4	.2874
1.0	75.5226	75.1768	4	.2733
0.5	77.9744	74.2917	5	.3485
0.0	35.0617	35.0596	8	.1183

and should be placed in `.../manifolds/stiefel` in your Manopt directory. For this purpose, consider the QR decomposition of a centred $p \times r$ matrix, i.e. replace `[Q,R]=qr(Y(:,:,i),0)` by `[Q,R]=qr(center(Y(:,:,i)),0)`.

The reader is encouraged to create such a new "factory" and repeat the numerical experiments in the previous section. The following code can be used, where (4.31) is utilized to approximate $\|A\|_{\ell_1}$, i.e.:

$$\|A\|_{\ell_1} \approx \operatorname{trace} A^\top \tanh(\gamma A) \, ,$$

for some large $\gamma > 0$ and tanh is the hyperbolic tangent.

```
1  function [A, more] = spca4coda(X, r, mu, varargin)
2
```

```
close all;

eps = .0000001;
penalty = 'TAN';%'N';%

%% Verify the specified solver
solver_list = { 'TR' 'SD' 'CG' 'PS' 'NM'};

if isempty(varargin)
    solver = 'TR';
elseif length(varargin) == 1
    if any(ismember(solver_list,varargin{1}))
        solver = varargin{1};
    else
        error('Unknown solver')
    end
else
    error('Incorrect number of input arguments')
end

%% initialization with original PCA
[n p] = size(X);
if n==p
    R = X;
else
    R = corrcoef(X);
end
[A,D] = eigs(R,r);A, sum(A)

% Create the problem structure.
%manifold = stiefelfactory(p, r);
manifold = stiefelcenteredfactory(p,r);
problem.M = manifold;

%options.maxiter = 5000;
%options.tolgradnorm = .01;
options.verbosity = 1; % or 0 for no output

% Define the problem cost function and its gradient.
problem.costgrad = @(A) mycostgrad(A);

% Numerically check gradient consistency.
% warning('off', 'manopt:stiefel:exp');
warning('off', 'manopt:getHessian:approx');
checkgradient(problem);

% Solve
if solver == 'TR'
    [A, xcost, info, options] = trustregions(problem,[],options);
elseif solver == 'SD'
    [A, xcost, info, options] = steepestdescent(problem,[],options);
elseif solver == 'CG'
    [A, xcost, info, options] = conjugategradient(problem,[],options);
elseif solver == 'PS'
    [A, xcost, info, options] = pso(problem,[]);
elseif solver == 'NM'
```

```
59        [A,  xcost ,  info ,  options ] = neldermead ( problem ,[]) ;
60   end
61
62   % Display  some  statistics .
63        plot_ok = 'N' %'Y' %
64        if  plot_ok == 'Y'
65             figure (2) ;
66             %semilogy ([ info . iter ] ,  [ info . gradnorm ] ,  'k.−') ;
67             plot ([ info . iter ] ,  [ info . gradnorm ] ,  'k.−') ;
68             hold  on;
69             %semilogy ([ info . iter ] ,  [ info . cost ] ,  'ro−') ;
70             plot ([ info . iter ] ,  [ info . cost ] ,  'ro−') ;
71             xlabel ('Iteration  #') ;  ylabel ('Gradient  norm/Cost  (red )') ;
72        end
73
74   %Subroutine  for  the  objective  function  f  and  its  gradient
75   function  [ f  g ] = mycostgrad (A)
76        W = R*A;
77        w = diag (A'*W) − diag (D) ;
78
79   % Penalty
80   if  penalty == 'TAN'
81        gamma = 1000;
82        Qga = gamma*A;
83        wt = tanh (Qga) ;
84        dth = Qga.*( ones ( size (wt)) − (wt.*wt)) ;
85        f1 = trace (A'*tanh (Qga)) ;
86        grad_f = wt + dth ;
87   else
88        W1 = sqrt (A.^2 + eps^2) ;
89        f1 = sum(sum(W1) − eps ) ;
90        grad_f = A./W1;
91   end
92
93        % objective  function
94        f = f1 + mu*(w'*w) ;
95
96        % gradient
97        if  nargout == 2
98             % Euclidean  gradient
99             g = grad_f + mu*2*(W.* repmat (w',p,1)) ;
100
101            % projected  gradient
102            W2 = A'*g;
103            g = .5*A*(W2 − W2') + ( eye (p) − A*A')*g;
104            %g = g − .5*A*(W2 + W2') ;
105        end
106  end
107
108  more . time = info (end) . time ;
109  more . iter = info (end) . iter ;
110  more . cost = info (end) . cost ;
111  %options
112
113  %disp ('Number  of  zeros  per  component ') ,  nz = sum( abs (A) <.0005)
114  %disp ('Total  number  of  zeros ') ,  sum(nz)
```

```
sum(A), disp('Compare with identity matrix!'), A'*A
% RS = A'*R*A; TV = trace(R);
% disp('Variances'), diag(RS)'/TV, sum(diag(RS))*100/TV
% disp('Adjusted Variances'), AdV = (diag(chol(RS)).^2)/TV, sum(AdV)
    *100
% disp('Correlations amomng sparse comonents'), corrcov(RS)

end
```

11.2.7 Conclusion

Section 11.2.1 briefly reviews the history of compositional data (CoDa) analysis. Section 11.2.2 gives definition of CoDa and lists main problems associated with them, as well as the most popular transformations of such data to make them suitable for standard data analysis. Section 11.2.3 concentrates on the clr CoDa transformation and shows that the interior of the simplex becomes a (real) Euclidean space. Section 11.2.4 considers the specific PCA application to CoDa and how it can be modified to produce sparse PCs for CoDa. These considerations are illustrated numerically in Section 11.2.5 on a small CoDa example. Finally, the Manopt codes for sparse PCA of CoDa are provided in Section 11.2.6.

11.2.8 Exercises

1. Check that $\mathcal{V}^{p-1}$ fulfils the conditions in Section 2.2 for a real vector space, with $\boxplus$ for "addition" and $\boxtimes$ for "multiplication".

2. For $\boxminus$ defined in (11.29), check that $v \boxminus w = v \boxplus [(-1) \boxtimes w] = v \boxplus w^{-1}$. Find $v \boxminus v$.

3. Show that

   (a) $\mathrm{clr}(v \boxplus w) = \mathrm{clr}(v) + \mathrm{clr}(w)$

   (b) (11.28) satisfies the axioms for inner product from Section 2.3.

4. Check that the function $d : \mathcal{V}^{p-1} \times \mathcal{V}^{p-1} \to [0, \infty)$ defined by

$$d(v, w) = \|\mathrm{clr}(v) - \mathrm{clr}(w)\|_2 = \left[\sum_{i=1}^{p} \left(\ln \frac{v_i}{g(v)} - \ln \frac{w_i}{g(w)} \right)^2 \right]^{1/2}$$

   is another expression for $d_{crl}(v, w)$.

5. Consider the following additive logratio transformation/operation alr :
$\mathring{\mathcal{V}}^{p-1} \to \mathbb{R}_+^{p-1}$ defined by:

$$\text{alr}(v) = \left[\ln \frac{v_1}{v_p}, \ldots, \ln \frac{v_{p-1}}{v_p}\right] \;,\quad \text{alr}^{-1}(x) = c(\exp([x,0])) = v \in \mathring{\mathcal{V}}^{p-1} \;.$$

For $v, w \in \mathring{\mathcal{V}}^{p-1}$:

(a) Show that $\text{alr}(v \boxplus w) = \text{alr}(v) + \text{alr}(w)$ and $\text{alr}(p^{-1}1_p) = 0_{p-1}$.

(b) Check that
$$\langle v, w \rangle_{alr} = \text{alr}(v)^\top J_{p-1}^a \text{alr}(w) \;,$$
defines an inner product in $\mathring{\mathcal{V}}^{p-1}$, where $J_{p-1}^a = I_{p-1} - \frac{E_{p-1}}{p}$.

(c) Write down the induced alr norm and distance in $\mathring{\mathcal{V}}^{p-1}$.

6. For $v, w \in \mathring{\mathcal{V}}^{p-1}$ and $\alpha > 0$:

(a) Show that $\text{ilr}(v \boxplus w) = \text{ilr}(v) + \text{ilr}(w)$ and $\text{ilr}(\alpha \boxtimes v) = \alpha \text{ilr}(v)$.

(b) Check that $\langle v, w \rangle_{clr} = \text{ilr}(v)^\top \text{ilr}(w)$.

(c) Show that $v = \left[\text{ilr}(v)_1 \boxtimes \text{clr}^{-1}(x_1)\right] \boxplus \ldots \boxplus \left[\text{ilr}(v)_{p-1} \boxtimes \text{clr}^{-1}(x_{p-1})\right]$ where $(x_1, \ldots, x_{p-1})$ is an orthonormal basis in $\mathbb{R}^{p-1}$.

7. (Vernicos, 2014) Consider the following (**Vernicos logratios**) transformation vlr : $\mathring{\mathcal{V}}^{p-1} \to \mathbb{R}^p$ defined by:

$$\text{vlr}(v) = \left[\ln \frac{v_1}{v_2}, \ln \frac{v_2}{v_3}, \ldots, \ln \frac{v_{p-1}}{v_p}, \ln \frac{v_p}{v_1}\right] \in \mathbb{H}^{p-1} \;.$$

Check that $\langle v, w \rangle_{vlr} = \langle \text{vlr}(v), \text{vlr}(w) \rangle_2$ defines a proper inner product in $\mathring{\mathcal{V}}^{p-1}$. Explore its properties and possible applications to CoDa.

8. Every composition $v \in \mathring{\mathcal{V}}^{p-1}$ can be viewed as a diagonal density matrix $D(v) = \text{diag}(v) \in \mathcal{DS}(p) \subset \mathcal{S}_+(p)$, i.e. a PD matrix with unit trace. For $v_1, \ldots, v_n \in \mathring{\mathcal{V}}^{p-1}$, check that their arithmetic mean $\frac{1}{n}\sum_{i=1}^n v_i \in \mathring{\mathcal{V}}^{p-1}$. Find their Riemannian mean composition v (Moakher, 2005):

$$\min_{D(v) \in \mathcal{D}\mathring{\mathcal{S}}(p)} \sum_{i=1}^n \left\|\log\left[D(v_i)^{-1}D(v)\right]\right\|_F^2 = \min_{v \in \mathring{\mathcal{V}}^{p-1}} \sum_{i=1}^n \|\ln v - \ln v_i\|_2^2 \;.$$

9. (Gonçalvesa et al., 2016) Check that the least squares projection of $A \in \mathcal{S}_+(p)$ onto $\mathcal{DS}(p)$, i.e.:

$$\min_{X \in \mathcal{D}\mathring{\mathcal{S}}(p)} \|A - X\|_F \;,$$

is, in fact, a projection of the vector of eigenvalues of A onto $\mathring{\mathcal{V}}^{p-1}$.

Bibliography

Absil, P.-A., & Gallivan, K. A. (2006). Joint diagonalization on the oblique manifold for independent component analysis. In *Proceedings of the IEEE International Conference on Acoustics, Speech, and Signal Processing (ICASSP)* (pp. V–945–V–948).

Absil, P.-A., & Kurdyka, K. (2006). On the stable equilibrium points of gradient systems. *Systems Control Letters, 55*, 573–577.

Absil, P.-A., Mahony, R., & Andrews, B. (2005). Convergence of the iterates of descent methods for analytic cost functions. *SIAM Journal on Optimization, 16*, 531–547.

Absil, P.-A., Mahony, R., & Sepulchre, R. (2008). *Optimization algorithms on matrix manifolds.* Princeton University Press.

Adachi, K. (2016). *Matrix-based introduction to multivariate data analysis.* Singapur: Springer.

Afsari, B. (2006). Simple LU and QR based non-orthogonal matrix joint diagonalization. In J. P. Rosca, D. Erdogmus, & S. Haykin (Eds.), *ICA*, volume 3889 of *Lecture notes in computer science* (pp. 1–7). Springer.

Aitchison, J. (1982). The statistical analysis of compositional data. *Journal of the Royal Statistical Society, B, 4*, 139–177.

Aitchison, J. (1983). Principal component analysis of compositional data. *Biometrika, 70*, 57–65.

Aitchison, J. (1986). *The statistical analysis of compositional data.* London: Chapman & Hall.

Aitchison, J., & Shen, S. M. (1980). Logistic-normal distributions. Some properties and uses. *Biometrika, 67*, 261–272.

Ammar, G., & Martin, C. (1986). The geometry of matrix eigenvalue methods. *Acta Applicandae Mathematcae, 5*, 239–278.

Anderson, T. W. (1984). *An introduction to multivariate statistical analysis.* New York, NY: Wiley.

Anderson, T. W., & Rubin, H. (1956). Statistical inference in factor analysis. In *Proceedings of the Third Berkeley Symposium on Mathematical Statistics and Probability* (Vol. 5, pp. 111–150). Berkeley, CA: University of California Press.

Aravkin, A., Becker, S., Cevher, V., & Olsen, P. (2014). A variational approach to stable principal component pursuit. In *Conference on Uncertainty in Artificial Intelligence (UAI)*.

Arnold, V. I. (1989). *Mathematical methods of classical mechanics* (2nd ed.). New York, NY: Springer.

Arrow, K. J., Azawa, H., Hurwicz, L., & Uzawa, H. (1958). *Studies in linear and nonlinear programming.* Stanford, CA: Stanford University Press.

Artin, M. (2011). *Algebra* (2nd ed.). London, UK: Pearson.

Athans, M., & Schweppe, F. C. (1965). *Gradient matrices and matrix calculations.* Technical Note 1965-53. Lexington, MA: MIT Lincoln Lab.

Bader, B. W., Harshman R. A., & Kolda, T. G. (2007). Temporal analysis of semantic graphs using ASALSAN. *Seventh IEEE International Conference on Data Mining (ICDM 2007)* (pp. 33–42). Omaha, NE.

Banfield, J. D., & Raftery, A. E. (1993). Model-based Gaussian and non-Gaussian clustering. *Biometrics, 49,* 803–821.

Barker, M., & Rayens, W. (2003). Partial least squares for discrimination. *Journal of Chemometrics, 17,* 166–173.

Barnett, V., & Lewis, T. (1978). *Outliers in statistical data.* Chichester, UK: Wiley.

Bartholomew, D., Knott, M., & Moustaki, I. (2011). *Latent variable models and factor analysis: A unified approach* (3rd ed.). Chichester, UK: Wiley.

Bartlett, M. S. (1938). Further aspects of the theory of multiple regression. *Proceeding of the Cambridge Philosophical Society, 34,* 33–40.

Barzilai, J., & Borwein, J. M. (1988). Two-point step size gradient methods. *IMA Journal on Numerical Analysis, 8,* 141–148.

Beaghen, M. (1997). Canonical variate analysis and related methods with longitudinal data. Ph.D. Thesis, Blacksburg, Virginia. Available from http://scholar.lib.vt.edu/theses/available/etd-?11997-?212717/.

Bellman, R. (1960). *Introduction to matrix analysis* (2nd ed.). New York, NY: McGraw – Hill.

Belouchrani, A., Abed-Meraim, K., Cardoso, J.-F., & Moulines, E. (1997). A blind source separation technique using second-order statistics. *IEEE Transactions of Signal Processing, 45*, 434–444.

Bensmail, H., & Celeux, G. (1996). Regularized Gaussian discriminant analysis through eigenvalue decomposition. *Journal of the American Statistical Association, 91*, 1743–1748.

Benzécri, J.-P. (1992). *Correspondence analysis handbook*. Boca Raton, FL: CRC Press.

Bergeron-Boucher, M. P., Simonacci, V., Oeppen, J., & Gallo, M. (2018). Coherent modeling and forecasting of mortality patterns for subpopulations using multiway analysis of compositions: An application to canadian provinces and territories. *North American Actuarial Journal, 22*, 92–118.

Bertsekas, D. P. (1999). *Nonlinear programming* (2nd ed.). Belmont, Massachusetts: Athena Scientific.

Bertsimas, D., Copenhaver, M. S., & Mazumder, R. (2017). Certifiably optimal low rank factor analysis. *Journal of Machine Learning Research, 18*, 1–53.

Bhatia, R. (1997). *Matrix analysis*. Berlin, GE: Springer.

Bhatia, R. (2001). Linear algebra to quantum cohomology: The story of Alfred Horn's inequalities. *American Mathematical Monthly, 37*, 165–191.

Bhatia, R. (2007). *Positive definite matrices*. Princeton, RI: Princeton University Press.

Bhatia, R., Jain, T., & Lim, Y. (2019). On the Bures–Wasserstein distance between positive definite matrices. *Expositiones Mathematicae, 108*, 289–318.

Bickel, P., & Levina, E. (2004). Some theory for Fisher's linear discriminant function, 'naive Bayes', and some alternatives when there are many more variables than observations. *Bernoulli, 10*, 989–1010.

Billheimer, D., Guttorp, P., & Fagan, W. (2001). Statistical interpretation of species composition. *Journal of the American Statistical Association, 456*, 1205–1214.

Borsuk, K. (1967). *Theory of retracts*. Warszawa, PO: PWN.

Boumal, N. (2020). *An introduction to optimization on smooth manifolds*. Available online http://www.nicolasboumal.net/book.

Boumal, N., Mishra, B., Absil, P.-A., & Sepulchre, R. (2014). Manopt: A Matlab toolbox for optimization on manifolds. *Journal of Machine Learning Research, 15,* 1455–1459.

Bourbaki, N. (1974). *Algebra I, Chapters 1–3*. Reading, MA: Addison-Wesley.

Boyd, S., & Vandenberghe, L. (2009). *Convex optimization*. Cambridge, UK: Cambridge University Press.

Boyd, S., & Vandenberghe, L. (2018). *Introduction to applied linear algebra - Vectors, matrices, and least squares*. Cambridge, UK: Cambridge University Press.

Bozma, H., Gillam, W., & Öztürk, F. (2019). Morse-Bott functions on orthogonal groups. *Topology and its Applications, 265,* 1–28.

Brickell, B., Dhillon, I. S., Sra, S., & Tropp, J. A. (2008). The metric nearness problem. *SIAM Journal on Matrix Analysis and Applications, 30,* 375–396.

Brockett, R. W. (1991). Dynamical systems that sort lists, diagonalize matrices, and solve linear programming problems. *Linear Algebra and its Applications, 146,* 79–91.

Browne, M. W. (1967). On oblique Procrustes rotation. *Psychometrika, 25,* 125–132.

Browne, M. W. (1972). Oblique rotation to a partially specified target. *British Journal of Mathematical and Statistical Psychology, 25,* 207–212.

Browne, M. W. (1987). The Young-Householder algorithm and the least squares multidimensional scaling of squared distances. *Journal of Classification, 4,* 175–219.

Browne, M. W. (2001). An overview of analytic rotation in exploratory factor analysis. *Multivariate Behavioral Research, 36,* 111–150.

Browne, M. W., & Kristof, W. (1969). On the oblique rotation of a factor matrix to a specified pattern. *Psychometrika, 34,* 237–248.

Browne, M. W., & Zhang, G. (2007). Developments in the factor analysis of individual time series. In R. Cudeck & R. C. MacCallum (Eds.), *Factor analysis at 100: Historical developments and future directions* (pp. 121–134). Mahway, New Jersey: Lawrence Erlbaum.

Cadima, J., & Jolliffe, I. T. (1995). Loadings and correlations in the interpretations of principal components. *Journal of Applied Statistics, 22*, 203–214.

Cai, J.-F., Candès, E. J., & Shen, Z. (2008). A singular value thresholding algorithm for matrix completion. *SIAM Journal on Optimization, 20*, 1956–1982.

Cai, T., & Liu, W. (2011). A direct estimation approach to sparse linear discriminant analysis. *Journal of the American Statistical Association, 106*, 1566–1577.

Campbell, N. A., & Reyment, R. A. (1981). Discriminant analysis of a cretaceous foraminifer using shrunken estimator. *Mathematical Geology, 10*, 347–359.

Candès, E. J., & Tao, T. (2007). The Dantzig selector: Statistical estimation when p is much larger than n. *Annals of Statistics, 35*, 2313–2351.

Candès, E. J., Li, X., Ma, Y., & Wright, J. (2009). Robust principal component analysis? *Journal of ACM, 58*, 1–37.

Cardoso, J.-F., & Souloumiac, A. (1993a). Blind beamforming for non gaussian signals. *Proc. Inst. Elect. Eng. F, 140*, 362–370.

Cardoso, J.-F., & Souloumiac, A. (1993b). Jacobi angles for simultaneous diagonalization. *SIAM Journal on Matrix Analysis and Applications, 17*, 161–162.

Cardoso, J. R., & Leite, F. S. (2010). Exponentials of skew-symmetric matrices and logarithms of orthogonal matrices. *Journal of Computational and Applied Mathematics, 11*, 2867–2875.

Carlen, E. (2010). Trace inequalities and quantum entropy: An introductory course. In R. Sims & D. Ueltschi (Eds.), *Entropy and the quantum* (pp. 73–140). Providence, RI: American Mathematical Society.

Carroll, J. D. (1968). Generalization of canonical correlation analysis to three or more sets of variables. In *Proceedings of the 76th Annual Convention of the American Psychological Association* (Vol. 3, pp. 227–228).

Carroll, J. D., & Chang, J. J. (1970). Analysis of individual differences in multidimensional scaling via an n—way generalization of "EckartYoung" decomposition. *Psychometrika, 103,* 283–319.

Casini, E., & Goebel, K. (2010). Why and how much brouwer's fixed point theorem fails in noncompact setting? *Milan Journal of Mathematics, 78,* 371–394.

Chandrasekaran, V., Sanghavi, S., Parrilo, P. A., & Willsky, A. S. (2011). Rank-sparsity incoherence for matrix decomposition. *SIAM Journal on Optimization, 21,* 572–596.

Charrad, M., Ghazzali, N., Boiteau, V., & Niknafs, A. (2014). NbClust: An R package for determining the relevant number of clusters in a data set. *Journal of Statistical Software, 61,* 1–36.

Chen, C., & Mangasarian, O. L. (1995). Smoothing methods for convex inequalities and linear complementarity problems. *Mathematical Programming, 71,* 51–69.

Chevalley, C. (1946). *Theory of Lie groups.* Princeton, NJ: Princeton University Press.

Chino, N. (1978). A graphical technique for representation asymmetric relationships between n objects. *Behaviormetrika, 5,* 23–40.

Chipman, H. A., & Gu, H. (2005). Interpretable dimension reduction. *Journal of Applied Statistics, 32,* 969–987.

Choi, J., Zou, H., & Oehlert, G. (2011). A penalized maximum likelihood approach to sparse factor analysis. *Statistics and Its Interface, 3,* 429–436.

Choulakian, V. (2003). The optimality of the centroid method. *Psychometrika, 68,* 473–475.

Choulakian, V. (2005). Transposition invariant principal component analysis in l_1 for long tailed data. *Statistics & Probability Letters, 71,* 23–31.

Chu, D., & Goh, S. T. (2010). A new and fast orthogonal linear discriminant analysis on undersampled problems. *SIAM Journal on Scientific Computing, 32,* 2274–2297.

Chu, M. T. (1988). On the continuous realization of iterative processes. *SIAM Review, 30,* 375–387.

Chu, M. T. (1991). A continuous-time jacobi-like approach to the simultaneous reduction of real matrices. *Linear Algebra and its Applications, 147,* 75–96.

Chu, M. T. (2008). Linear algebra algorithms as dynamical systems. *Acta Numerica, 17,* 1–86.

Chu, M. T., & Driessel, K. R. (1990). The projected gradient method for least squares matrix approximations with spectral constraints. *SIAM Journal on Numerical Analysis, 27*(4), 1050–1060.

Chu, M. T., & Funderlic, R. E. (2001). The centroid decomposition: Relationships between discrete variational decompositions and SVD. *SIAM Journal on Matrix Analysis and Applications, 23,* 1025–1044.

Chu, M. T., & Trendafilov, N. T. (1998a). On a differential equation approach to the weighted orthogonal Procrustes problem. *Statistics and Computing, 8,* 125–133.

Chu, M. T., & Trendafilov, N. T. (1998b). ORTHOMAX rotation problem. A differential equation approach. *Behaviormetrika, 25,* 13–23.

Chu, M. T., & Trendafilov N. T. (2001). The orthogonally constrained regression revisited. *Journal of Computational and Graphical Statistics, 10,* 746–771.

Clark, D. I. (1985). The mathematical structure of huber's-estimator. *SIAM Journal on Scientific and Statistical Computing, 6,* 209–219.

Clemmensen, L., Hastie, T., Witten, D., & Ersbøll, B. (2011). Sparse discriminant analysis. *Technometrics, 53,* 406–413.

Comon, P. (1994). Independent component analysis - a new concept? *Signal Processing, 36,* 287–314.

Conlon, L. (2001). *Differentiable manifolds* (2nd ed.). Basel, SU: Birkhäuser.

Constantine, A. G., & Gower, J. C. (1978). Graphical representation of asymmetric matrices. *Applied Statistics, 27,* 297–304.

Courcoux, P., Devaux, M.-F., & Bouchet, B. (2002). Simultaneous decomposition of multivariate images using three-way data analysis. Application to the comparison of cereal grains by confocal laser scanning microscopy. *Chemometrics and Intelligent Laboratory Systems, 62,* 103–113.

Cox, T. F., & Cox, M. A. A. (2001). *Multidimensional scaling* (2nd ed.). London: Chapman.

Cramer, E. (1974). On Browne's solution for oblique Procrustes rotation. *Psychometrika, 39,* 139–163.

Cramér, H. (1945). *Mathematical methods of statistics.* Princeton, RI: Princeton University Press.

Culver, W. J. (1966). On the existence and uniqueness of the real logarithm of a matrix. *Proceedings of the American Mathematical Society, 17,* 1146–1151.

Curtis, M. L. (1984). *Matrix groups* (2nd ed.). New York, NY: Springer.

Cutler, A., & Breiman, L. (1994). Archetypal analysis. *Technometrics, 36,* 338–347.

Cuturi, M., & Avis, D. (2014). Ground metric learning. *Journal of Machine Learning Research, 15,* 533–564.

d'Aspremont, A., Ghaoui, L., Jordan, M., & Lanckriet, G. (2007). A direct formulation for sparse PCA using semidefinite programming. *SIAM Review, 49,* 434–448.

d'Aspremont, A., Bach, F., & Ghaoui, L. (2008). Optimal solutions for sparse principal component analysis. *Journal of Machine Learning Research, 9,* 1269–1294.

de la Harpe, P. (1993). On Hilbert's metric for simplices. In G. A. Niblo & M. A. Roller (Eds.), *Geometric group theory* (pp. 97–119). Cambridge, UK: Cambridge University Press.

De Lathauwer, L. (1997). Signal processing based on multilinear algebra. Ph.D. Thesis, Katholieke Universiteit Leuven, http://www.esat.kuleuven. ac.be/sista/members/delathau.html.

De Lathauwer, L., De Moor, B., & Vandewalle, J. (2000). A multilinear singular value decomposition. *SIAM Journal on Matrix Analysis and Applications, 21,* 1253–1278.

De Leeuw, J., & Pruzansky, S. (1978). A new computational method to fit the weighted euclidean distance model. *Psychometrika, 43,* 479–490.

de Silva, V., & Lim, L. H. (2008). Tensor rank and the ill-posedness of the best low-rank approximation problem. *SIAM Journal on Matrix Analysis and Applications, 20,* 1084–1127.

De Soete, G., Carroll, J. D., & Chaturvedi, A. D. (1993). A modified CANDECOMP method for fitting the extended INDSCAL model. *Journal of Classification, 10,* 75–92.

Dean, E. B. (1988). Continuous optimization on constraint manifolds. In *Presented at the TIMS/ORSA Joint National Meeting,* Washington D.C.

Deift, P., Nanda, T., & Tomei, C. (1983). Differential equations for the symmetric eigenvalue problem. *SIAM Journal on Numerical Analysis, 20,* 1–22.

Del Buono, N., & Lopez, L. (2001). Runge-Kutta type methods based on geodesics for systems of ODEs on the Stiefel manifold. *BIT Numerical Mathematics, 41*(5), 912–923.

Del Buono, N., & Lopez, L. (2002). Geometric integration on manifold of square oblique rotation matrices. *SIAM Journal on Matrix Analysis and Applications, 23*(4), 974–989.

Demmel, J. (1997). *Applied numerical linear algebra.* Philadelphia, PA: SIAM.

Deshpande, A., Rademacher, L., Vempala, S., & Wang, G. (2006). Matrix approximation and projective clustering via volume sampling. *Theory of Computing, 2,* 225–247.

Deshpande, Y., & Montanari, A. (2016). Sparse PCA via covariance thresholding. *Journal of Machine Learning Research, 17,* 1–41.

Dhillon, I. S., Modha, D. S., & Spangler, W. S. (2002). Class visualization of high-dimensional data with applications. *Computational Statistics and Data Analysis, 41,* 59–90.

Diele, F., Lopez, L., & Peluso, R. (1998). The Cayley transform in the numerical solution of unitary differential systems. *Advances in Computational Mathematics, 8,* 317–334.

Dieudonné, J. (1969a). *Foundations of modern analysis.* New York, NY: Academic Press.

Dieudonné, J. (1969b). *Linear algebra and geometry*. Paris, FR: Hermann.

Ding, X., He, L., & Carin, L. (2011). Bayesian robust principal component analysis. *IEEE Transactions on Image Processing, 20,* 3419–3430.

Do Carmo, M. P. (1992). *Riemannian geometry*. Boston, MA: Birkhäuser.

Donoho, D. L., & Johnstone, I. M. (1994). Ideal spatial adaptation via wavelet shrinkage. *Biometrika, 81,* 425–455.

Drineas, P., & Mahoney, M. (2005). On the Nyström method for approximating a Gram matrix for improved kernel-based learning. *Journal of Machine Learning Research, 6,* 2153–2175.

Dryden, I. L., & Mardia, K. V. (1998). *Statistical shape analysis*. Chichester, UK: Wiley.

Duchene, J., & Leclercq, S. (1988). An optimal transformation for discriminant and principal component analysis. *IEEE Transactions on Pattern Analysis and Machine Intelligence, 10,* 978–983.

Duda, R., Hart, P., & Stork, D. (2001). *Pattern classification* (2nd ed.). New York, N.Y.: Wiley.

Duintjer Tebbens, J., & Schlesinger, P. (2007). Improving implementation of linear discriminant analysis for the high dimension/small sample size problem. *Computational Statistics and Data Analysis, 52,* 423–437.

Eckart, C., & Young, G. (1936). The approximation of one matrix by another of lower rank. *Psychometrika, 1,* 211–218.

Edelman, A., Arias, T. A., & Smith, S. T. (1998). The geometry of algorithms with orthogonality constraints. *SIAM Journal on Matrix Analysis and Applications, 20,* 303–353.

Egozcue, J., Pawlowsky-Glahn, V., Mateu-Figueras, G., & Barceló-Vidal, C. (2003). Isometric logratio transformations for compositional data analysis. *Mathematical Geology, 35,* 279–300.

Egozcue, J., Barcelo-Vidal, C., Martín-Fernández, J., Jarauta-Bragulat, E., Díaz-Barrero, J., & Mateu-Figueras, G. (2011). *Elements of simplicial linear algebra and geometry, compositional data analysis: Theory and applications*. Chichester: Wiley.

Eldén, L. (1977). Algorithms for the regularization of ill-conditioned least squares problems. *BIT, 17,* 134–145.

Eldén, L. (2004). Partial least-squares vs. Lanczos bidiagonalization - I: Analysis of a projection method for multiple regression. *Computational Statistics & Data Analysis, 46,* 11–31.

Eldén, L. (2007). *Matrix methods in data mining and pattern recognition.* Philadelphia: SIAM.

Eldén, L., & Trendafilov, N. (2019). Semi-sparse PCA. *Psychometrika, 84,* 164–185.

Elkan, C. (2003). Using the triangle inequality to accelerate k-means. In *Proceedings of the Twentieth International Conference on Machine Learning,* Washington, DC. ICML-2003.

Engle, M. A., Gallo, M., Schroeder, K. T., Geboy, N. J., & Zupancic, J. W. (2014). Three-way compositional analysis of water quality monitoring data. *Environmental and Ecological Statistics, 21,* 565–581.

Engø, K., Marthinsen, A., & Munthe-Kaas, H. (1999). DiffMan: An object oriented MATLAB toolbox for solving differential equations on manifolds. http://www.math.ntnu.no/num/.

Enki, D., & Trendafilov, N. T. (2012). Sparse principal components by semi-partition clustering. *Computational Statistics, 4,* 605–626.

Enki, D., Trendafilov, N. T., & Jolliffe, I. T. (2013). A clustering approach to interpretable principal components. *Journal of Applied Statistics, 3,* 583–599.

Escande, A. (2016). Fast closest logarithm algorithm in the special orthogonal group. *IMA Journal of Numerical Analysis, 36,* 675–687.

Escofier, B., & Pages, J. (1983). Méthode pour l'analyse de plusieurs groups de variables - Application a la caracterisation de vins rouge du Val de Loire. *Revue de Statistique Appliquée, 31,* 43–59.

Everitt, B. S., & Dunn, G. M. (2001). *Applied multivariate data analysis* (2nd ed.). London: Arnold.

Fan, J., & Li, R. (2001). Variable selection via nonconcave penalized likelihood and its oracle properties. *Journal of the American Statistical Association, 96,* 1348–1360.

Fan, J., Feng, Y., & Tong, X. (2012). A road to classification in high dimensional space: the regularized optimal affine discriminant. *Journal of the Royal Statistical Society, B, 74*, 745–771.

Fang, H. R., & O'Leary, D. P. (2012). Euclidean distance matrix completion problems. *Optimization Methods and Software, 27*, 695–717.

Farebrother, R. (1987). The historical development of the l_1 and l_∞ estimation procedures, 1793–1930. In Y. Dodge (Ed.), *Statistical data analysis based on the L_1 norm and related methods* (pp. 37–63). Amsterdam, NL: North Holland.

Farebrother, R. (1999). *Fitting linear relationships: A history of the calculus of observations 1750–1900*. Berlin, DE: Springer.

Faybusovich, L. E. (1991). Dynamical systems which solve optimization problems with linear constraints. *IMA Journal of Mathematical Control and Information, 8*, 135–149.

Faybusovich, L. E. (2003). Dynamical systems that solve linear programming problems. In *Proceedings of the IEEE Conference on Decision and Control* (pp. 1626–1631). Tuscon, Arizona.

Fisher, R. A. (1936). The use of multiple measurements in taxonomic problems. *Annals of Eugenics, 7*, 179–184.

Flaschka, H. (1974). The Toda lattice. *Physical Review, 9*, 1924–1925.

Flury, B. (1986). Proportionality of k covariance matrices. *Statistics & Probability Letters, 4*, 29–33.

Flury, B. (1988). *Common principal components and related multivariate models*. New York: Wiley.

Flury, B., & Gautschi, W. (1986). An algorithm for simultaneous orthogonal transformation of several positive definite symmetric matrices to nearly diagonal form. *Journal on Scientific and Statistical Computing, 7*, 169–184.

Flury, B., Schmid, M. J., & Narayanan, A. (1994). Error rates in quadratic discrimination with constraints on the covariance matrices. *Journal of Classification, 11*, 101–120.

Friedland, S. (2015). *MATRICES: Algebra, analysis and applications*. Singapore: World Scientific.

Friedman, H. P., & Rubin, J. (1967). On some invariant criteria for grouping data. *Journal of the American Statistical Association, 62,* 1159–1178.

Friedman, J. H. (1989). Regularized discriminant analysis. *Journal of the American Statistical Association, 84,* 165–175.

Fukunaga, K. (1990). *Introduction to statistical pattern recognition* (2nd ed.). San Diego, CA: Academic Press.

Furuta, T. (2005). Two reverse inequalities associated with Tsallis relative operator entropy via generalized Kantorovich constant and their applications. *Linear Algebra and its Applications, 412.*

Gabay, D. (1982). Minimizing a differentiable function over a differential manifold. *Journal of Optimization Theory and Applications, 37,* 177–219.

Gan, G., Ma, C., & Wu, J. (2007). *Data clustering.* Philadelphia, PA: SIAM.

Gander, W., Golub, G., & von Matt, U. (1989). A constrained eigenvalue problem. *Linear Algebra and its Applications, 114,* 815–839.

Gantmacher, F. R. (1966). *Teoriya matrits* (2nd ed.). Moskwa, RU: Nauka.

Gel'fand, I. M. (1961). *Lectures on linear algebra.* New York, NY: Interscience Publishers Inc.

Genicot, M., Huang, W., & Trendafilov, N. (2015). Weakly correlated sparse components with nearly orthonormal loadings. In F. Nielsen & F. Barbaresco (Eds.), *Proceedings of the 2nd Conference on Geometric Science of Information, École Polytechnique, Palaiseau, France, October 28–30, 2015,* Lecture Notes in Computer Science (Vol. 9389, pp. 484–490). Berlin, GE: Springer.

Gifi, A. (1990). *Nonlinear multivariate analysis.* Chichester, UK: Wiley.

Gilbert, A. C., & Jain, L. (2017). If it ain't broke, don't fix it: Sparse metric repair. In *55th Annual Allerton Conference on Communication, Control, and Computing.* IEEE, https://doi.org/10.1109/ALLERTON.2017.8262793, Monticello, IL, USA. 3–6 Oct. 2017.

Giraud, C. (2015). *Introduction to high-dimensional statistics.* Boca Raton, FL: CRC Press, Taylor & Francis Group.

Golub, G. H., & Van Loan, C. F. (2013). *Matrix computations* (4th ed.). Baltimore, MD: Johns Hopkins University Press.

Gonçalvesa, D., Gomes-Ruggierob, M., & Lavor, C. (2016). A projected gradient method for optimization over density matrices. *Optimization Methods & Software, 31*, 328–341.

Goreinov, S. A., Tyrtyshnikov, E. E., & Zamarashkin, N. L. (1997). A theory of pseudoskeleton approximations. *Linear Algebra and its Applications, 261*, 1–21.

Goria, M. N., & Flury, B. D. (1996). Common canonical variates in k independent groups. *Journal of the American Statistical Association, 91*, 1735–1742.

Gower, J. C. (1966). Some distance properties of latent root and vector methods used in multivariate analysis. *Biometrika, 53*, 325–338.

Gower, J. C. (1977). The analysis of asymmetry and orthogonality. In J. R. Barra, F. Brodeau, G. Romier, & B. van Cutsem (Eds.), *Recent developments in statistics* (pp. 109–123). Amsterdam, The Netherlands: North Holland.

Gower, J. C. (1985). Multivariate analysis: Ordination, multidimensional scaling and allied topics. In E. Lloyd & W. Ledermann (Eds.), *Handbook of Applicable Mathematics* (v. 6, Part B, pp. 727–781). Chichester, UK: Wiley.

Gower, J. C. (1989). Generalised canonical analysis. In R. Coppi & S. Bolasko (Eds.), *Multiway data analysis* (pp. 221–232). Amsterdam, NL: North Holland.

Gower, J. C., & Dijksterhuis, G. B. (2004). *Procrustes problems*. Oxford, UK: Oxford University Press.

Gower, J. C., & Krzanowski, W. J. (1999). Analysis of distance for structured multivariate data and extensions to multivariate analysis of variance. *Journal of the Royal Statistical Society: Series C (Applied Statistics), 48*, 505–519.

Grant, M., & Boyd, S. (2016). CVX: Matlab software for disciplined convex programming, version 2.1. http://cvxr.com/cvx.

Grasedyck, L., Kressner, D., & Tobler, C. (2013). A literature survey of low-rank tensor approximation techniques. *GAMM-Mitteilungen, 36*, 53–78.

Greenacre, M. J. (1984). *Theory and applications of correspondence analysis.* New York, NY: Academic Press.

Greenacre, M. J. (2018a). *Compositional data analysis in practice.* Boca Raton, FL: Chapman & Hall / CRC.

Greenacre, M. J. (2018b). Variable selection in compositional data analysis, using pairwise logratio. *Mathematical Geoscience.*

Gruvaeus, G. T. (1970). A general approach to Procrustes pattern rotation. *Psychometrika, 35,* 493–505.

Guan, Y., & Dy, J. (2009). Sparse probabilistic principal component analysis. *Proceedings of the Twelfth International Conference on Artificial Intelligence and Statistics, 5,* 185–192.

Guillemin, V., & Pollack, A. (1974). *Differential topology.* Providence, RI: AMS Chelsea Publishing.

Guo, F., Gareth, J., Levina, E., Michailidis, G., & Zhu, J. (2010). Principal component analysis with sparse fused loadings. *Journal of Computational and Graphical Statistics, 19,* 947–962.

Guo, Y., Hastie, T., & Tibshirani, R. (2007). Regularized linear discriminant analysis and its application in microarrays. *Biostatistics, 8,* 86–100.

Hackbusch, W. (2012). *Tensor spaces and numerical tensor calculus.* Heidelberg, GE: Springer.

Hage, C., & Kleinsteuber, M. (2014). Robust PCA and subspace tracking from incomplete observations using ℓ_0-surrogates. *Computational Statistics, 29,* 467–487.

Halko, E., Martinsson, P. G., & Tropp, J. A. (2011). Finding structure with randomness: Probabilistic algorithms for matrix decompositions. *SIAM Review, 53,* 217–288.

Hall, B. C. (2015). *Lie groups, Lie algebras, and representations: An elementary introduction* (2nd ed.). Berlin, GE: Springer.

Halmosh, P. R. (1987). *Finite-dimensional vector spaces* (3rd ed.). New York, NY: Springer.

Halmosh, P. R. (1995). *Linear algebra problem book.* Washington DC: Mathematical Association of America.

Hanafi, M., & Kiers, H. A. L. (2006). Analysis of k sets of data, with differential emphasis on agreement between and within sets. *Computational Statistics and Data Analysis, 51,* 1491–1508.

Hannachi, A., & Trendafilov, N. T. (2017). Archetypal analysis: Mining weather and climate extremes. *Journal of Climate, 30,* 6927–6944.

Hao, N., Dong, B., & Fan, J. (2015). Sparsifying the Fisher linear discriminant by rotation. *Journal of the Royal Statistical Society, B, 77,* 827–851.

Harman, H. H. (1976). *Modern factor analysis* (3rd ed.). Chicago, IL: University of Chicago Press.

Harshman, R. A. (1970). Foundations of the PARAFAC procedure: Models and conditions for an 'explanatory' multi-modal factor analysis. *UCLA Working Papers in Phonetics, 16,* 1–84.

Harshman, R. A. (1972). Determination and proof of minimum uniqueness conditions for PARAFAC1. *UCLA Working Papers in Phonetics, 22,* 111–117.

Harshman, R. A. (1978). Models for analysis of asymmetrical relationships among n objects or stimuli. In *The First Joint Meeting of the Psychometric Society and the Society of Mathematical Psychology,* Hamilton, Ontario, Canada.

Harshman, R. A., & Kiers, H. A. L. (1987). Algorithms for DEDICOM analysis of asymmetric data. In *The European Meeting of the Psychometric Society.* Enschede, The Netherlands.

Hartigan, J. A. (1975). *Clustering algorithms.* New York, NY: Wiley.

Hartley, R., Aftab, K., & Trumpf, J. (2011). L1 rotation averaging using the Weiszfeld algorithm. In *Proceedings of the 2011 IEEE Conference on Computer Vision and Pattern Recognition* (pp. 3041–3048). Washington, DC: IEEE Computer Society.

Hastie, T., Tibshirani, R., & Buja, A. (1994). Flexible discriminant analysis by optimal scoring. *Journal of the American Statistical Association, 89,* 1255–1270.

Hastie, T., Buja, A., & Tibshirani, R. (1995). Penalized discriminant analysis. *The Annals of Statistics, 23,* 73–102.

Hastie, T., Tibshirani, R., & Friedman, J. H. (2009). *The elements of statistical learning: Data mining, inference, and prediction* (3rd ed.). New York, NY: Springer.

Hausman, R. E. (1982). Constrained multivariate analysis. In S. H. Zanakis & J. S. Rustagi (Eds.), *Optimization in statistics* (pp. 137–151). Amsterdam, NL: North-Holland.

Helmke, U., & Moore, J. B. (1994). *Optimization and dynamical systems.* London, UK: Springer.

Hiai, F., & Petz, D. (2014). *Introduction to matrix analysis and applications.* Heidelberg, GE: Springer.

Higham, N. J. (1989). Matrix nearness problems and applications. In M. J. C. Gover & S. Barnett (Eds.), *Applications of matrix theory* (pp. 1–27). Oxford, UK: Oxford University Press.

Hinton, G., & Roweis, S. (2002). Stochastic neighbor embedding. In *Advances in neural information processing systems* (Vol. 15, pp. 833–840). Cambridge, MA: The MIT Press.

Hirose, K., & Yamamoto, M. (2014). Estimation of an oblique structure via penalized likelihood factor analysis. *Computational Statistics and Data Analysis, 79,* 120–132.

Hirose, K., & Yamamoto, M. (2015). Sparse estimation via nonconcave penalized likelihood in a factor analysis model. *Statistics and Computing, 25,* 863–875.

Hirsh, M. W. (1976). *Differential topology.* New York, NY: Springer.

Hirsh, M. W., & Smale, S. (1974). *Differential equations, dynamical systems, and linear algebra.* San Diego, CA: Academic Press.

Hitchcock, F. (1927a). The expression of a tensor of a polyadic as a sum of products. *Journal of Mathematical Physics Analysis, 6,* 164–189.

Hitchcock, F. (1927b). Multiple invariants and generalized rank of a p-way matrix or tensor. *Journal of Mathematical Physics Analysis, 7,* 39–79.

Horn, R. A., & Johnson, C. A. (1985). *Matrix analysis.* Cambridge, UK: Cambridge University Press.

Horn, R. A., & Johnson, C. A. (1994). *Topics in matrix analysis*. Cambridge, UK: Cambridge University Press.

Horn, R. A., & Johnson, C. A. (2013). *Matrix analysis* (2nd ed.). Cambridge, UK: Cambridge University Press.

Horst, P. (1961). Relations among m sets of measures. *Psychometrika, 26,* 129–149.

Horst, P. (1965). *Factor analysis of data matrices*. New York, NY: Holt, Rinehart and Winston.

Hotelling, H. (1933). Analysis of a complex of statistical variables into principal components. *Journal of Educational Psychology, 24*(417–441), 498–520.

Hotelling, H. (1936). Relations between two sets of variates. *Biometrtka, 28,* 321–377.

Howe, W. G. (1955). Some contributions to factor analysis. Internal Report. Oak Ridge National Laboratory.

Hron, K., Filzmoser, P., de Caritat, P., Fišerová, E., & Gardlo, A. (2017a). Weighted pivot coordinates for compositional data and their application to geochemical mapping. *Mathematical Geoscience, 49,* 777–796.

Hron, K., Filzmoser, P., de Caritat, P., Fišerová, E., & Gardlo, A. (2017b). Weighted pivot coordinates for compositional data and their application to geochemical mapping. *Mathematical Geoscience, 49,* 797–814.

Huang, W., Absil, P.-A., Gallivan, K., & Hand, P. (2015). ROPTLIB: an object–oriented C++ library for optimization on Riemannian manifolds. https://www.math.fsu.edu/~whuang2.

Huang W, Absil, P.-A., & Gallivan, K. A. (2017). Intrinsic representation of tangent vectors and vector transports on matrix manifolds. *Numerische Mathematik, 136,* 523–543.

Huber, P. J. (1981). *Robust statistical procedures*. SIAM Series in Applied Mathematics, # 27, Philadelphia, PA.

Hunter, D. R., & Lange, K. (2004). A tutorial on MM algorithms. *The American Statistician, 58,* 30–37.

Hyvärinen, A., Karhunen, J., & Oja, E. (2001). *Independent component analysis*. New York, NY: Wiley.

Ikemoto, H., & Adachi, K. (2016). Sparse tucker2 analysis of three-way data subject to a constrained number of zero elements in a core array. *Computational Statistics & Data Analysis, 98,* 1–18.

Iserles, A., Munthe-Kaas, H., Norset, S. P., & Zanna, A. (2000). Lie group methods. In A. Iserles (Ed.), *Acta Numerica 2000* (pp. 1–151). Cambridge, UK: Cambridge University Press.

Izenman, A. J. (2008). *Modern multivariate statistical techniques*. New York, NY: Springer.

Jeffers, J. N. R. (1967). Two case studies in the application of principal component analysis. *Applied Statistics, 16,* 225–236.

Jennrich, R. I. (2001). A simple general procedure for orthogonal rotation. *Psychometrika, 66,* 289–306.

Jennrich, R. I. (2002). A simple general method for oblique rotation. *Psychometrika, 67,* 7–20.

Jennrich, R. I. (2004). Rotation to simple loadings using component loss functions: The orthogonal case. *Psychometrika, 69,* 257–273.

Jennrich, R. I. (2006). Rotation to simple loadings using component loss functions: The oblique case. *Psychometrika, 71,* 173–191.

Jennrich, R. I. (2007). Rotation methods, algorithms, and standard errors. In R. Cudeck & R. C. MacCallum (Eds.), *Factor analysis at 100* (pp. 315–335). NJ, Mahwah: Lawrens Erlbaum Associates.

Jennrich, R. I., & Trendafilov, N. T. (2005). Independent component analysis as a rotation method: A very different solution to Thurstone's box problem. *British Journal of Mathematical and Statistical Psychology, 58,* 199–208.

Johnstone, I. M., & Lu, A. Y. (2009). On consistency and sparsity for principal components analysis in high dimensions. *Journal of the American Statistical Association, 104,* 682–693.

Joho, M. (2008). Newton method for joint approximate diagonalization of positive definite Hermitian matrices. *SIAM Journal on Matrix Analysis and Applications, 30,* 1205–1218.

Jolliffe, I. T. (2002). *Principal component analysis* (2nd ed.). New York, NY: Springer.

Jolliffe, I. T., & Uddin, M. (2000). The simplified component technique: An alternative to rotated principal components. *Journal of Computational and Graphical Statistics, 9,* 689–710.

Jolliffe, I. T., Trendafilov, N. T., & Uddin, M. (2003). A modified principal component technique based on the LASSO. *Journal of Computational and Graphical Statistics, 12,* 531–547.

Jones, M. C., & Sibson, R. (1987). What is projection pursuit? *Journal of Royal Statistical Society A, 150,* 1–36.

Jöreskog, K. G. (1977). Factor analysis by least-squares and maximum likelihood methods. In K. Enslein, A. Ralston, & H. S. Wilf (Eds.), *Mathematical methods for digital computers* (pp. 125–153). New York, NY: Wiley.

Journée, M., Nesterov, Y., Richtárik, P., & Sepulchre, R. (2010). Generalized power method for sparse principal component analysis. *Journal of Machine Learning Research, 11,* 517–553.

Kaiser, H. F. (1958). The varimax criterion for analytic rotation in factor analysis. *Psychometrika, 23,* 187–200.

Kannan, R., & Vempala, S. (2017). Randomized algorithms in numerical linear algebra. *Acta Numerica, 26,* 95–135.

Kantorovich, L. V., & Akilov, G. P. (1982). *Functional analysis* (2nd ed.). Oxford, UK: Pergamon Press.

Karmarkar, N. (1984). A new polynomial time algorithm for linear programming. *Combinatorica, 4,* 373–395.

Karmarkar, N. (1990). Riemannian geometry underlying interior-point methods for linear programming. *Contemporary Mathematics, 114,* 51–75.

Kass, R. E. (1989). The geometry of asymptotic inference. *Statistical Science, 4,* 188–219.

Kettenring, J. R. (1971). Canonical analysis of several sets of variables. *Biometrika, 58,* 433–451.

Kiers, H. A. L. (1989a). An alternating least squares algorithm for fitting the two- and three-way DEDICOM model and the IDIOSCAL model. *Psychometrika, 55*, 515–521.

Kiers, H. A. L. (1989b). Majorization as a tool for optimizing a class of matrix functions. *Psychometrika, 55*, 417–428.

Kiers, H. A. L. (1993). An alternating least squares algorithm for PARAFAC2 and three-way DEDICOM. *Computational Statistics & Data Analysis, 16*, 103–118.

Kiers, H. A. L. (1997). Three-mode orthomax rotation. *Psychometrika, 62*, 579–598.

Kiers, H. A. L. (1998). Joint orthomax rotation of the core and component matrices resulting from three-mode principal component analysis. *Journal of Classification, 15*, 245–263.

Kiers, H. A. L., & Takane, Y. (1994). A generalization of GIPSCAL for the analysis of nonsymmetric data. *Journal of Classification, 11*, 79–99.

Kiers, H. A. L., ten Berge, J. M. F., Takane, Y., & de Leeuw, J. (1990). A generalization of Takane's algorithm for DEDICOM. *Psychometrika, 55*, 151–158.

Knüsel, L. (2008). Chisquare as a rotation criterion in factor analysis. *Computational Statistics and Data Analysis, 52*, 4243–4252.

Kolda, T. G., & Bader, B. W. (2009). Tensor decompositions and applications. *SIAM REVIEW, 51*, 455–500.

Kolda, T. G., & O'Leary, D. P. (1998). A semi-discrete matrix decomposition for latent semantic indexing in information retrieval. *ACM Transactions on Information Systems, 16*, 322–346.

Kosinski, A. (1993). *Differential manifolds*. Boston, MA: Academic Press.

Kostrikin, A. I., & Manin, Y. I. (1997). *Linear algebra and geometry*. New York, NY: Gordon and Breach Science Publishers.

Krijnen, W. P., Dijkstra, T. K., & Stegeman, A. (2008). On the non-existence of optimal solutions and the occurence of "degeneracy" in the candecomp/parafac model. *Psychometrika, 73*, 431–439.

Kroonenberg, P. M., & De Leeuw, J. (1980). Principal component analysis of three-mode data by means of alternating least squares algorithms. *Psychometrika, 45,* 69–97.

Kruskal, J. B. (1977). Three-way arrays: Rank and uniqueness of trilinear decompositions, with application to arithmetic complexity and statistics. *Linear Algebra and its Applications, 18,* 95–138.

Kruskal, J. B. (1993). Rank, decomposition, and uniqueness for 3-way and N-way arrays, in multiway data analysis. In R. Coppi & S. Bolasco (Eds.), *Multiway data analysis* (pp. 7–18). Amsterdam, NL: North-Holland.

Krzanowski, W. J. (1984). Principal component analysis in the presence of group structure. *Applied Statistics, 33,* 164–168.

Krzanowski, W. J. (1990). Between-group analysis with heterogeneous covariance matrices: The common principal components model. *Journal of Classification, 7,* 81–98.

Krzanowski, W. J. (1995). Orthogonal canonical variates for discrimination and classification. *Journal of Chemometrics, 9,* 509–520.

Krzanowski, W. J. (2003). *Principles of multivariate analysis: A user's perspective* (revised edition). Oxford, UK: Oxford University Press.

Krzanowski, W. J., Jonathan, P., McCarthy, W. V., & Thomas, M. R. (1995). Discriminant analysis with singular covariance matrices: Methods and applications to spectroscopic data. *Journal of the Royal Statistical Society, C, 44,* 101–115.

Kumar, N. K., & Schneider, J. (2017). Literature survey on low rank approximation of matrices. *Linear and Multilinear Algebra, 65,* 2212–2244.

Kwak, N. (2008). Principal component analysis based on L1-norm maximization. *IEEE Transactions on Pattern Analysis and Machine Intelligence, 30,* 1672–1680.

Lai, Z., Lim, L.-H., & Ye, K. (2020). Simpler Grassmannian optimization. https://www.stat.uchicago.edu/~lekheng/work/simpler.pdf.

Lancaster, P., & Tismenetsky, M. (1985). *The theory of matrices, with applications* (2nd ed.). San Diego, CA: Academic Press.

Landsberg, J. (2012). *Tensors: Geometry and applications.* Providence, RI: American Mathematical Society.

Lang, S. (1962). *Introduction to differentiable manifolds.* New York, NY: Springer.

Lang, S. (1987). *Linear algebra.* New York, NY: Springer.

Lang, S. (1993). *Real and functional analysis.* New York, NY: Springer.

Lavit, C., Escoufier, Y., Sabatier, R., & Traissac, P. (1994). The ACT (STATIS method). *Computational Statistics & Data Analysis, 18,* 97–119.

Lawley, D. N., & Maxwell, A. E. (1971). *Factor analysis as a statistical method.* London, UK: Butterworth.

Lawson, C. L., & Hanson, R. J. (1974). *Solving least squares problems.* Englewood Cliffs, NJ: Prentice-Hall.

Lax, P. (2007). *Linear algebra and its applications.* Hoboken, NJ: Wiley.

Lee, J. M. (2000). *Introduction to topological manifolds.* New York, NY: Springer.

Lee, J. M. (2003). *Introduction to smooth manifolds.* New York, NY: Springer.

Levi-Civita, T. (1927). *The absolute differential calculus.* London and Glasgow, UK: Blackie and Son.

Li, W., & Swetits, J. (1998). Linear ℓ_1 estimator and Huber M-estimator. *SIAM Journal on Optimization, 8,* 457–475.

Lim, L.-H. (2013). Tensors and hypermatrices. In L. Hogben (Ed.), *Handbook of linear algebra* (p. 960, 2nd ed.). Boca Raton, FL: Chapman and Hall/CRC.

Liu, X., Wang, X., Wang, W. G. (2015). Maximization of matrix trace function of product Stiefel manifolds. *SIAM Journal on Matrix Analysis and Applications, 36,* 1489–1506.

Lock, E., Hoadley, K., Marron, J., & Nobel, A. (2013). Joint and individual variation explained (jive) for integrated analysis of multiple data types. *Annals of Applied Statistics, 7,* 523–542.

Lu, Z., & Zhang, Y. (2012). An augmented Lagrangian approach for sparse principal component analysis. *Mathematical Programming. Ser. A, 135,* 149–193.

Luenberger, D. G. (1972). The gradient projection method along geodesics. *Management Science, 18,* 620–630.

Luenberger, D. G., & Ye, Y. (2008). *Linear and nonlinear programming* (3rd ed.). New York, NY: Springer.

Luo, D., Ding, C., & Huang, H. (2011). Linear discriminant analysis: New formulations and overfit analysis. In *Proceedings of the Twenty-Fifth AAAI Conference on Artificial Intelligence* (pp. 417–422).

Luss, R., & Teboulle, M. (2013). Conditional gradient algorithms for rank-one matrix approximations with a sparsity constraint. *SIAM Review, 55,* 65–98.

Lykou, A., & Whittaker, J. (2010). Sparse cca using a lasso with positivity constraints. *Computational Statistics and Data Analysis, 54,* 3144–3157.

Mackey, L. (2009). Deflation methods for sparse PCA. In D. Koller, D. Schuurmans, Y. Bengio, & L. Bottou (Eds.), *Advances in neural information processing systems* (Vol. 21, pp. 1017–1024).

Madsen, K., & Nielsen, H. (1993). A finite smoothing algorithm for linear ℓ_1 estimation. *SIAM Journal on Optimization, 3,* 223–235.

Madsen, K., Nielsen, H., & Pinar, M. (1994). New characterizations of ℓ_1 solutions to overdetermined systems of linear equations. *Operation Research Letters, 16,* 159–166.

Magnus, J., & Neudecker, H. (1988). *Matrix differential calculus with application in statistics and econometrics.* New York, NY: Wiley.

Mahoney, M. W. (2011). Randomized algorithms for matrices and data. *Foundations and Trends in Machine Learning, 3,* 123–224.

Mahoney, M. W., & Drineas, P. (2009). Cur matrix decompositions for improved data analysis. *The National Academy of Sciences of the USA, 106,* 697–702.

Mai, Q., Yang, Y., & Zou, H. (2016). Multiclass sparse discriminant analysis. *Statistica Sinica.* https://doi.org/10.5705/ss.202016.0117.

Mardia, K. V., Kent, J. T., & Bibby, J. M. (1979). *Multivariate analysis.* London, UK: Academic Press.

Marsden, J., & Ratiu, T. (1999). *Introduction to mechanics and symmetry* (2nd ed.). New York, NY: Springer.

Marshall, A., & Olkin, I. (1979). *Inequalities: Theory of majorization and its applications.* London, UK: Academic Press.

Marshall, A., Olkin, I., & Arnold, B. (2011). *Inequalities: Theory of majorization and its applications* (2nd ed.). New York, NY: Springer.

Martin, S., Raim, A. M., Huang, W., & Adragni, K. P. (2020). ManifoldOptim: An R interface to the ROPTLIB library for Riemannian manifold optimization. *Journal of Statistical Software, 93.*

MATLAB. (2019). *MATLAB R2019a.* New York, NY: The MathWorks Inc.

McLachlan, G. J. (2004). *Discriminant analysis and statistical pattern recognition* (2nd ed.). New Jersey, NJ: Wiley.

Merchante, L., Grandvalet, Y., & Govaert, G. (2012). An efficient approach to sparse linear discriminant analysis. In *Proceedings of the 29 th International Conference on Machine Learning*, Edinburgh, Scotland, UK.

Michal, A. D. (1940). Differentials of functions with arguments and values in topological abelian groups. *Proceedings of the National Academy of Sciences of the USA, 26,* 356–359.

Milnor, J. W. (1958). *Differential topology.* Lectures Fall term. Princeton University.

Milnor, J. W. (1965). *Topology from the differential viewpoint.* Charlottesville, VA: The University Press of Virginia.

Mirsky, L. (1955). *An introduction to linear algebra.* Oxford, UK: Oxford University Press.

Mishra, B., Meyer, G., & Sepulchre, R. (2011). Low-rank optimization for distance matrix completion. In *Proceedings of the 50th IEEE Conference on Decision and Control and European Control Conference.* IEEE, Orlando, FL, USA. 12–15 Dec. 2011.

Moakher, M. (2002). Means and averaging in the group of rotations. *SIAM Journal on Matrix Analysis and Applications, 24,* 1–16.

Moakher, M. (2005). A differential geometric approach to the geometric mean of symmetric positive-definite matrices. *SIAM Journal on Matrix Analysis and Applications, 26*, 735–747.

Moghaddam, B., Weiss, Y., & Avidan, S. (2006). Spectral bounds for sparse PCA: Exact and greedy algorithms. *Advances in Neural Information Processing Systems, 18*, 915–922.

Mosier, C. I. (1939). Determining a simple structure when loadings for certain tests are known. *Psychometrika, 4*, 149–162.

Mulaik, S. A. (1972). *The foundations of factor analysis*. New York, NY: McGraw-Hill.

Mulaik, S. A. (2010). *The foundations of factor analysis* (2nd ed.). Boca Raton, FL: Chapman and Hall/CRC.

Munkres, R. J. (1966). *Elementary differential topology*. Princeton, NJ: Princeton University Press.

Nashed, M. Z. (1966). Some remarks on variations and differentials. *American Mathematical Monthly, 73*, 63–76.

Neudecker, H. (1981). On the matrix formulation of kaiser's varimax criterion. *Psychometrika, 46*, 343–345.

Neuenschwander, B. E., & Flury, B. D. (1995). Common canonical variates. *Biometrika, 82*, 553–560.

Ng, M., Li-Zhi, L., & Zhang, L. (2011). On sparse linear discriminant analysis algorithm for high-dimensional data classification. *Numerical Linear Algebra with Applications, 18*, 223–235.

Nielsen, F., & Sun, K. (2019). Clustering in Hilbert's projective geometry: The case studies of the probability simplex and the elliptope of correlation matrices. In F. Nielsen (Ed.), *Geometric structures of information* (pp. 297–331). Cham: Springer.

Ning, N., & Georgiou, T. T. (2011). Sparse factor analysis via likelihood and ℓ_1-regularization. In *50th IEEE Conference on Decision and Control and European Control Conference (CDC-ECC) Orlando*, FL, USA, December 12–15, 2011.

O'Hagan, A. (1984). Motivating principal components, and a stronger optimality result. *The Statistician, 33*, 313–315.

Oseledets, I. (2011). Tensor-train decomposition. *SIAM Journal on Scientific Computing, 33,* 2295–2317.

Oseledets, I., & Tyrtyshnikov, E. (2010). Tt-cross approximation for multidimensional arrays. *Linear Algebra and its Applications, 432,* 70–88.

Owrem, B., & Welfert, B. (2000). The Newton iteration on lie groups. *BIT, 40,* 121–145.

Pawlowsky-Glahn, V., & Egozcue, J. (2001). Geometric approach to statistical analysis on the simplex. *Stochastic Environmental Research and Risk Assessment, 15,* 384–398.

Pearson, K. (1901). On lines and planes of closest fit to systems of points in space. *Philosophical Magazine, 2,* 559–572.

Pedhazur, E. (1982). *Multiple regression in behavioral research* (2nd ed.). Fort Worth, TX: Holt, Rinehart and Winston, Inc.

Petz, D. (2008). *Quantum information theory and quantum statistics.* Heidelberg, GE: Springer.

Pham, D. T. (2001). Joint approximate diagonalization of positive definite Hermitian matrices. *SIAM Journal on Matrix Analysis and Applications, 22,* 1136–1152.

Platt, J. (2005). Fastmap, MetricMap, and Landmark MDS are all Nyström algorithms. In *10th International Workshop on Artificial Intelligence and Statistics,* (pp. 261–268).

Qi, X., Luo, R., & Zhao, H. (2013). Sparse principal component analysis by choice of norm. *Journal of Multivariate Analysis, 114,* 127–160.

Ramsay, J. O. (1977). Maximum likelihood estimation in multidimensional scaling. *Psychometrika, 42,* 241–266.

Recht, B., Fazel, M., & Parrilo, P. (2007). Guaranteed minimum-rank solutions of linear matrix equations via nuclear norm minimization. In *Allerton Conference,* Allerton House, Illinois.

Rencher, A. (1992). Interpretation of canonical discriminant functions, canonical variates, and principal components. *The American Statistician, 46,* 217–225.

Rencher, A. C. (2002). *Methods of multivariate analysis*. New York, NY: Wiley.

Riccia, G., & Shapiro, A. (1982). Minimum rank and minimum trace of covariance matrices. *Psychometrika, 47*, 443–448.

Rinehart, R. F. (1960). Skew matrices as square roots. *The American Mathematical Monthly, 67*, 157–161.

Rockova, V., & George, E. I. (2016). Fast Bayesian factor analysis via automatic rotations to sparsity. *Journal of the American Statistical Association, 111*, 1608–1622.

Rohlf, F. J., & Slice, D. (1990). Extensions of the Procrustes method for the optimal superimposition of landmarcs. *Systematic Zoology, 39*, 40–59.

Rosen, J. B. (1960). The gradient projection method for nonlinear programming. part i. linear constraints. *Journal of the Society for Industrial and Applied Mathematics, 8*, 181–217.

Rosen, J. B. (1961). The gradient projection method for nonlinear programming. part ii. nonlinear constraints. *Journal of the Society for Industrial and Applied Mathematics, 9*, 514–532.

Rosipal, R., & Krämer, N. (2006). Overview and recent advances in partial least squares. In C. Saunders, M. Grobelnik, S. Gunn, & J. Shawe-Taylor (Eds.), *Subspace, latent structure and feature selection. SLSFS 2005. Lecture notes in computer science* (Vol. 3940, pp. 34–51). Berlin, GE: Springer.

Rousseeuw, P., & Leroy, A. (1987). *Robust regression and outlier detection*. New York, NY: Wiley.

Rousson, V., & Gasser, T. (2004). Simple component analysis. *Applied Statistics, 53*, 539–555.

Rudin, W. (1976). *Principles of mathematical analysis* (3rd ed.). New York, NY: McGraw-Hill, Inc.

Sabatier, R., & Escoufier, Y. (1976). A unifying tool for linear multivariate statistical methods: The RV coefficient. *Applied Statistics C, 25*, 257–265.

Sagle, A., & Walde, R. (1973). *Introduction to Lie groups and Lie algebra*. Ney Work, NY: Academic Press.

Saunderson, J., Chandrasekaran, V., Parrilo, P. A., & Willsky, A. S. (2012). Diagonal and low-rank matrix decompositions, correlation matrices, and ellipsoid fitting. *SIAM Journal on Matrix Analysis and Applications, 33,* 1395–1416.

Savas, B., & Dhillon, I. S. (2016). Clustered matrix approximation. *SIAM Journal on Matrix Analysis and Applications, 37,* 1531–1555.

Schechter, M. (1984). Differentiation in abstract spaces. *Journal of Differential Equations, 55,* 330–345.

Scott, A. J., & Symons, M. J. (1971). Clustering methods based on likelihood ratio criteria. *Biometrics, 27,* 387–397.

Seber, G. A. F. (2004). *Multivariate observations* (2nd ed.). New Jersey, NJ: Wiley.

Shampine, L. F., & Reichelt, M. W. (1997). The MATLAB ode suite. *SIAM Journal on Scientific Computing, 18,* 1–22.

Shen, H., & Huang, J. Z. (2008). Sparse principal component analysis via regularized low-rank matrix approximation. *Journal of Multivariate Analysis, 99,* 1015–1034.

Shin, H., & Eubank, R. (2011). Unit canonical correlations and high-dimensional discriminant analysis. *Journal of Statistical Computation and Simulation, 81,* 167–178.

Shingel, T. (2009). Interpolation in special orthogonal groups. *IMA Journal of Numerical Analysis, 29,* 731–745.

Sidiropoulos, N. D., & Bro, R. (2000). On the uniqueness of multilinear decomposition of N-way arrays. *Journal of Chemometrics, 14,* 229–239.

Siegel, A., & Benson, R. (1982). A robust comparison of biological shapes. *Biometrics, 38,* 341–350.

Siegel, A., & Pinkerton, J. (1982). Robust comparison of three-dimensional shapes with an application to protein molecule configurations. Technical report. Technical Report # 217, Series 2, Department of Statistics, Princeton University.

Simonacci, V., & Gallo, M. (2017). Statistical tools for student evaluation of academic educational quality. *Quality and Quantity, 51,* 565–579.

Spearman, C. (1904). "general intelligence", objectively determined and measured. *The American Journal of Psychology, 15,* 201–292.

Spivak, M. (1993). *Calculus on manifolds.* New York, NY: Benjamin.

SPSS. (2001). *SPSS 11.0.* Chicago, IL: SPSS Inc.

Sriperumbudur, B. K., Torres, D. A., & Lanckriet, G. R. G. (2011). A majorization-minimization approach to the sparse generalized eigenvalue problem. *Machine Learning, 85,* 3–39.

Stegeman, A. (2007). Degeneracy in CANDECOMP/PARAFAC explained for $p \times p \times 2$ arrays of rank $p+1$ or higher. *Psychometrika, 71,* 483–501.

Steiger, J. H. (1979). Factor indeterminacy in the 1930's and the 1970's: Some interesting parallels. *Psychometrika, 44,* 157–166.

Stiefel, E. (1935). Richtungsfelder und fernparallelismus in n-dimensionalel manning faltigkeiten.

Stigler, S. (1986). *The history of statistics: The measurement of uncertainty before 1900.* Cambridge, MA: Harvard University Press.

Stuart, A. M., & Humphries, A. R. (1996). *Dynamical systems and numerical analysis.* Cambridge, UK: Cambridge University Press.

Sun, Y., Gao, J., Hong, X., Mishra, B., & Yin, B. (2016). Heterogeneous tensor decomposition for clustering via manifold optimization. *IEEE Transactions on Pattern Analysis and Machine Intelligence, 38,* 476–489.

Sundberg, R., & Feldmann, U. (2016). Exploratory factor analysis - parameter estimation and scoresprediction with high-dimensional data. *Journal of Multivariate Analysis, 148,* 49–59.

Takane, Y., Young, F. W., & Leeuw, J. D. (1977). Nonmetric individual differences multidimensional scaling: Alternating least squares with optimal scaling features. *Psychometrika, 42,* 7–67.

Takane, Y., Jung, K., & Hwang, H. (2010). An acceleration method for Ten Berge et al'.s algorithm for orthogonal INDSCAL. *Computational Statistics, 25,* 409–428.

Takeuchi, K., Yanai, H., & Mukherjee, B. (1982). *The foundations of mutivariate analysis.* New York, NY: Wiley.

ten Berge, J. (1977). Orthogonal procrustes rotation for two or more matrices. *Psychometrika, 42,* 267–276.

ten Berge, J. (1984). A joint treatment of varimax rotation and the problem of diagonalizing symmetric matrices simultaniously in the least-squares sense. *Psychometrika, 49,* 347–358.

ten Berge, J., & Nevels, K. (1977). A general solution to Mosier's oblique Procrestes problem. *Psychometrika, 42,* 593–600.

ten Berge, J. M. F. (1991). A general solution for a class of weakly constrained linear regression problems. *Psychometrika, 56,* 601–609.

ten Berge, J. M. F. (1993). *Least squares optimization in multivariate analysis.* Leiden, NL: DSWO Press.

ten Berge, J. M. F., & Kiers, H. A. L. (1981). Computational aspects of the greatest lower bound to the reliability and constrained minimum trace factor analysis. *Psychometrika, 46,* 201–213.

ten Berge, J. M. F., & Kiers, H. A. L. (1991). Some clarifications of the CANDECOMP algorithm applied to INDSCAL. *Psychometrika, 56,* 317–326.

ten Berge, J. M. F., Kiers, H. A. L., & Krijnen, W. P. (1993). Computational solutions for the problem of negative saliences and nonsymmetry in INDSCAL. *Journal of Classification, 10,* 115–124.

Thurstone, L. L. (1935). *The vectors of mind.* Chicago, IL: University of Chicago Press.

Thurstone, L. L. (1947). *Multiple factor analysis.* Chicago, IL: University of Chicago Press.

Tibshirani, R. (1996). Regression shrinkage and selection via the LASSO. *Journal of the Royal Statistical Society, Ser B, 58,* 267–288.

Tibshirani, R., Saunders, M., Rosset, S., Zhu, J., & Knight, K. (2005). Sparsity and smoothness via the fused lasso. *Journal of the Royal Statistical Society, Ser B, 67,* 91–108.

Timmerman, M. E., Ceulemans, E., Kiers, H. A. L., & Vichi, M. (2010). Factorial and reduced k-means reconsidered. *Computational Statistics and Data Analysis, 54,* 1858–1871.

Torgerson, W. S. (1957). *Theory and methods of scaling*. New York, NY: Wiley.

Trefethen, L. N., & Bau, D., III. (1997). *Numerical linear algebra*. Philadelphia, PA: Society for Industrial and Applied Mathematics.

Trendafilov, N., & Fontanella, S. (2019). Exploratory factor analysis of large data matrices. *Statistical Analysis and Data Mining, 12*, 5–11.

Trendafilov, N., Fontanella, S., & Adachi, K. (2017). Sparse exploratory factor analysis. *Psychometrika, 82*, 778–794.

Trendafilov, N. T. (1994). A simple method for Procrustean rotation in factor analysis using majorization theory. *Multivariate Behavioral Research, 29*, 385–408.

Trendafilov, N. T. (1999). A continuous-time approach to the oblique Procrustes problem. *Behaviormetrika, 26*, 167–181.

Trendafilov, N. T. (2002). GIPSCAL revisited. A projected gradient approach. *Statistics and Computing, 12*, 135–145.

Trendafilov, N. T. (2003). Dynamical system approach to factor analysis parameter estimation. *British Journal of Mathematical and Statistical Psychology, 56*, 27–46.

Trendafilov, N. T. (2004). Orthonormality-constrained INDSCAL with non-negative saliences. In A. Laganà et al. (Eds.), *Computational science and its applications (ICCSA 2004), Lecture notes in computer science series 3044, Part II* (pp. 952–960). Berlin, GE: Springer.

Trendafilov, N. T. (2006). The dynamical system approach to multivariate data analysis, a review. *Journal of Computational and Graphical Statistics, 50*, 628–650.

Trendafilov, N. T. (2010). Stepwise estimation of common principal components. *Computational Statistics and Data Analysis, 54*, 3446–3457.

Trendafilov, N. T. (2012). DINDSCAL: Direct INDSCAL. *Statistics and Computing, 22*, 445–454.

Trendafilov, N. T. (2014). From simple structure to sparse components: A review. *Computational Statistics, 29*, 431–454.

Trendafilov, N. T., & Adachi, K. (2015). Sparse versus simple structure load-ings. *Psychometrika, 80,* 776–790.

Trendafilov, N. T., & Jolliffe, I. T. (2006). Projected gradient approach to the numerical solution of the SCoTLASS. *Computational Statistics and Data Analysis, 50,* 242–253.

Trendafilov, N. T., & Jolliffe, I. T. (2007). DALASS: Variable selection in discriminant analysis via the LASSO. *Computational Statistics and Data Analysis, 51,* 3718–3736.

Trendafilov, N. T., & Unkel, S. (2011). Exploratory factor analysis of data matrices with more variables than observations. *Journal of Computational and Graphical Statistics, 20,* 874–891.

Trendafilov, N. T., & Vines, K. (2009). Simple and interpretable discrimina-tion. *Computational Statistics and Data Analysis, 53,* 979–989.

Trendafilov, N. T., & Watson, G. A. (2004). The ℓ_1 oblique Procrustes prob-lem. *Statistics and Computing, 14,* 39–51.

Tropp, J. A. (2015). An introduction to matrix concentration inequalities. In *Foundations and trends in machine learning* (Vol. 8, Issues 1-2). Now Publishers.

Troyanov, M. (2014). On the origin of Hilbert geometry. In A. Papadopoulos & M. Troyanov (Eds.), *Handbook of Hilbert geometry*. IRMA lectures in mathematics and theoretical physics (Vol. 22, pp. 383–390). EMS Pub-lishing House.

Truesdell, C. A. (1991). *A fist course in rational continuum mechanics* (2nd ed.). Boston, MA: Academic Press.

Tu, L. W. (2010). *An introduction to manifolds* (2nd ed). New York, NY: Springer.

Tucker, L. R. (1966). Some mathematical notes on three-mode factor anal-ysis. *Psychometrika, 31,* 279–311.

Turk, M., & Pentland, A. (1991). Eigenfaces for recognition. *Journal of Cog-nitive Neuroscience, 3,* 71–86.

Unkel, S., & Trendafilov, N. T. (2010). Simultaneous parameter estimation in exploratory factor analysis: An expository review. *International Statistical Review, 78,* 363–382.

Unkel, S., & Trendafilov, N. T. (2013). Zig-zag routine for exploratory factor analysis of data matrices with more variables than observations. *Computational Statistics, 28,* 107–125.

Uno, K., Adachi, K., & Trendafilov, N. T. (2019). Clustered common factor exploration in factor analysis. *Psychometrika, 84,* 1048–1067.

van de Velden, M. (2011). On generalized canonical correlation analysis. In *Proceedings of the 58th World Statistical Congress* (pp. 758–765). Dublin, IR.

van der Maaten, L., & Hinton, H. (2008). Visualizing data using t-SNE. *Journal of Machine Learning Research, 9,* 2579–2605.

Verboon, P. (1994). *A robust approach to nonlinear multivariate analysis.* Leiden, NL: DSWO Press.

Vernicos, C. (2014). On the Hilbert geometry of convex polytopes. In A. Papadopoulos & M. Troyanov (Eds.), *Handbook of Hilbert geometry.* IRMA lectures in mathematics and theoretical physics (Vol. 22, pp. 111–126). EMS Publishing House.

Vichi, M., & Saporta, G. (2009). Clustering and disjoint principal component analysis. *Computational Statistics and Data Analysis, 53,* 3194–3208.

Vines, S. K. (2000). Simple principal components. *Applied Statistics, 49,* 441–451.

Wald, R. (1984). *General relativity.* Chicago, IL: Chicago University Press.

Wall, M. M., & Amemiya, Y. (2007). A review of nonlinear factor analysis and nonlinear structural equation modeling. In R. Cudeck & R. C. MacCallum (Eds.), *Factor analysis at 100: Historical developments and future directions* (pp. 337–361). Mahway, New Jersey: Lawrence Erlbaum.

Wang, S., & Zhang, Z. (2013). Improving CUR matrix decomposition and the Nyström approximation via adaptive sampling. *Journal of Machine Learning Research, 14,* 2729–2769.

Watkins, D. S. (1983). Understanding the QR algorithm. *SIAM Review, 24,* 427–440.

Watson, G. A. (1992). Algorithms for minimum trace factor analysis. *SIAM Journal on Matrix Analysis and Applications, 13,* 1039–1053.

Wen, Z., & Yin, W. (2013). A feasible method for optimization with orthogonality constraints. *Mathematical Programming, 142,* 397–434.

West, M. (2003). Bayesian factor regression models in the "large p, small n" paradigm. In *Bayesian statistics* (pp. 723–732). Oxford University Press.

Westerhuis, J. A., Kourti, T., & Macgregor, J. F. (1998). Analysis of multiblock and hierarchical PCA and PLS models. *Journal of Chemometrics, 12,* 301–321.

Witten, D. M., & Tibshirani, R. (2011). Penalized classification using Fisher's linear discriminant. *Journal of the Royal Statistical Society, B, 73,* 753–772.

Witten, D. M., Tibshirani, R., & Hastie, T. (2009). A penalized matrix decomposition, with applications to sparse principal components and canonical correlation. *Biostatistics, 10,* 515–534.

Wold, H. (1975). Path models with latent variables: The NIPALS approach. In H. Blalock (Ed.), *Quantitative sociology: International perspectives on mathematical and statistical model building* (pp. 307–357). Academic Press.

Wright, S. (2011). Gradient algorithms for regularized optimization. SPARS11, Edinburgh, Scotland, http://pages.cs.wisc.edu/~swright.

Yanai, H., & Takane, Y. (2007). Matrix methods and their applications to factor analysis. In S.-Y. Lee (Ed.), *Handbook of latent variable and related models* (pp. 345–366). NE: Elsevier.

Ye, J. (2007). Least squares linear discriminant analysis. In *Proceedings of the 24th International Conference on Machine Learning* (pp. 1087–1094). Corvallis, OR.

Ye, J., & Xiong, T. (2006). Computational and theoretical analysis of null space and orthogonal linear discriminant analysis. *Journal of Machine Learning Research, 7,* 1183–1204.

Yeredor, A. (2002). Non-orthogonal joint diagonalization in the least-squares sense with application in blind source separation. *IEEE Transactions on Signal Processing, 50,* 1545–1553.

Young, G., & Householder, A. S. (1938). Discussion of a set of points in terms of their mutual distances. *Psychometrika, 3,* 19–22.

Zhai, Y., Yang, Z., Liao, Z., Wright, J., & Ma, Y. (2020). Complete dictionary learning via l4-norm maximization over the orthogonal group. *Journal of Machine Learning Research, 21*, 1–68.

Zhang, C. (2010). Nearly unbiased variable selection under minimax concave penalty. *Annals of Statistics, 38*, 894–942.

Zhang, H., & Hager, W. W. (2004). A nonmonotone line search technique and its application to unconstrained optimization. *SIAM Journal on Optimization, 14*, 1043–1056.

Zhang, L.-H., Liao, L.-Z., & Ng, M. K. (2006). Fast algorithms for the generalized Foley-Sammon discriminant analysis. *SIAM Journal on Matrix Analysis and Applications, 31*, 1584–1605.

Zhang, T., Fang, B., Tang, Y. Y., Shang, Z., & Xu, B. (2010). Generalized discriminant analysis: A matrix exponential approach. *IEEE Transaction on Systems, Man, and Cybernetics-Part B: Cybernetics, 93*, 186–197.

Zhu, X. (2017). A Riemannian conjugate gradient method for optimization on the Stiefel manifold. *Computational Optimization and Applications, 67*, 73–110.

Zimmermann, R. (2017). A matrix-algebraic algorithm for the Riemannian logarithm on the Stiefel manifold under the canonical metric. *SIAM Journal on Matrix Analysis and Applications, 38*, 322–342.

Zou, H., & Hastie, T. (2005). Regularization and variable selection via the elastic net. *Journal of the Royal Statistical Society Ser. B, 67*, 301–320.

Zou, H., Hastie, T., & Tibshirani, R. (2006). Sparse principal component analysis. *Journal of Computational and Graphical Statistics, 15*, 265–286.

Zou, H., Hastie, T., & Tibshirani, R. (2007). On the degrees of freedom of the LASSO. *Annals of Statistics, 35*, 2173–2192.

Zou, M. (2006). Discriminant analysis with common principal components. *Biometrika, 93*, 1018–1024.

Index

© Springer Nature Switzerland AG 2021

N. Trendafilov and M. Gallo, *Multivariate Data Analysis on Matrix Manifolds*,

Springer Series in the Data Sciences, https://doi.org/10.1007/978-3-030-76974-1

Printed in the United States
by Baker & Taylor Publisher Services